Application of Remote Sensing Approaches in Geohazard Risk

Application of Remote Sensing Approaches in Geohazard Risk

Guest Editors

Chao Zhou

Kunlong Yin

Federico Raspini

MDPI

Basel • Beijing • Wuhan • Barcelona • Belgrade • Novi Sad • Cluj • Manchester

Guest Editors
Chao Zhou
School of Geography and Information Engineering
China University of Geosciences
Wuhan
China

Kunlong Yin
Faculty of Engineering
China University of Geosciences
Wuhan
China

Federico Raspini
Department of Earth Sciences
University of Florence
Firenze
Italy

Editorial Office
MDPI AG
Grosspeteranlage 5
4052 Basel, Switzerland

This is a reprint of the Special Issue, published open access by the journal *Remote Sensing* (ISSN 2072-4292), freely accessible at: https://www.mdpi.com/journal/remotesensing/special_issues/N216U627PD.

For citation purposes, cite each article independently as indicated on the article page online and as indicated below:

Lastname, A.A.; Lastname, B.B. Article Title. *Journal Name* **Year**, *Volume Number*, Page Range.

ISBN 978-3-7258-3433-4 (Hbk)
ISBN 978-3-7258-3434-1 (PDF)
https://doi.org/10.3390/books978-3-7258-3434-1

Contents

remote sensing

Article

Insights into Deformation and Mechanism of a Reactivated Landslide Occurrence from Multi-Source Data: A Case Study in Li County, China

Yingjin Du [1], Kun He [1,*], Xiewen Hu [1] and Hongsheng Ma [2]

[1] Faculty of Geosciences and Engineering, Southwest Jiaotong University, Chengdu 610031, China; dyj@swjtu.edu.cn (Y.D.); huxiewen@swjtu.edu.cn (X.H.)
[2] Sichuan Highway Planning, Survey, Design and Research Institute Ltd., Chengdu 610041, China
* Correspondence: hekun@swjtu.edu.cn

Abstract: The investigation of reactivated landslides in the alpine-canyon areas suffers the difficult accessibility of precipitous terrain. In particular, when reactivated landslides occur along the major roads, efforts are focused on measuring ground surface displacements during road construction. Nevertheless, the ancient landslide deposits may reactivate after several years of road operation, while they show a stable state during the road construction. The characterization of this type of reactivated landslides is challenging, due to their complex mechanism and the limited monitoring data. Appropriate multi-source data can provide insights into deformation fields and enhance the understanding of landslide mechanisms, ensuring the outperformance of remedial works. This paper reports a recent Tangjiawan reactivated landslide along the Wenchuan-Maerkang Highway in Li County, China. The outcomes, including satellite InSAR, in situ real-time monitoring, and detailed ground and UAV investigation, conducted at this landslide are presented. Early deformation of the reactivated landslide began from 2019, with an InSAR-derived velocity of −11.7 mm/year, furthermore, a significant subsidence of about 21.2 mm, which occurred within a span of only 12 days from 3 June 2020 to 15 June 2020, was observed. The deformation characteristics derived from in situ monitoring during the remedial works were likely firstly associated with the initial unreinforced slope condition and the heavy rainfall. Subsequently, the displacement evolution transformed into deformation induced by time-dependent reduction in slope strength under rainfall conditions. The existing of unconsolidated deposits derived from ancient landslides, along with a fragile geo-structure consisting of rock blocks and gravels interlayered with breccias, exacerbated by large relief created a predisposition for landslide reactivation. Furthermore, 13 days of antecedent cumulative rainfall totaling 224.5 mm directly triggered the occurrence of a landslide event. The significance and implications of integrating multiple monitoring techniques are emphasized.

Keywords: reactivated landslide; deformation; mechanism; integrated monitoring

Citation: Du, Y.; He, K.; Hu, X.; Ma, H. Insights into Deformation and Mechanism of a Reactivated Landslide Occurrence from Multi-Source Data: A Case Study in Li County, China. *Remote Sens.* **2024**, *16*, 1317. https://doi.org/10.3390/rs16081317

Academic Editor: Sandro Moretti

Received: 20 February 2024
Revised: 2 April 2024
Accepted: 5 April 2024
Published: 9 April 2024

1. Introduction

Landslides are a significant concern in the field of geomorphology for road planning, design, construction, and maintenance [1–5]. Particularly in the areas characterized by alpine-canyon terrain, active tectonism, intense seismic activities, and weak rock mass, landslides not only occur in the form of new landslide, but they also exhibit as the reactivation of ancient landslide deposits [6–8].

The construction of major roads and highways on mountain slopes necessitates attention to anthropogenic activities such as slope excavating, which can significantly increase the likelihood of reactivated landslide occurrences [9–11]. However, little emphasis has been placed on monitoring slope activity for landslides that do not reactivate due to excavation but fail sometime after the completion of construction. Once a reactivated landslide is detected along the road, it is important to investigate its deformation characteristics, reveal the prerequisite and triggering factors, and indicate whether this landslide could reactivate catastrophically and induce devastating aftermath [12–14]. Enhanced efforts have been made to understand the characteristics and mechanisms of reactivated landslides, the deformations of which are mainly induced by precipitation, water level fluctuation and anthropogenic activity [15–17]. For instance, Li et al. [18] reported the 2019 Leidashi reactivated landslide with a volume of 2.2×10^6 m^3 triggered by short-term but continuous heavy rain, which damaged two main roads and 44 anti-sliding piles. Zhang et al. [19] investigated the 2019 Yahuokou landslide reactivation induced by rainfall, with a volume of 3.92×10^6 m^3 debris that slid from the slope, and causing road disruption at the slope toe. Notti et al. [20] studied a rotational landslide in November 2016 in the upper Tanarello stream basin (NW Italy) reactivated by an extreme rainfall event, resulting in the complete collapse of two buildings and a section of the main road. In some of these cases, many attempts focus on the landslide reactivated mechanisms using pre-event rainfall data and post-event geomorphologic features [6,20,21]. Post-event deformation evolutions, in particular, of partially reactivated landslides which are subjected to remedial works in order to explore the deformation behaviors during the construction period, are often not considered. Consequently, the reconstruction of landslide evolution cannot be solely based on in situ measurement data acquired during the events. Instead, a comprehensive strategy that incorporates the entire duration of pre-event and post-event monitoring, as well as field survey data, should be adopted.

Relevant to the absence of in situ monitoring data, remote sensing techniques present significant advantages for landslide analysis [22–24]. Particularly, they are an excellent tool to reveal landslide deformation during the pre-event phase. For decades, among various remote sensing techniques, increasing contributions have been provided by Interferometric Synthetic Aperture Radar (InSAR), which exploits the phase difference between two SAR images acquired at different times over the same area, offering measurements of ground surface displacement and velocity along the radar line of sight (LOS) direction; it is currently one of the foremost methods for landslide deformation monitoring [25–28]. Numerous successful cases evidence the noteworthy preponderances of satellite InSAR in monitoring daily deformation [29,30], examining precursory movements [31,32], detecting wide-region slope deformations [33,34], and assisting the analysis of landslide mechanisms [35].

For the post-reactivation phase, especially when remedial works are performed, it is of utmost importance to center on the continuous monitoring of the slope deformations [36–39]. In situ real-time monitoring strategies are often employed to analyze landslide behaviors and to follow the remedial work performance, during and after its implementation [40–42]. The prime objective of in situ real-time monitoring is to evaluate the efficacy of the intervention by means of ground surface, crack, deep displacement and associated triggering factors analysis, which can provide essential information on the dynamics and evolution of the landslide during and after remedial works. The integration of data from various sources can effectively tackle these challenges and uncover properties of the phenomenon that would otherwise remain unknown.

This paper aims to analyze the Tangjiawan reactivated landslide by means of a combined analysis of InSAR-based and in situ real-time monitoring data, which were acquired using an integrated remote sensing monitoring system for this case. Coupled with geological and geomorphological observations from detailed and immediate field investigations, the integration of the monitoring datasets has yielded a comprehensive perspective on the deformation field of the landslide before reactivation, during and following remedial works, and enabled the unravelling of key features concerning its mechanisms and behavior. This study emphasizes the benefits and implications of integrating monitoring data from diverse sources to enhance the comprehension of reactivated landslides along major road infrastructure. This approach offers a more comprehensive understanding of the dynamics and behavior of reactivated landslides, thereby contributing to improved road engineering practices.

2. Study Area

2.1. Geological Background

The ancient landslide falls in a complex shear tectonic system, involving the "S"-shaped tectonic structure of Xuecheng and the northwest-strike structure of Maerkang, dominated by fold structures of reverse anticlinorium and synclinorium (Figure 1c). A great deal of "S"-shape-like folds and faults are observed in the study area, attributed to the "S"-shaped tectonic structure of Xuecheng, with strikes of 60° and 220° in the northeast and southwest direction, respectively. The bedrock in the landslide zone belongs to dark gray phyllite, interlayered with thin-bedded quartz sandstone, with cleavage with an occurrence of 265°∠70°~85°, corresponding to the lower Weiguan formation of the Devonian period of the Paleozoic era. Two sets of discontinuities are observed with occurrences of 345°∠45° and 200°∠65°, respectively, which result in the rock masses presenting a fractured structure. The main loose deposits of the study area are collapsed deposits, landslide deposits and alluvium deposits of the Quaternary Holocene. The study area was characterized by active tectonic movements, with frequent earthquakes caused by the collision between the Indian Ocean Plate and the Eurasian Plate, including the 2008 Wenchuan earthquake (Ms 8.0), 66 km away from the study site. Severe earthquakes could influence the stability of steep slopes within the area and cause geological disasters. Nevertheless, only two small earthquakes with a magnitude of less than 3.0 were captured 35 km away from the landslide site (https://news.ceic.ac.cn/, accessed on 20 March 2024.).

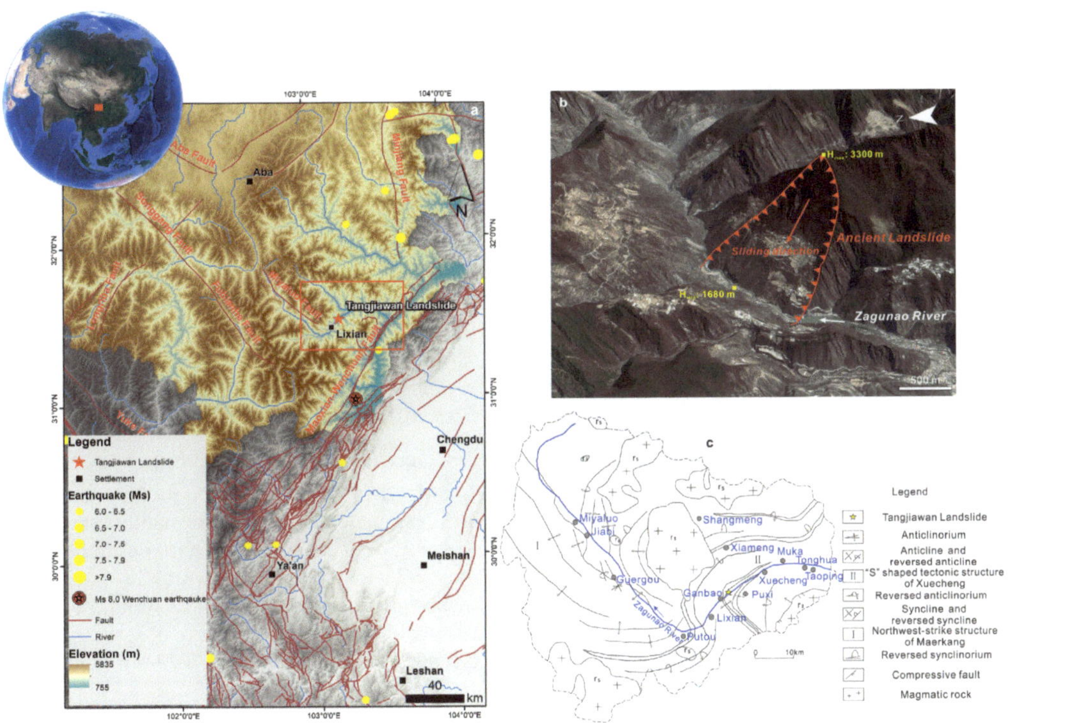

Figure 1. (**a**) Regional geological map of the study area. (**b**) Plan view of the ancient landslide. (**c**) Tectonic map of the study area.

2.2. Geomorphology

The study area is situated on the southeastern margin of the Tibet Plateau, characterized by its lofty elevation and rugged terrain. The landscape is defined by deep canyons surrounded by high mountains (Figure 1a). The elevation range of the ancient landslide area spans from 1600 m to 3300 m, with the peak of the slope positioned at an approximate elevation of 3300 m and the toe situated in the ravine bed of the Zagunao River at around 1680 m, resulting in a vertical variance of approximately 1620 m (Figure 1b).

2.3. Climate and Hydrogeology

The research area is located within a subtropical plateau monsoon climate zone. The average annual precipitation from 2005 to 2022 ranged between 650 and 1000 mm, with a peak daily rainfall of 55.9 mm. The wet season spans from May to September, during which time 80% of the yearly precipitation occurs, often manifesting as rainstorms or heavy downpours. A rain gauge positioned on the pre-existing anti-slide piles recorded a cumulative rainfall of 224.5 mm from 10 May to 22 June, in the year 2022 (Figure 2). The groundwater in the study area is predominantly stored within unconsolidated soil pores and fractured rock fissures.

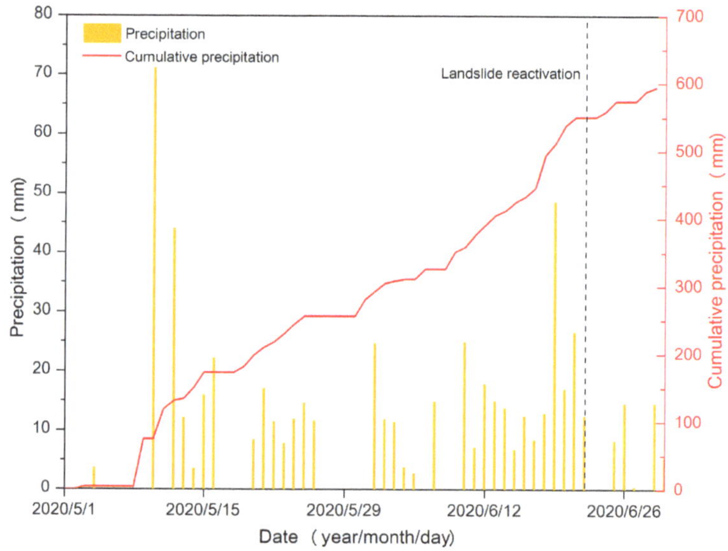

Figure 2. Daily and cumulative precipitation of May to June 2020 in the study area (Dash line means the occurrence of the landslide reactivation).

2.4. General Description of the Ancient Landslide and Recent Reactivation

The Tangjiawan reactivated landslide occurred in a slope with a historical event of ancient landslide occurrence, presenting a chair-like shape in its planform morphology, with a length along the sliding direction of 2120 m, an average lateral width of 1600 m, with a total area of approximately 2.96 km^2, and a total volume of about 1.18×10^8 m^3 (Figure 1b). The local collapse of the Tangjiawan landslide began from 22 June 2020; however, the deformation signs including road subsidence and cracks were observed since June 2019. The remedial work of concrete grouting was carried out in 2019 but failed in stabilizing the slope.

3. Methodology

After the severe deformation and failure of the slope, emergency and permanent remedial works were performed, through a construction period of November 2020 to July 2022 (Figure 3), with the construction work conducted by Sichuan Highway Planning, Survey, Design and Research Institute Ltd. The InSAR technique was deployed to reveal the pre-reactivation landslide evolution; then, ground and UAV investigations were carried out to identify the landslide characteristics, and the in situ real-time monitoring system was conducted to achieve the outperformance of the remedial works (Figure 4).

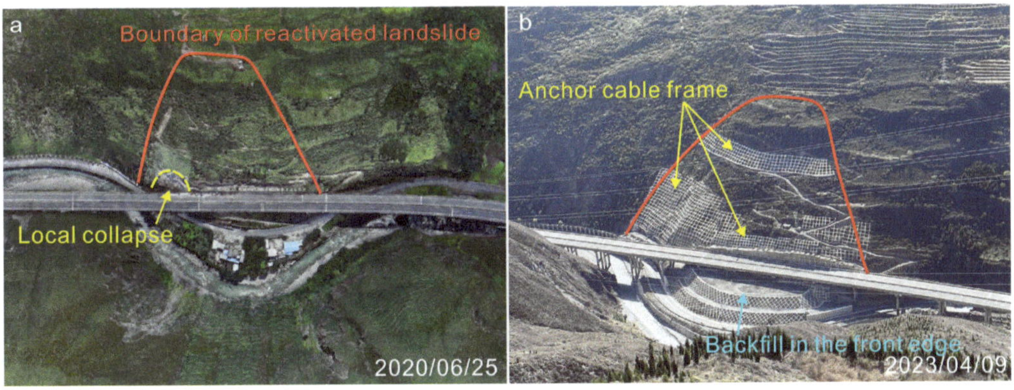

Figure 3. Geomorphologic characteristics of (**a**) pre-remediation and (**b**) post-remediation.

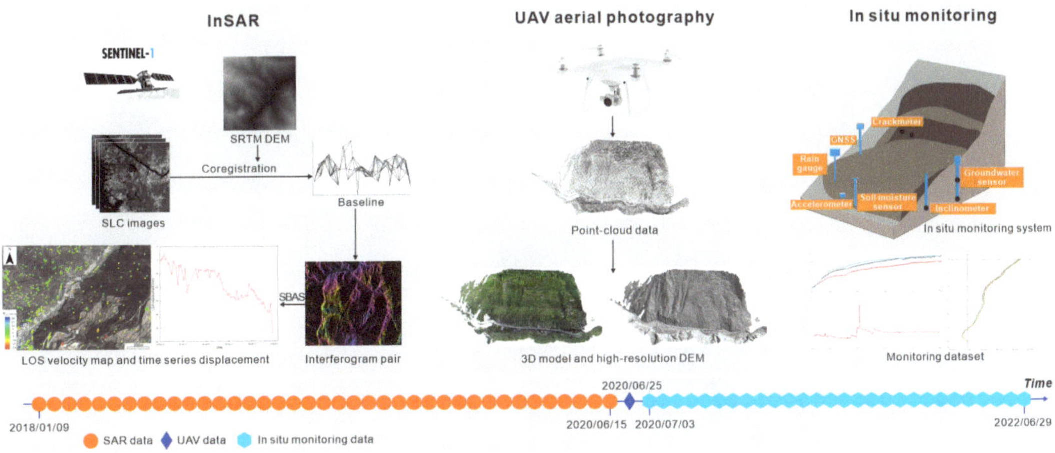

Figure 4. Integrated monitoring dataset and temporal coverage from different sources.

3.1. Ground and UAV Investigations

The detailed field investigations focusing on the Tangjiawan landslide were conducted from 22 June 2020 to 31 December 2020. Firstly, the location and occurrence of all the cracks, the deformation signs of infrastructures, and the boundaries of the reactivated landslide were identified. Six boreholes were drilled and cored to explore the geo-structures of the landslide. In total, 1568 orthoimages were acquired using an unmanned aerial vehicle, DJI Phantom 4 Pro (SZ DJI Technology Co Ltd, Shenzhen, China), and then a stereo aerial map and an orthophoto map were created using ContextCapture software (version 10.20).

3.2. InSAR Technique

Sentinel-1 Single Look Complex (SLC) products in Interferometric Wide (IW) mode were collected to derive the surface deformation in the study area. The Sentinel-1 satellite performs C-band SAR imaging and provides long-term monitoring with a revisit cycle of a minimum of 6 days. The delivered images have 5 (in range) $\times$ 20 (in azimuth) m spatial resolution. As indicated from the optical images, the main direction of the slope is toward the northwest, which causes a foreshortening effect in ascending data. Therefore, 72 Sentinel-1 images (Track 62) in descending orbital direction spanning from 9 January 2018 to 15 June 2020 were collected to generate a stack of 237 interferograms (Figure 5). The stacks of SAR images were processed using ISCE [43]. The 30 m Shuttle Radar Topography

Mission (SRTM) digital elevation model was used to remove topographic phases and geocode interferograms, and the Generic Atmospheric Correction Online Service (GACOS) was used to remove the atmospheric phase [44], and displacement and time series were derived using the StaMPS package [45].

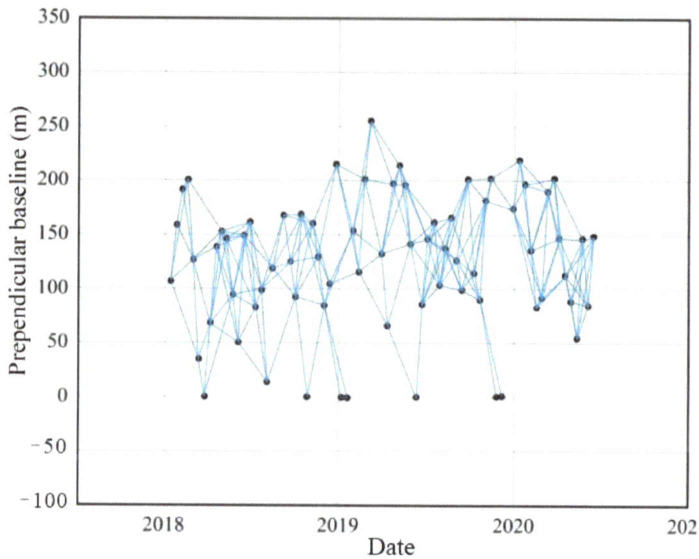

Figure 5. Temporal and perpendicular baselines of the InSAR processing.

3.3. In Situ Real-Time Monitoring

A monitor network of 12 permanent Differential Global Navigation Satellite System (GNSS) stations, 4 crack meters, 3 soil moisture sensors, 4 inclinometers, 3 groundwater monitoring stations and 2 accelerometers was deployed on the landslide in the aftermath of the June 2020 reactivation (Figure 6). The GNSS stations were set on the slope surface to obtain the ground surface displacement, and the displacements in the landslide area during the monitoring period were analyzed through the cumulative displacement and deformation velocity changes of each monitoring point. The cracks (including cracks on the infrastructures) were monitored by implementing crack meters to unravel the development tendency of cracks. The utilization of inclinometers was imperative to ascertaining the depth of the sliding surface and in monitoring the magnitude, velocity, and direction of landslide displacement. The soil moisture meters were set up to monitor the soil moisture content on the slope surface, and the groundwater monitoring gauges were installed to obtain the groundwater level of the study area. The accelerometers were installed on the anti-slide piles to measure the tilt of the structures. Moreover, near the boundary of the landslide, a rain gauge for monitoring precipitation was installed in an unobstructed and stable area. The deadline for all data statistics presented in this study is 29 June 2022.

3.4. Geotechnical Test

The physical mechanic properties of sliding zone soils are vital for unveiling the mechanism of landslides. In order to determine the properties of the soils, geotechnical tests were performed in a laboratory, including index properties, such as specific gravity, density, and Atterberg limits. In addition, a direct shear test was conducted to obtain the shear strength of the soils within the sliding zone. Ten samples were collected from drilling holes ZK01 and ZK02 in the front and middle parts of the study area.

Figure 6. In situ real-time monitoring network and the locations of different instruments.

4. Characteristics of the Reactivated Landslide

Based on the field investigation, borehole drilling and UAV photogrammetry, reactivated area and depth were identified. The reactivated landslide shows a chair shape in the plane, spanning a length along the sliding direction of 100 m and reaching a maximum width of 100 m. The landslide covers an area of about 10×10^3 m^2, with a summit altitude of the area of approximately 1855 m a.s.l and a minimum altitude near the foot of approximately 1685 m a.s.l. The toe of reactivated area falls into the K93 + 900~K94 + 200 section of the Wenchuan-Maerkang Highway. The gradient of front edge is about 20~30°, while the middle and rear parts have a slope gradient of 30~40°, and the main sliding direction is 300° (Figure 7). The determination of the thickness of the sliding mass is achieved through the process of borehole drilling, which is about 25.3~40.1 m, with an average thickness of 35.0 m; thus, the reactivated landslide has a volume of about 350×10^3 m^3 (Figure 8).

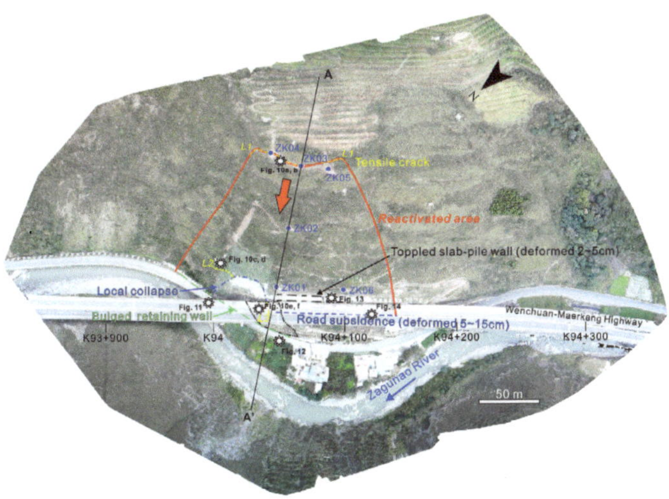

Figure 7. The geomorphological map of the reactivated landslide and the distribution of in situ investigation points. The base map is a UAV orthophoto from 25 June 2020.

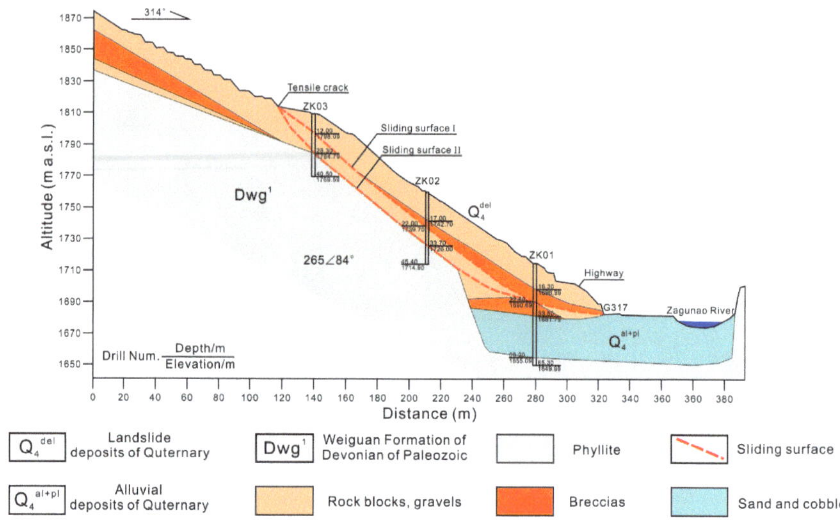

Figure 8. Longitudinal profile of the reactivated landslide (section A–A′ in Figure 7).

4.1. Structural and Material Characteristics

Based on the findings from borehole drilling, exploratory trench testing, and field outcrop investigations, it has been determined that the sliding mass primarily consists of rock blocks, gravels, and breccias (Figure 9). The rock blocks and gravels are characterized by loose structuring with a medium density and a grayish-brown color, and good permeability but a limited water retention capacity. Similarly, the breccias also present a loose structure, mainly medium density, with a gray-brown color, good permeability and water retention ability. The soil composition within the sliding masses is heterogeneous; while rock blocks and gravels have high porosity and strong permeability characteristics, breccias exhibit low porosity and poor permeability properties. Drillings indicate that the clay particles are concentrated in the breccias locally, where the smooth mirrors are also observed with a dip angle of 10~25° (Figure 9c,d).

Multiple slip zones were observed within the reactivated landslide, which is generally steep in the upper and middle parts and gentle in the lower part longitudinally. The middle part of the landslide mainly deforms and slides along the soil–rock interface, with a dip angle of about 36°, while the front edge primarily passes through the gravel and breccia layers, with an inclination of about 5° (Figure 8). The material in the slip zone predominantly contains grayish-brown breccias evolved from phyllite, containing clay, with a particle size of 20~200 mm for about 10% and 2~20 mm for about 50%. The sliding bed is phyllite, belonging to the Devonian Weiguan formation, preponderantly including fractured phyllite. Owing to the regional tectonic influence, the rock mass exhibits a relatively fractured state, characterized by the development of numerous joints and fissures.

4.2. Ground Deformation

A detailed field investigation revealed that the deformation and reactivation signs began on 22 June 2020, which included tensile cracks in the ground surface and a local collapse at the front edge of the slope.

Figure 9. Borehole drillings reveal the structure of the landslide. (**a**) Angular revealed by ZK04. (**b**) Rock blocks and debris from ZK03. (**c**,**d**) Breccia soil from ZK01 and ZK02, respectively.

The ground surface tensile cracks involved three major cracks, mostly located along the boundary of the landslide (rose diagram of the crack strike in Figure 10). The crack L1 had developed at the rear edge, comprising several small cracks, with a length of approximately 100 m, a width of about 1~10 cm and a maximum drop of 180 cm. The strike of L1 was 210~230°, and was nearly parallel to the road direction (Figure 10a,b). The crack L2 was located at the right part of the landslide, belonging to the extension of the rear boundary of the local collapse, with a length of about 30 m, a width of about 1~10 cm, a drop of approximately 20~100 cm, and a strike of 210~230° (Figure 10c,d). The crack L3 was near the front edge of the local collapse, mainly composed of cracks parallel and oblique to the road. The width of the crack was about 0.2~2 cm, the length was about 5~20 m, and the strikes were 210~230° and 350°, respectively Figure 10e,f).

The local collapse had occurred on 22 June 2020, which was located in the right-lower part of the reactivated landslide (Figure 7). This local collapse had a length along the sliding direction of 30 m, and a width of 20 m, covering an area of 600 m^2. Based on the geometry measurement of the surrounding slope, the collapse had a thickness of about 2 m, corresponding to a shallow landslide (Figure 11). The volume of this area was approximately 1200 m^3. The slide direction of this local collapse was approximately 30°, and several tensile cracks and fissures formed in its rear part.

The 210~230° orientation and progressive widening of these cracks, as well as the local collapse, are indicative of the overall shear strength reduction of the slope. The development of tensile cracks probably marked the pre- and entire failure creep deformation. Such occurrences afford a reasonable lead time, representing the temporal interval between the initial observations of precursors and the occurrence of a major failure event, with the aim of providing a timely early warning.

Figure 10. Cracks developed in the landslide area and their rose diagram of strike. (**a**,**b**) Tensile crack L1 at the crown of the reactivated landslide. (**c**,**d**) Tensile crack L2 observed in the vicinity of the local failure. (**e**,**f**) Tensile cracks at the ground surface at the slope toe.

4.3. Infrastructure Deformation

The deformation of infrastructure began from 2019 with slight signs of deformation on the road surface; due to the conducting of countermeasures, the deformation was not recorded. Since June 2020, the infrastructures within the landslide area were seriously damaged, which mainly included retaining walls, ditches, slab-pile walls and roads. National road G317 passes through the front edge in the form of excavation, with an excavated height of about 2~5 m, and a retaining wall with a slope ratio of about 1:0.3~1:0.5 was constructed around 2010 and has been used for nearly 10 years. On 22 June 2020, after heavy rainfall, the slope began to deform, and a bulging crack on the wall and a bulged deformation were observed (Figure 12a). Moreover, the boundary of left side of the road was a ditch with a concrete cover plate. The soil mass of this location was uplifted due to the extrusion, with a height of about 2–20 cm, and the cover plate was damaged (Figure 12b,c).

Figure 11. Local collapse occurred at the lower section of the slope. (**a**) The collapse reached the bridge. (**b**) Rock blocks distributed within the residual slope. (**c**) The failure of the slope damaged the pier of the bridge. (**d**) The rear edge of the local collapse.

Figure 12. (**a**) Bulging deformation of the retaining wall. (**b**,**c**) Uplift deformation of the cover plate of the road ditch.

A 160 m long slab-pile wall was constructed at the front edge of slope. The investigation revealed that cracks appeared on both sides of the slab-pile wall. The tensile crack appeared at the junction of the pile and lattice frame, with a width of about 5~10 cm, showing severe damage (Figure 13a). About 11 vertical cracks were observed in front of the slab-pile wall, with widths ranging from 0.1 to 0.8 cm. It was noticed that the cracks began to gradually extend from 22 June 2020 (Figure 13b). The rear edge of the slab-pile wall was a ditch constructed using concrete, where the cracks began to develop from 22 June, with a width of 1~8 cm, and progressively extended (Figure 13c,d).

Figure 13. Crack deformations of the slab-pile wall in the front edge of slope. (**a**) The crack developed in the junction of pile and lattice frame. (**b**) Crack developed on the slab-pile wall. Crack observed on different dates: (**c**) 7 July and (**d**) 5 September 2020.

A concrete retaining wall was set on the right side of the subgrade. The subsidence and tensile deformation of the retaining wall began to appear after June 2020, with the maximum subsidence reaching 12 cm; in the entire retaining wall, some sections were cracked and damaged (Figure 14).

In late June 2019, the road ground surface of the left Wen-Ma highway exhibited local subsiding, and the subsidence height reached 2~13 cm. The subgrade of this section was grouting treatment. However, since June 2020, the subgrade continued to subside (Figure 14e), and several transverse cracks appeared, with the crack width ranging from 0.2 to 2 cm (Figure 14d).

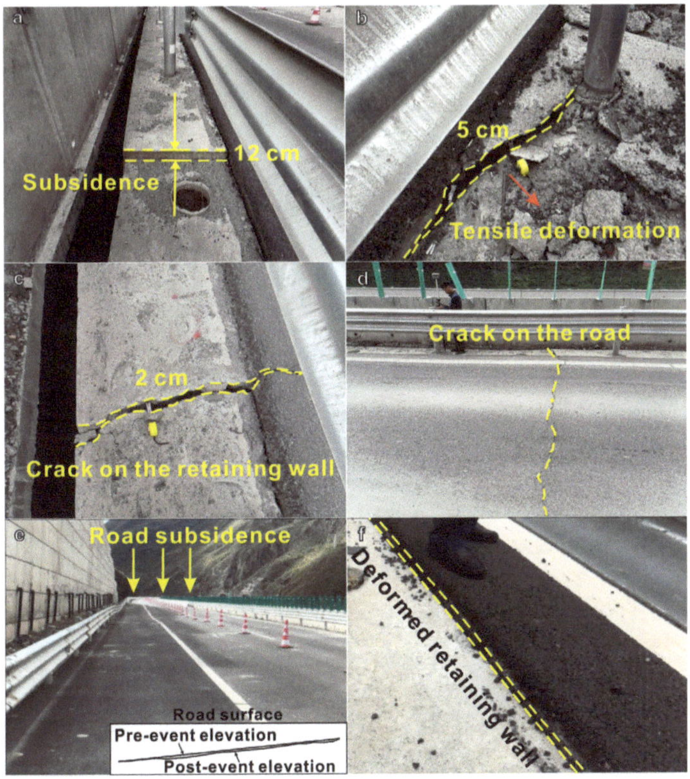

Figure 14. Deformation signs of highway and retaining wall. (**a**) Subsidence of retaining wall. (**b**) Tensile crack in the inside of road. (**c**) Crack on the retaining wall. (**d**) Crack observed on the road. (**e**) Subsidence of road after the reactivation in 2020 while the profile showing the elevation change of the road surface. (**f**) Tensile crack developed in the junction of road and retaining wall.

5. InSAR-Derived History Deformation

Figure 15 depicts the mean velocity of deformation in the LOS direction of the study area between 9 January 2018 and 15 June 2020, which is derived from the SBAS-InSAR technique. The monitored time stack has a mean LOS velocity of −18.9~6.5 mm/yr. Two monitoring points of interest are obtained near the ground surface of the Wen-Ma highway, and the mean LOS velocity of point A was about −11.7 mm/year. The subsidence area was mainly distributed along K94 + 000~K94 + 200. The landslide area suffered uneven subsidence, with a mean LOS velocity of −18~−5 mm/year.

The cumulative displacement curve of InSAR-derived monitoring points in time series can be used to identify the evolution stage of the reactivated landslide. Cumulative displacement time tendency analysis of two monitoring points in three deformation regions was conducted. Figure 16 presents the time series LOS displacement of two PS monitoring points on the front edge of the slope. The ground surface represented by point A has shown a seasonal subsidence tendency since January 2018, but a large subsidence of about 21.2 mm occurred over 12 days from 3 June 2020 to 15 June 2020. The ground surface represented by point B also has a seasonal subsidence trend during the monitoring period, with a mean LOS velocity of about −5.9 mm/year and a total displacement of 50 mm from January 2018 to June 2020. The seasonal deformation indicates that the precipitation plays an important role on the slope stability.

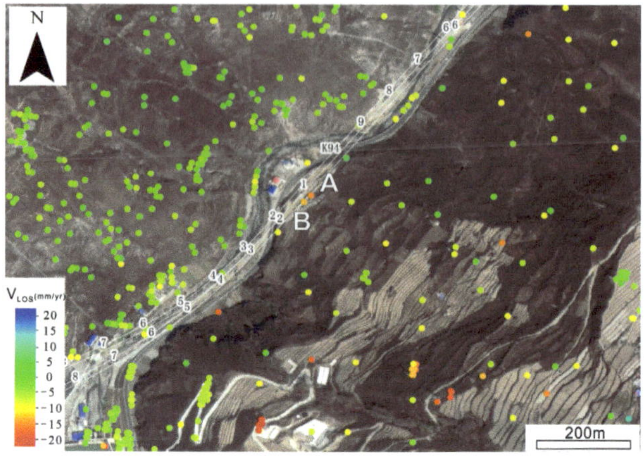

Figure 15. LOS Velocity of study area based on SBAS-InSAR (Points A and B are the selected monitoring points, while the numbers mean the mileage of the highway).

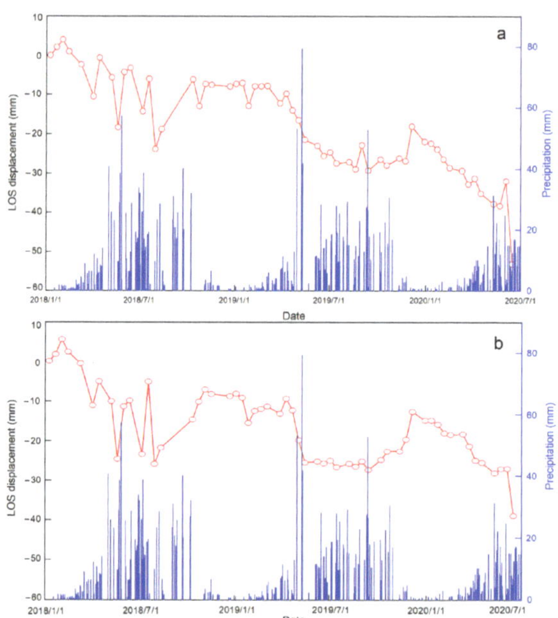

Figure 16. LOS displacement versus time of point (**a**) A and (**b**) B.

6. In Situ Real-Time Monitoring Results

6.1. The Monitoring Results of the Surface Displacement

Figure 17 depicts the monitoring results of the surface deformation of the reactivated landslide obtained from the GNSS stations. Among the 12 GNSS stations, GNSS01-03 stations are installed on the original anti-slide piles at the toe of the slope. In terms of deformation varying with time, a period of rapid deformation was recorded in the three monitoring points from 17 July to 13 November 2020, with horizontal displacement increments of 56.4 mm, 77.3 mm and 92.6 mm for GNSS01-03, respectively. Such deformation characteristics are likely attributed to the unreinforced slope condition and heavy rainfall. Subsequently, the displacement evolution transforms into that of deformation induced by time-dependent slope strength reduction under rainfall and anthropogenic activities.

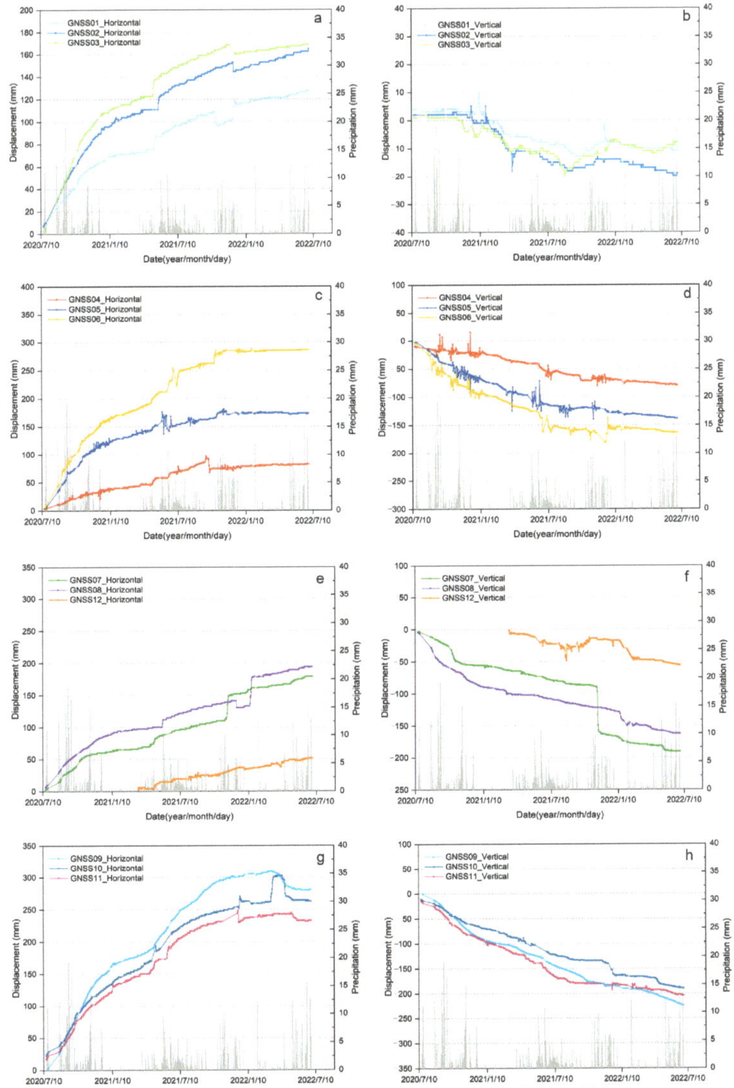

Figure 17. Displacement and precipitation versus time curves: (**a**) horizontal and (**b**) vertical displacement versus time monitored by GNSS01-03. (**c**) Horizontal and (**d**) vertical displacement versus time monitored by GNSS04-06. (**e**) Horizontal and (**f**) vertical displacement versus time monitored by GNSS07, 08 and 12. (**g**) Horizontal and (**h**) vertical displacement versus time monitored by GNSS09-11.

GNSS04-06 stations were installed on the lower part of the slope, which had a similar deformation response to GNSS01-03 after implementation until January 2021. A rapid deformation period can be observed at the onset of deformation monitoring, with horizontal displacements of 29.9 mm, 102.2 mm and 135.7 mm for GNSS04-06, respectively, but it disappeared after the countermeasures that were conducted on December 2020, and the deformation tended to be stable since December 2021.

For the ground surface of Wen-Ma Highway and the National Road G317 with GNSS07-08 and GNSS12 stations installed (GNSS12 installed on 20 March 2021), an increasing displacement over time is observed, which consists of initially rapid deformation followed by slow increasing, due to the effective countermeasures. Notably, GNSS07 and GNSS08 present a sudden increase in both horizontal and vertical displacement in the period of 15 November 2021 to 18 January 2022, with displacement increments of 10~40 mm, indicating that the road surface deforms due to the cumulative precipitation of 95 mm in this period.

GNSS09-11 stations are installed at the right lateral boundary of the reactivated landslide, where the maximum displacement is recorded in the monitoring period, reaching 280.7 mm, 262.6 mm and 232.6 mm, respectively. Such deformation features corresponding to the countermeasures are conducted first at the sections in the vicinity of the Wenchuan-Maerkang Highway, followed by the right boundary; thus, these deformations become stable after November 2021.

6.2. The Monitoring Results of the Deep Displacement

Borehole inclinometers are the most direct way to reveal the sliding zone and to understand the subsurface deformation of the landslide. Figure 18 depicts the deep displacement data measured by four inclinometers. It can be seen the cumulative displacements of inclinometers IN-1, IN-3 and IN-4 have been increasing during the selected monitoring period, suggesting that landslide deformation is still ongoing but at a low deformation rate. Due to the IN-2 being installed on 20 April 2021, the monitoring period of 1 October to 20 October 2021 was selected, when the deep displacement was constant, showing no increasing deformation in IN-2. In addition, Figure 18 illustrates that inclinometers IN-1, IN-3 and IN-4 show obvious shear displacement at different depths, that is, the depth of the sliding zone is 10~12 m and 28~30 m in borehole IN-1, 20~22 m in borehole IN-3, and 20~22 m in borehole IN-4.

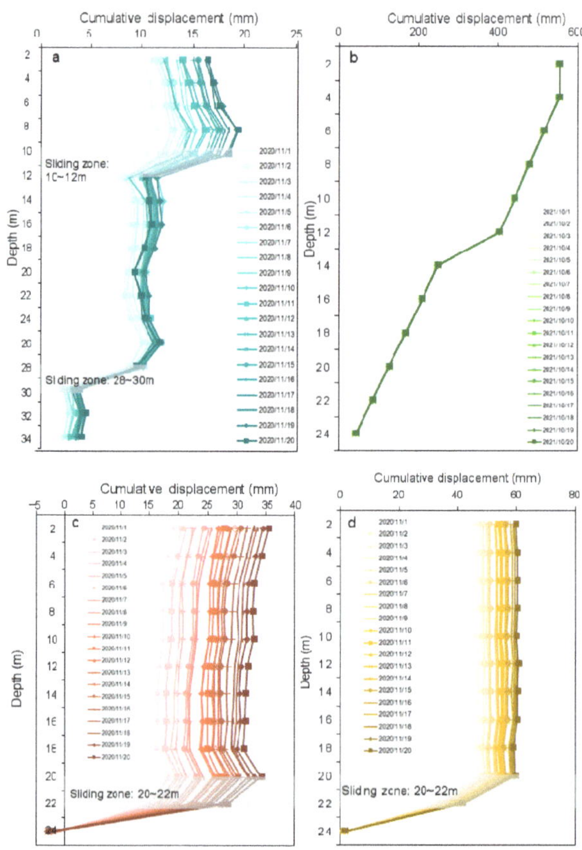

Figure 18. (**a–d**) Cumulative displacement of inclinometers IN01-04.

6.3. The Monitoring Results of the Crack Deformation

Crack meter 01 (CM01) and CM02 are installed on the anti-slide piles, and CM03 and CM04 are located at the rear edge of the local collapse. As shown in Figure 19, all cracks showed coherent opening rates at the early stage of the countermeasure constructing, reached displacements of 16.7 mm in crack meter 01, 51.9 mm in in CM02, 111.2 mm in CM03, and 82.9 mm in CM04, which were overprinted by short-term accelerations triggered by heavy rainfall. Hereafter, the deformation rates showed a low level, except for several intense jump points, which may be influenced by the construction of the anchor cable frame. The ground cracks (CM03 and CM04) present a tendency for high rates and greater displacements compared to the infrastructure cracks (CM01 and CM02).

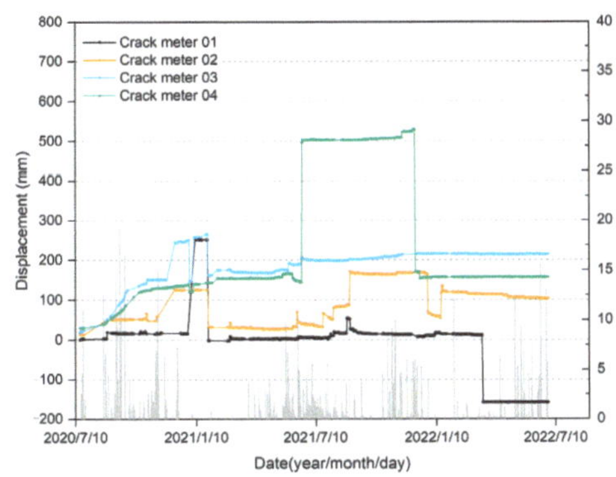

Figure 19. Displacement and precipitation versus time curves on cracks monitored by crack meters 01-04 (Grey bar is the precipitation).

6.4. The Monitoring Results of the Soil Moisture

Figure 20 depicts the change in soil moisture within the landslide area. During the monitoring period, rainy seasons were accompanied by the marked increase in the soil moisture, with soil moisture peaking in the rainy season in 2022 for SF1 and SF2 at the left boundary and middle parts of the landslide, at 19.2% and 24.8% soil moisture content, respectively, as rainfall continued to occur with heavy precipitation in this period. SF3 at the right boundary of the landslide was characterized by stable soil moisture, changing from around 5 to 10%, and the variation mode was also related to the onset of the precipitation event. The deformation of the slope was related to the increase in soil moisture. Before the main countermeasures were conducted and completed in December 2021, the increase in soil moisture induced significant deformations of the slope; nevertheless, after the remediation, the slope tended to be stable, even though soil moisture increased.

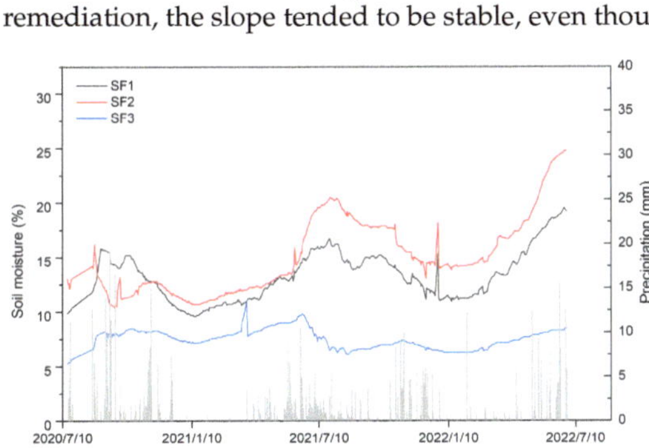

Figure 20. Soil moisture versus time curves monitored by sensors SF1-3 in Figure 6 (Grey bar is the precipitation).

7. Discussion

7.1. Factors Contributing to the Reactivated Landslide

7.1.1. Predisposing Factors

- Unconsolidated deposits derived from ancient landslide at the base of the slope

Surrounding mountainous landscape and the chair-like shape of the studied slope shows that an ancient landslide had occurred. Based on the materials indicated by the borehole drillings, the ancient landslide deposits overlie the alluvium deposits formed by the Zagunao River, indicating that a river-damming event had occurred by the ancient landslide. Moreover, the ancient accumulation layer of the landslide exhibited characteristics of high porosity, elevated water content, and a rich water medium. As previously mentioned, the sliding mass possessed a loose structure, primarily comprised of rock blocks, gravels, and breccias, with coarse particles accounting for 50~60% of its composition. The formation of a skeletal framework between the rock blocks created favorable conditions for predominant seepage. This intricate network facilitated the dominant flow of water within the landslide accumulation layer. Thus, the concentrated seepage is prone to occur then, causing the finer soil particles to be washed away, forming macropores within the rock blocks when the deposits are subjected to high-intensity precipitation, ultimately decreasing the stability of the unconsolidated landslide deposits. The residual rock blocks and gravels could be observed in the failed local collapse, which is attributed to the concentrated seepage.

- Fragile geo-structure of rock blocks and gravels interlayered with breccias

The ancient landslide deposits consist of rock blocks, gravels interlayered with breccias, in which the breccias with poor permeability are located between the rock blocks, and gravels with high permeability. A large amount of precipitation in this area is transferred to groundwater through the rock blocks and gravels and is discharged along the interlayer of breccias and bedrock. However, due to the limited drainage points, part of the groundwater is retained in the breccias layer. Under the action of the high overlying sliding mass weight pressure, the water and the breccia are strongly hydrated, that is, the breccia is argillated, and the strength of the breccia is greatly reduced and it becomes a potential slip zone.

Ten soil samples in the sliding zone were collected in the landslide area (Figure 7). Through laboratory experiments, the geotechnical properties are obtained. Table 1 lists the physical mechanics indexes of the sliding zone soils. The measured natural and saturated water content of the sliding zone soil is 7.1% and 21.3%, respectively. Regarding the peak strength to residual strength, the strength parameter of sliding zone soils decreases significantly, while cohesion decreases by 67% and the internal friction angle decreases by 33% in the natural water content condition of soils. The same shear test was conducted for the saturated condition of the samples. The cohesion decreases from 6.5 kPa to 2.0 kPa, and the internal friction angle decreases from 15.5° to 11.2° between the peak strength and residual strength. The results indicate that the strength of the sliding zone soil will be significantly weakened under the action of water infiltration. The liquid limit is 35.1%, and the plastic limit is 16.2%. The soil has a plasticity index of 18.9. When the water content reaches the plastic limit, any further increase in water content will precipitate a marked deterioration in the strength of the sliding zone soils, thereby compromising the stability of the slope. The inherent natural water content of sliding zone soils is measured at 7.1%, that is, the soils require the infiltration from rainwater to be saturated. Hence, the role of rainwater infiltration is considered as the primary factor this reactivation.

Table 1. Physical and mechanics parameters of sliding zone soils.

Natural Density (g/cm³)	Water Content (%)		Liquid Limit (%)	Plastic Limit (%)	Plasticity Index	Peak Strength				Residual Strength			
	Natural	Saturated				Cohesion (kPa)		Internal Friction Angle (°)		Cohesion (kPa)		Internal Friction Angle (°)	
						Natural	Saturated	Natural	Saturated	Natural	Saturated	Natural	Saturated
1.94	7.1	21.3	35.1	16.2	18.9	16.5	6.5	22.6	15.5	5.5	2.0	15.1	11.2

- Large relief of the original slope

The ancient landslide has an area of 2.96 km^2, with an elevation difference of 1620 m, and an average slope of 36°, presenting a steep-gentle-steep terrain longitudinally. The rear edge of the reactivated landslide is located at the slope break zone, which belongs to the concentrated region of tension stress. The Zagunao river valley is characterized by a deep incision, showing a "V" shape, with 80 m in width near the Tangjiawan ancient landslide at the bottom of the valley and less than 20 m wide at the narrowest section. Unraveled by the borehole drilling, the alluvium deposits of the Zagunao river are overlayed by ancient landslide deposits, causing a river-damming event (Figure 8). Due to the strong incising of the river, the landslide dam was eroded and a high-steep free slope surface formed in the toe of the ancient landslide deposits with a height of more than 130 m and an average slope of 40°, resulting in the re-distribution of stress and providing the potential energy of the deformation or even failure. Therefore, the large relief and unique topography facilitated the development of the reactivated landslide and promoted a retrogressive mode for failure.

7.1.2. Triggering Factors and Reactivation Mechanism

- A total of 13 days of heavy precipitation preceding the reactivation

The intense deformation of the reactivated landslide began on 22 June 2020, corresponding to 13 consecutive days of precipitation of 224.5 mm prior to the event. The cumulative rainfall from 10 May to 22 June 2022 reached 546.8 mm, which exceeded 55% of the total annual rainfall. Thus, the long-term rainfall, particularly the last 13 days of the antecedent rainfall period, was the primary triggering factor of this reactivated landslide. The effects of early-stage rainfall include: (1) weakening the mechanical parameters (cohesion and internal friction angle) of the loose ancient landslide deposits; (2) the rise in the river water level reduces the hydraulic gradient of the slope, reduces the amount of groundwater excretion on the slope, and strands a great deal of rainwater on the slope for a long time, thus increasing the self-weight and sliding force of the slope; (3) elevating the groundwater level, leading to a significant reduction in the mechanical parameters at the interface of the foundation cover, causing a sharp drop in the strength of the sliding zone; (4) the dynamic water pressure of long-term rainfall causes the loss of fine-grained soils, enabling a resultant loose structure and poor cohesion, which is more conducive to rainfall infiltration.

- The contribution of anthropogenic activity in changing the strength of the slope

National Road G317 and Wenchuan-Maerkang Highway pass through the slope front in the form of excavation. The highest excavated slope height is about 11 m. The construction of National Road G317 and Wen-Ma Highway reduced the anti-slip section of the slope, forming the steep free face on the slope front, and damaging the strength of the slope. Therefore, anthropogenic activity is the secondary triggering factor of landslide reactivation.

7.2. Integrated Monitor Network on Ancient Landslides in Mountainous Area

The reactivation of the Tangjiawan ancient landslide event was disastrous for road construction and operation in mountainous areas, causing damage to infrastructures and interrupting road operation, consequently leading to severe economic losses. Based on our field investigation, compared with other rainfall- and road excavation-related reactivated landslides in China, such as the Leibo landslide [12,46], Baiyangwan landslide [47], Maobazi landslide [48], and the Dongla landslide [49], the occurrence of such reactivated landslides is particularly related to their complex geological environments [50]. Hence, this event serves as a poignant reminder to diligently consider the spatial distribution and characteristics of active ancient landslides in main road engineering.

The mountainous area in western Sichuan experiences extremely active neotectonics and many strong earthquakes. Numerous ancient landslide deposits are distributed within this area. In order to achieve early detection, real-time monitoring and early warning of reactivated landslides possible along the Wenchuan to Li County section of the Wenchuan-Maerkang Highway. Investigations of these landslides were carried out through employing optical images, the SBAS-InSAR technique, and field investigations. As depicted in Figure 21, twenty-one ancient landslides are distributed along the highway, which pose a severe threat to the road and surrounding settlements. Efforts are concentrated on the remote measurement of ground surface deformations, in order to define the slope dynamics and landslide activities. Characterization of these ancient landslide deposits remains challenging because of their complex environments and the limitations of monitoring techniques. This paper utilizes an integrated monitor network, appropriately using the outcomes of satellite InSAR, UAV, and in situ monitoring instruments, such as GNSS. These monitoring data from different sources can help identify potentially active ancient landslide deposits preceding road construction and monitor real-time deformation during and after construction. Moreover, for an ancient landslide, which was stable during the construction stage but may be partially revived again due to triggering factors during the operation, the joint analysis could provide a comprehensive overview of the deformation of the landslide, and could evaluate the instability of the landslide in a timely manner, reducing the uncertainties of qualitative analysis.

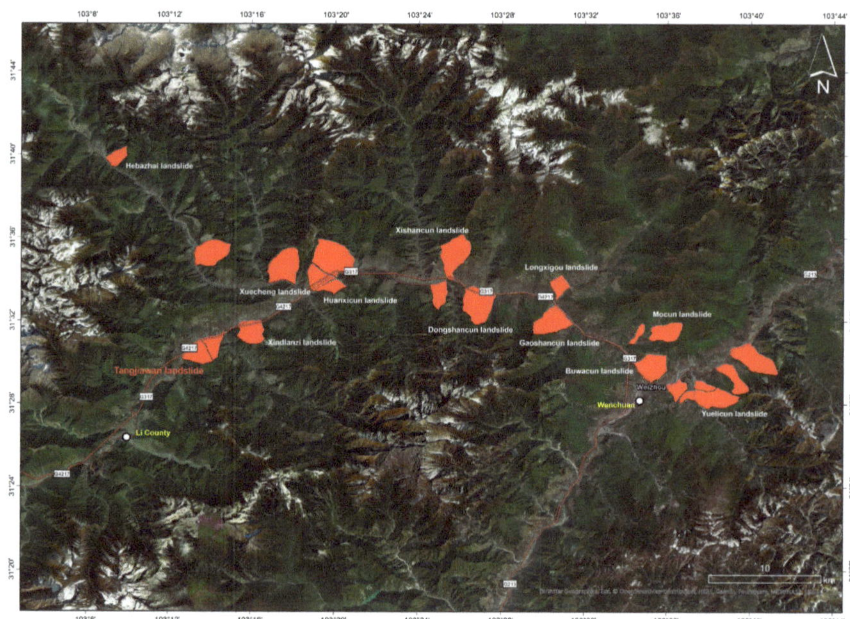

Figure 21. Distribution of ancient landslides on the Wenchuan-Li County section of Wen-Ma Highway from a comprehensive investigation (Red polygons are the ancient landslide areas).

8. Conclusions

This study conducts a detailed investigation on the deformation and mechanisms of a recently reactivated landslide, Tangjiawan landslide in Li County, Sichuan, China. This landslide is located on a steep alpine slope and experienced a large-scale reactivation on 22 June 2020. This event was triggered by a period of prolonged and intense precipitation. By employing an integrated monitoring network, the deformation pre- and post-event, including during the construction of countermeasures, is addressed.

The joint analysis of satellite InSAR, UAV and in situ monitoring data, spanning a time period of four years in total, provided essential insights into the deformation and mechanisms of the reactivated landslide. The landslide activity has been in a state of creep since 2019, with a mean LOS velocity of $-18\sim-5$ mm/year. Once the countermeasures were implemented, the deformation rates of the slope were reduced based on the data measured from the GNSS stations, inclinometers and crack meters. The deformation rates were also closely associated with the precipitation.

The reactivation of the landslide is attributed to the loose ancient landslide deposits, fragile geo-structure, and large relief, and it was directly triggered by precipitation, while anthropogenic activity plays a secondary role in its reactivation. Shearing of the sliding mass occurred at a depth of 25.3~40.0 m, exploiting a heavily deformed weak zone composed of breccias with poor permeability.

The distribution and characteristics of ancient landslides along the Wenchuan-Maerkang Highway can be explored using the integrated monitor network, which enables the safe construction and operation of such traffic corridors.

The innovation of this study lies in its comprehensive depiction of the entire spatial-temporal evolution process, from the initiation to the significant deformation, of a reactivated landslide along a major road. This case serves as an excellent reference for analyzing and monitoring similar landslide evolution processes before, during, and after remedial measures are implemented. Furthermore, it demonstrates the integration of remote sensing and continuous surveillance and the establishment of a warning system at potential landslide sites. Consequently, it allows for tracing evolutionary history, generating activity maps, and analyzing development trends with remarkable efficacy in reducing landslide risks.

Author Contributions: Y.D. and K.H. framed the study plan and wrote the paper; Y.D., X.H. and H.M. carried out field surveys; Y.D. conducted the InSAR data treatment; X.H. provided feedback on the manuscript. All authors have read and agreed to the published version of the manuscript.

Funding: This research was funded by the National Natural Science Foundation of China (41907225, 41731285) and the National-funded Postdoctoral Researcher Program of China (GZB20230606).

Data Availability Statement: The original contributions presented in the study are included in the article, further inquiries can be directed to the corresponding author.

Acknowledgments: The authors would like to acknowledge the reviewers for their comments and suggestions, which helped in improving the quality of the manuscript.

Conflicts of Interest: Hongsheng Ma is employed by Sichuan Highway Planning, Survey, Design and Research Institute Ltd. The authors declare no conflict of interest.

References

1. Donnini, M.; Napolitano, E.; Salvati, P.; Ardizzone, F.; Bucci, F.; Fiorucci, F.; Santangelo, M.; Cardinali, M.; Guzzetti, F. Impact of event landslides on road networks: A statistical analysis of two Italian case studies. *Landslides* **2017**, *14*, 1521–1535. [CrossRef]
2. Jaiswal, P.; van Westen, C.J.; Jetten, V. Quantitative assessment of landslide hazard along transportation lines using historical records. *Landslides* **2011**, *8*, 279–291. [CrossRef]
3. Penna, D.; Borga, M.; Aronica, G.T.; Brigandì, G.; Tarolli, P. The influence of grid resolution on the prediction of natural and road-related shallow landslides. *Hydrol. Earth Syst. Sci.* **2014**, *18*, 2127–2139. [CrossRef]
4. Yan, Y.; Cui, Y.; Huang, X.; Zhou, J.; Zhang, W.; Yin, S.; Guo, J.; Hu, S. Combining seismic signal dynamic inversion and numerical modeling improves landslide process reconstruction. *Earth Surf. Dyn.* **2022**, *10*, 1233–1252. [CrossRef]
5. Zhang, Y.; Ayyub, B.M.; Gong, W.; Tang, H. Risk assessment of roadway networks exposed to landslides in mountainous regions—A case study in Fengjie County, China. *Landslides* **2023**, *20*, 1419–1431. [CrossRef]
6. Gu, D.M.; Huang, D.; Yang, W.D.; Zhu, J.L.; Fu, G.Y. Understanding the triggering mechanism and possible kinematic evolution of a reactivated landslide in the Three Gorges Reservoir. *Landslides* **2017**, *14*, 2073–2087. [CrossRef]
7. Lu, N.; Wayllace, A.; Oh, S. Infiltration-induced seasonally reactivated instability of a highway embankment near the Eisenhower Tunnel, Colorado, USA. *Eng. Geol.* **2013**, *162*, 22–32. [CrossRef]
8. Massey, C.I.; Petley, D.N.; McSaveney, M.J. Patterns of movement in reactivated landslides. *Eng. Geol.* **2013**, *159*, 1–19. [CrossRef]
9. Jaafari, A.; Najafi, A.; Rezaeian, J.; Sattarian, A.; Ghajar, I. Planning road networks in landslide-prone areas: A case study from the northern forests of Iran. *Land Use Policy* **2015**, *47*, 198–208. [CrossRef]
10. Sun, D.; Gu, Q.; Wen, H.; Xu, J.; Zhang, Y.; Shi, S.; Xue, M.; Zhou, X. Assessment of landslide susceptibility along mountain highways based on different machine learning algorithms and mapping units by hybrid factors screening and sample optimization. *Gondwana Res.* **2023**, *123*, 89–106. [CrossRef]
11. Tanyaş, H.; Görüm, T.; Kirschbaum, D.; Lombardo, L. Could road constructions be more hazardous than an earthquake in terms of mass movement? *Nat. Hazards* **2022**, *112*, 639–663. [CrossRef]
12. He, K.; Ma, G.; Hu, X.; Luo, G.; Mei, X.; Liu, B.; He, X. Characteristics and mechanisms of coupled road and rainfall-induced landslide in Sichuan China. *Geomat. Nat. Hazards Risk* **2019**, *10*, 2313–2329. [CrossRef]

13. Jiang, N.; Li, H.B.; Hu, Y.X.; Zhang, J.Y.; Dai, W.; Li, C.J.; Zhou, J.W. Dynamic evolution mechanism and subsequent reactivated ancient landslide analyses of the "6.17" Danba sequential disasters. *Bull. Eng. Geol. Environ.* **2022**, *81*, 149. [CrossRef]
14. Lee, C.F.; Huang, W.K.; Chang, Y.L.; Chi, S.Y.; Liao, W.C. Regional landslide susceptibility assessment using multi-stage remote sensing data along the coastal range highway in northeastern Taiwan. *Geomorphology* **2018**, *300*, 113–127. [CrossRef]
15. Barla, G.; Debernardi, D.; Perino, A. Lessons learned from deep-seated landslides activated by tunnel excavation. *Geomech. Tunn.* **2015**, *8*, 394–401. [CrossRef]
16. Carlà, T.; Macciotta, R.; Hendry, M.; Martin, D.; Edwards, T.; Evans, T.; Farina, P.; Intrieri, E.; Casagli, N. Displacement of a landslide retaining wall and application of an enhanced failure forecasting approach. *Landslides* **2018**, *15*, 489–505. [CrossRef]
17. Zhang, Y.; Ren, S.; Liu, X.; Guo, C.; Li, J.; Bi, J.; Ran, L. Reactivation mechanism of old landslide triggered by coupling of fault creep and water infiltration: A case study from the east Tibetan Plateau. *Bull. Eng. Geol. Environ.* **2023**, *82*, 291. [CrossRef]
18. Li, W.; Zhao, B.; Xu, Q.; Yang, F.; Fu, H.; Dai, C.; Wu, X. Deformation characteristics and failure mechanism of a reactivated landslide in Leidashi, Sichuan, China, on August 6, 2019: An emergency investigation report. *Landslides* **2020**, *17*, 1405–1413. [CrossRef]
19. Zhang, Z. Mechanism of the 2019 Yahuokou landslide reactivation in Gansu, China and its causes. *Landslides* **2020**, *17*, 1429–1440. [CrossRef]
20. He, K.; Ma, G.; Hu, X.; Liu, B. Failure mechanism and stability analysis of a reactivated landslide occurrence in Yanyuan City, China. *Landslides* **2021**, *18*, 1097–1114. [CrossRef]
21. Nguyen, L.C.; Tien, P.V.; Do, T.N. Deep-seated rainfall-induced landslides on a new expressway: A case study in Vietnam. *Landslides* **2020**, *17*, 395–407. [CrossRef]
22. Carlà, T.; Tofani, V.; Lombardi, L.; Raspini, F.; Bianchini, S.; Bertolo, D.; Thuegaz, P.; Casagli, N. Combination of GNSS, satellite InSAR, and GBInSAR remote sensing monitoring to improve the understanding of a large landslide in high alpine environment. *Geomorphology* **2019**, *335*, 62–75. [CrossRef]
23. Casagli, N.; Intrieri, E.; Tofani, V.; Gigli, G.; Raspini, F. Landslide detection, monitoring and prediction with remote-sensing techniques. *Nat. Rev. Earth Environ.* **2023**, *4*, 51–64. [CrossRef]
24. Wang, W.; Motagh, M.; Mirzaee, S.; Li, T.; Zhou, C.; Tang, H.; Roessner, S. The 21 July 2020 Shaziba landslide in China: Results from multi-source satellite remote sensing. *Remote Sens. Environ.* **2023**, *295*, 113669. [CrossRef]
25. He, K.; Tanyas, H.; Chang, L.; Hu, X.; Luo, G.; Lombardo, L. Modelling InSAR-derived hillslope velocities with multivariate statistics: A first attempt to generate interpretable predictions. *Remote Sens. Environ.* **2023**, *289*, 113518. [CrossRef]
26. He, K.; Lombardo, L.; Chang, L.; Sadhasivam, N.; Hu, X.; Fang, Z.; Dahal, A.; Fadel, I.; Luo, G.; Tanyas, H. Investigating earthquake legacy effect on hillslope deformation using InSAR-derived time series. *Earth Surf. Process. Landf.* **2024**, *49*, 980–990. [CrossRef]
27. Lacroix, P.; Bièvre, G.; Pathier, E.; Kniess, U.; Jongmans, D. Use of Sentinel-2 images for the detection of precursory motions before landslide failures. *Remote Sens. Environ.* **2018**, *215*, 507–516. [CrossRef]
28. Xie, M.; Zhao, W.; Ju, N.; He, C.; Huang, H.; Cui, Q. Landslide evolution assessment based on InSAR and real-time monitoring of a large reactivated landslide, Wenchuan, China. *Eng. Geol.* **2020**, *277*, 105781. [CrossRef]
29. Bian, S.; Chen, G.; Zeng, R.; Meng, X.; Jin, J.; Lin, L.; Zhang, Y.; Shi, W. Post-failure evolution analysis of an irrigation-induced loess landslide using multiple remote sensing approaches integrated with time-lapse ERT imaging: Lessons from Heifangtai, China. *Landslides* **2022**, *19*, 1179–1197. [CrossRef]
30. Shi, X.; Wang, J.; Jiang, M.; Zhang, S.; Wu, Y.; Zhong, Y. Extreme rainfall-related accelerations in landslides in Danba County, Sichuan Province, as detected by InSAR. *Int. J. Appl. Earth Obs.* **2022**, *115*, 103109. [CrossRef]
31. Cook, M.E.; Brook, M.S.; Hamling, I.J.; Cave, M.; Tunnicliffe, J.F.; Holley, R. Investigating slow-moving shallow soil landslides using Sentinel-1 InSAR data in Gisborne, New Zealand. *Landslides* **2023**, *20*, 427–446. [CrossRef]
32. Dong, J.; Zhang, L.; Li, M.; Yu, Y.; Liao, M.; Gong, J.; Luo, H. Measuring precursory movements of the recent Xinmo landslide in Mao County, China with Sentinel-1 and ALOS-2 PALSAR-2 datasets. *Landslides* **2018**, *15*, 135–144. [CrossRef]
33. Cai, J.; Zhang, L.; Dong, J.; Guo, J.; Wang, Y.; Liao, M. Automatic identification of active landslides over wide areas from time-series InSAR measurements using Faster RCNN. *Int. J. Appl. Earth Obs.* **2023**, *124*, 103516. [CrossRef]
34. van Natijne, A.L.; Bogaard, T.A.; van Leijen, F.J.; Hanssen, R.F.; Lindenbergh, R.C. World-wide InSAR sensitivity index for landslide deformation tracking. *Int. J. Appl. Earth Obs.* **2022**, *111*, 102829. [CrossRef]
35. Xia, Z.; Motagh, M.; Li, T.; Peng, M.; Roessner, S. A methodology to characterize 4D post-failure slope instability dynamics using remote sensing measurements: A case study of the Aniangzhai landslide in Sichuan, Southwest China. *ISPRS J. Photogramm. Remote Sens.* **2023**, *196*, 402–414. [CrossRef]
36. Confuorto, P.; Di Martire, D.; Infante, D.; Novellino, A.; Papa, R.; Calcaterra, D.; Ramondini, M. Monitoring of remedial works performance on landslide-affected areas through ground- and satellite-based techniques. *Catena* **2019**, *178*, 77–89. [CrossRef]
37. Di Maio, C.; Fornaro, G.; Gioia, D.; Reale, D.; Schiattarella, M.; Vassallo, R. In situ and satellite long-term monitoring of the Latronico landslide, Italy: Displacement evolution, damage to buildings, and effectiveness of remedial works. *Eng. Geol.* **2018**, *245*, 218–235. [CrossRef]
38. Nappo, N.; Peduto, D.; Mavrouli, O.; van Westen, C.J.; Gullà, G. Slow-moving landslides interacting with the road network: Analysis of damage using ancillary data, in situ surveys and multi-source monitoring data. *Eng. Geol.* **2019**, *260*, 105244. [CrossRef]

39. Stevens, W.R.; Zehrbach, B.E. Inclinometer Data Analysis for Remediated Landslides. In *Geotechnical Measurements: Lab and Field, Proceedings of Sessions of Geo-Denver 2000, Denver, CO, USA, 5–8 August 2000*; American Society of Civil Engineers: Reston, VA, USA, 2000; Volume 294, pp. 126–137.
40. Lollino, P.; Giordan, D.; Allasia, P.; Fazio, N.L.; Perrotti, M.; Cafaro, F. Assessment of post-failure evolution of a large earthflow through field monitoring and numerical modelling. *Landslides* **2020**, *17*, 2013–2026. [CrossRef]
41. Tang, H.; Li, C.; Hu, X.; Su, A.; Wang, L.; Wu, Y.; Criss, R.; Xiong, C.; Li, Y. Evolution characteristics of the Huangtupo landslide based on in situ tunneling and monitoring. *Landslides* **2015**, *12*, 511–521. [CrossRef]
42. Tildy, P.; Neducza, B.; Nagy, P.; Kanli, A.I.; Hegymegi, C. Time lapse 3D geoelectric measurements for monitoring of in-situ remediation. *J. Appl. Geophys.* **2017**, *136*, 99–113. [CrossRef]
43. Rosen, P.A.; Gurrola, E.; Sacco, G.F.; Zebker, H. The InSAR scientific computing environment. In Proceedings of the EUSAR 2012; 9th European Conference on Synthetic Aperture Radar, Nuremberg, Germany, 23–26 April 2012.
44. Yu, C.; Li, Z.; Penna, N.T.; Crippa, P. Generic Atmospheric Correction Model for Interferometric Synthetic Aperture Radar Observations. *J. Geophys. Res. Solid Earth* **2018**, *123*, 9202–9222. [CrossRef]
45. Hooper, A.; Bekaert, D.; Hussain, E.; Spaans, K. *StaMPS/MTI Manual: Version 4.1b*; School of Earth and Environment, University of Leeds: Leeds, UK, 2018.
46. He, K.; Ma, G.; Hu, X.; Liu, B.; Han, M. The July 2, 2017, Lantian landslide in Leibo, China: Mechanisms and mitigation measures. *Geomech. Eng.* **2022**, *28*, 283–298.
47. Zhang, C.; Yin, Y.; Dai, Z.; Huang, B.; Zhang, Z.; Jiang, X.; Tan, W.; Wang, L. Reactivation mechanism of a large-scale ancient landslide. *Landslides* **2021**, *18*, 397–407. [CrossRef]
48. Zhao, B.; Yuan, L.; Geng, X.; Su, L.; Qian, J.; Wu, H.; Liu, M.; Li, J. Deformation characteristics of a large landslide reactivated by human activity in Wanyuan city, Sichuan Province, China. *Landslides* **2022**, *19*, 1131–1141. [CrossRef]
49. Luo, G.; Hu, X.; Bowman, E.T.; Liang, J. Stability evaluation and prediction of the Dongla reactivated ancient landslide as well as emergency mitigation for the Dongla Bridge. *Landslides* **2017**, *14*, 1403–1418. [CrossRef]
50. Wen, H.; Zhao, S.; Liang, Y.; Wang, S.; Tao, L.; Xie, J. Landslide development and susceptibility along the Yunling–Yanjing segment of the Lancang River using grid and slope units. *Nat. Hazards* **2024**. [CrossRef]

remote sensing

MDPI

Article

Landslide Hazard Assessment for Wanzhou Considering the Correlation of Rainfall and Surface Deformation

Xiangjie She, Deying Li *, Shuo Yang, Xiaoxu Xie, Yiqing Sun and Wenjie Zhao

Faculty of Engineering, China University of Geosciences, Wuhan 430074, China; shexiangjie@cug.edu.cn (X.S.); yang_shuo@cug.edu.cn (S.Y.); xiexx@cug.edu.cn (X.X.); syiqing@cug.edu.cn (Y.S.); zhaowenjie@cug.edu.cn (W.Z.)
* Correspondence: lideying@cug.edu.cn

Abstract: The landslide hazard assessment plays a crucial role in landslide risk mitigation and land use planning. The result of landslide hazard assessment corrected by surface deformation, obtained through time-series InSAR, has usually proven to have good application capabilities. However, the issue lies in the uncertainty of InSAR results, where some deformations cannot be calculated, and some are not true deformations. This uncertainty of InSAR results will lead to errors in landslide hazard assessment. Here, we attempt to evaluate landslide hazards by considering combined rainfall and surface deformation. The main objective of this research was to mitigate the impact of bias and explore the accurate landslide hazard assessment method. A total of 201 landslides and 11 geo-environment factors were utilized for landslide susceptibility assessment by support vector machine (SVM) model in Wanzhou District, Three Gorges Reservoir Area (TGRA). The preliminary hazard is obtained by analyzing the statistical data of landslides and rainfall. Based on the SAR image data of Sentinel-1A satellites from September 2019 to October 2021, the SBAS-InSAR method was used to analyze surface deformation. The correlation between surface deformation and rainfall was analyzed, and the deformation factor variables were applied to landslide hazard assessment. The research results demonstrate that the error caused by the uncertainty of InSAR results can be effectively avoided by analyzing the relationship between rainfall and surface deformation. Our results can effectively adjust and correct the hazard results and eliminate the errors in the general hazard assessment. Our proposed method can be used to assess the landslide hazard in more detail and provide a reference for fine risk management and control.

Keywords: landslide hazard assessment; SBAS InSAR; correlation; deformation factor; Three Gorges Reservoir Area

Citation: She, X.; Li, D.; Yang, S.; Xie, X.; Sun, Y.; Zhao, W. Landslide Hazard Assessment for Wanzhou Considering the Correlation of Rainfall and Surface Deformation. *Remote Sens.* **2024**, *16*, 1587. https://doi.org/10.3390/rs16091587

Academic Editor: Sandro Moretti

Received: 13 March 2024
Revised: 14 April 2024
Accepted: 21 April 2024
Published: 29 April 2024

1. Introduction

Landslide disasters are common geological disasters, occurring widely in strong earthquake areas, reservoir areas and mountainous areas [1–4]. Due to the suddenness, concealment and frequency characteristics, the infrastructure and live of the residents are seriously threatened by landslides. Therefore, landslide hazard evaluation plays an important role in land use planning, engineering site selection, and landslide risk mitigation [5,6].

Landslide hazard assessment is a critical part of risk assessment, aiming to evaluate the spatial and temporal probability of landslide occurrence. Generally, the spatial probability of landslide occurrence is delineated by the susceptibility assessment based on the historical and present landslide events, and the temporal probability is analyzed by a physical model or statistical model of previous landslides [7,8]. However, for the large study area, landslide hazard is affected by geo-environment, rainfall, human activities, etc., which means that the precision of the physical model is limited by the parameter uncertainty [9–11]. Therefore, statistical analysis has wider application scenarios. In recent years, the difference and dynamic change in landslide hazard at different periods was noticed by researchers. For instance, the influence of rainfall intensities and meteorological conditions in landslide hazard assessment was studied [12,13]. With the development and maturity of InSAR technology, it has been widely used in the study of monitoring and landslide identification [14–20]. The surface deformation based on InSAR has also been introduced into dynamic landslide hazard assessment [14–16].

The surface deformation of the landslide at a certain time period can reflect the current situation of the landslide hazard. As is well known, time-series InSAR technology has the characteristics of wide coverage, high precision and short periods [17–20]. Therefore, the surface deformation data obtained from time-series InSAR are not only useful for landslide monitoring and identification [21–24] suit for dynamic landslide hazard assessment [25–28]. Yi Zhang et al. used the information on surface deformation extracted by SBAS-InSAR as an internal index to establish a machine-learning model for landslide hazard analysis [16]. Guo Zhang et al. used the WOE model and static factor data for the LS map. They used the contribution matrix to grade and assign values to susceptibility and surface deformation and produced a more accurate and reasonable LS distribution [25]. Yunfeng Shan et al. propose a novel method for assessing landslide hazards by combining deformation results from ascending and descending SAR data images. They constructed hazard matrices for all InSAR surface deformation results for different orbits, thus carrying out a more temporal hazard evaluation [26]. Chao Zhou et al. considered the seasonality of rainfall in the Three Gorges Reservoir Area, used the ground deformation velocity extracted from MT-InSAR, and used an empirical matrix to evaluate the dynamic landslide hazard in two periods of a water year [27]. It demonstrates that it is achievable to incorporate time-series InSAR into the landslide hazard assessment for long-term sustainable assessment and dynamic adjustment.

However, the surface deformation obtained by InSAR technology is uncertain, some of them cannot be calculated, and some are not real landslide deformations. It means that the landslide hazard assessment result will be induced bias by the direct introduction of surface deformation obtained from InSAR. For instance, the surface deformation is only used as an intrinsic index to be added directly into the evaluation model [16] or simply superimposed with other evaluation indicators (spatial probability, rainfall) [25–29]. Especially in the hazard assessment of rainfall-induced landslides, most landslide surface deformations are caused by rainfall. The current research severed the relationship between surface deformation and rainfall events. No attention was paid to the correspondence between surface deformation and rainfall events, which resulted in some errors in the evaluation results. This type of deformation should be effectually introduced in dynamic and precise hazard evaluations rather than ignore the relationship between surface deformation and rainfall events.

Therefore, the intent of this paper is to alleviate the bias caused by directly introducing surface deformation obtained by InSAR in landslide risk assessment by analyzing the correlation between rainfall and surface deformation. The Wanzhou District of the Three Gorges Reservoir Area was selected as the study area. The preliminary landslide hazard assessment result was constructed by the spatial probability (analyzed by landslide susceptibility) and the temporal probability (obtained by rainfall events). The surface deformation from 2019 to 2021 was extracted by PS+SBAS InSAR, and the correlation between surface deformation and rainfall was analyzed. The deformation factor was applied to the preliminary assessment results. Through this assessment framework, more accurate and more practical dynamic assessment results of landslide hazards were provided.

2. Study Area

The study area in Wanzhou is located in the northeastern part of Chongqing, China. The area belongs to the Three Gorges Reservoir Area (TGRA), and the Yangtze River passes through the area from southwest to northeast (Figure 1). It is located in a humid subtropical monsoon climate area with abundant rainfall. The average annual rainfall is 1200 mm, and the maximum annual rainfall is 1641 mm. The maximum monthly rainfall is 683 mm. The rainy season is from April to October, and the cumulative rainfall accounts for about 80–90% of the total annual rainfall. The water level in the Yangtze River has been raised from the original level to 175 m since 2003 due to the construction of the Three Gorges Dam. The reservoir water level fluctuates from 145 m to 175 m every year. The big rise and periodic fluctuations in reservoir water level significantly change the local groundwater flow in the reservoir banks, which often results in landslide occurrence. The landslides in the reservoir banks are influenced by reservoir water level and/or heavy rainfall.

The geology is an important factor affecting landslide occurrence and evolution. It is located on the eastern edge of the Sichuan Basin, where the geological structure is complex with dense fold structure. However, there is no large-scale fracture tectonic movement. The Jurassic and Triassic sedimentary rocks are widely distributed in the area overlain by Quaternary soils, mainly including sandstones and mudstones. The landslides often occur in the mudstone layers, where the soft rocks are easily transferred to soil in heavy rainfall and reservoir water levels. The area is located in a high landslide susceptibility zone. The inventory of 201 landslides in the area was obtained from remote sensing, field investigations and landslide reports, as shown in Figure 1. Detailed landslide records span a period of 20 years from 1997 to 2017, including landslide type, failure time, location, landslide process, loss, etc. The landslides are mainly distributed in the reservoir banks and most of them are colluvial landslides triggered by rainfall and reservoir water level.

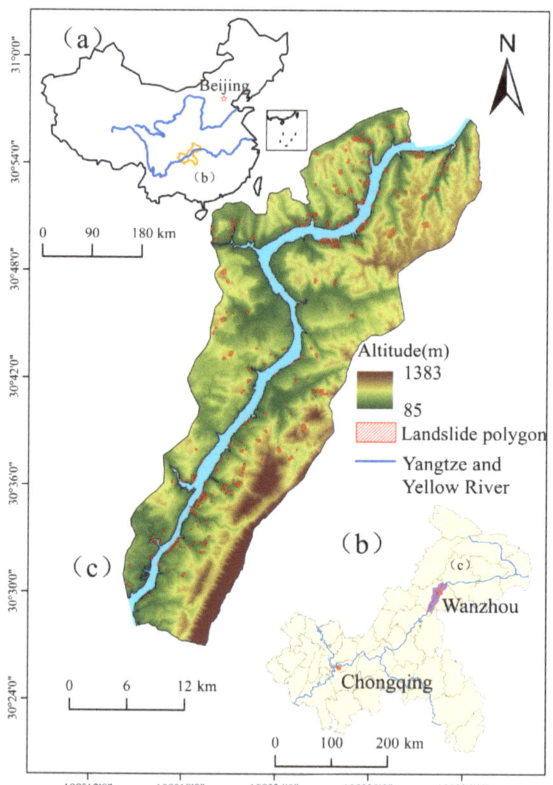

Figure 1. (**a**) Geographic location of the study area. (**b**) Chongqing, China; (**c**) landslides map in the study area.

3. Methods

Rainfall is the main trigger of landslide deformation and landslide failure. For single landslides, the relationship between landslide deformation and rainfall was widely studied based on long-term monitoring data [30]. However, landslide deformation was seldom used for regional landslide hazard assessment. In the paper, the regional landslide deformation obtained from the time-series InSAR method was attempted for landslide hazard analysis. The preliminary landslide hazard assessment considering rainfall trigger factors was first obtained by probabilistic analysis. The final result of the landslide hazard assessment was corrected through the regional landslide surface deformation. The flowchart of landslide hazard assessment is shown in Figure 2. It is mainly divided into four steps: (1) Landslide inventory in the study area was obtained by remote sensing interpretation, field investigation and data collection. The predisposing factors of landslides were determined. The machine learning method (SVM) was adopted to predict the regional landslide susceptibility. (2) The historical rainfall data and landslide records were collected to analyze the temporal probability of landslide occurrence under different rainfall intensities. The preliminary result of landslide hazard evaluation was gained through the combination with the landslide susceptibility result. (3) The Sentinel-1 provides free SAR images for regional image interpretation. The integrated PS+SBAS time-series InSAR method was used to research the surface deformation. The correlation between surface deformation and rainfall was analyzed to provide an effective deformation factor. (4) The empirical matrix considering surface deformation and the preliminary landslide hazard result was constructed to get the final landslide hazard assessment result. In the study, the key is to find out the relationship between surface deformation and rainfall in a certain period and apply it to the landslide hazard assessment.

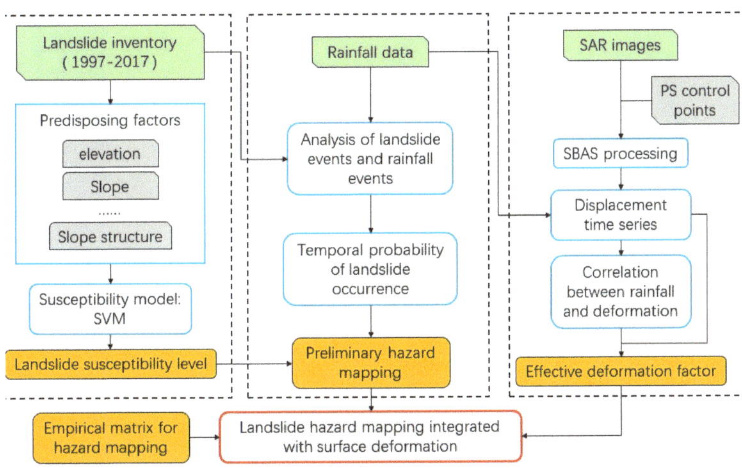

Figure 2. Flowchart of landslide hazard assessment considering surface deformation.

3.1. Landslide Susceptibility Assessment

Landslide susceptibility assessment is the quantification of the spatial probability of landslide occurrence. The basic assumption of landslide susceptibility assessment is that "Future landslide events are more likely to occur under conditions similar to those of previous landslide events" [31]. Therefore, the landslide possibility of the study area was predicted by analyzing the environmental conditions where the landslide events had occurred. Statistical methods and machine learning models are often used in the landslide susceptibility assessment, which are data-driven methods. Among them, support vector machine (SVM) has been widely used in landslide susceptibility evaluation due to its excellent classification ability [27,32,33].

In this paper, the support vector machine was adopted to evaluate the landslide susceptibility. Through the analysis of landslide occurrence and landslide triggers, the 14 variables were extracted from DEM and geology map as landslide susceptibility evaluation factors (Figure 3), including elevation, slope, aspect, topographic wetness index (TWI), topographic roughness index (TRI), distance from gullies, distance from roads, topographic curvature, plane curvature, land use, distance from to rivers, slope structure, lithology and normalized difference vegetation index (NDVI). After multiple covariance analyses and importance calculations, 11 evaluation factors are finally selected, as shown in Table 1. In the study, landslide data and evaluation factors have been rasterized and resampled to 20 m × 20 m pixel grid units. A total of 70% of the known landslides were used for training, and the remaining 30% was for validation. Finally, the ROC index was used to evaluate the landslide susceptibility result.

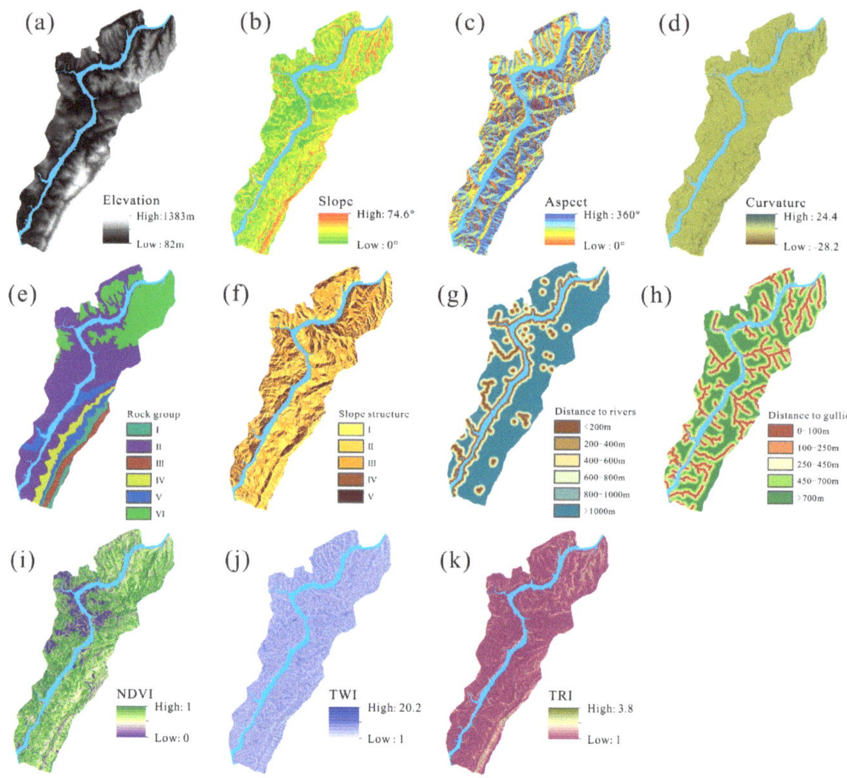

Figure 3. Landslide susceptibility assessment factors. (**a**) Elevation; (**b**) slope; (**c**) aspect; (**d**) curvature; (**e**) engineering geological rock group; (**f**) slope structure; (**g**) distance to rivers; (**h**) distance to gullies; (**i**) NDVI; (**j**) TWI; (**k**) TRI.

Table 1. Predictor factors.

Factors	Class
Elevation	137–350 m, 350–550 m, 550–780 m, 780–1141 m
Slope	0–5°, 5–15°, 15–30°, 30–45°, >45°
Aspect	0–45°, 45–90°, 90–135°, 135–180°, 180–225°, 225–270°, 270–315°, 315–360°
Curvature	<−2, −2–0, 0–3, >3
Engineering geological rock group	I (hard thick limestone rock group), II (hard thick sandstone intercalated with shale rock group), III (hard middle argillaceous limestone rock group), IV (soft thin mudstone shale rock group), V (hard thick lithic sandstone rock group), VI (harder middle sandstone rock group)

Table 1. *Cont.*

Factors	Class
Slope structure	I: dip slope, II: dip-diagonal slope, III: transverse slope, IV: reverse-diagonal slope, V: reverse slope
Topographic wetness index (TWI)	<5.9, 5.9–8.3, 8.3–12, >12
Topographic roughness index (TRI)	1–1.2, 1.2–1.4, 1.4–3.7
Normalized difference vegetation index (NDVI)	Bare land (0–0.3), low vegetation cover (0.3–0.5), medium vegetation coverage (0.5–0.7), high coverage vegetation (0.7–1.0)
Distance to rivers	<200 m, 200–400 m, 400–600 m, 600–800 m, 800–1000 m, >1000 m
Distance to gullies	0–100 m, 100–250 m, 250–450 m, 450–700 m, >700 m

3.2. Landslide Hazard Analysis Based on Rainfall

Landslide hazard assessment (LHA) aims to determine the spatial-temporal probability of landslide occurrence in the target area within a specified period of time. The frequency or return period of landslide events under specific triggering conditions of rainfall or earthquake was considered in the common practice [27,34,35]. The landslide hazard index H in relation to the triggering factor of rainfall can be expressed as follows:

$$H = P(s) \times P(t) \times P(e) \tag{1}$$

where P(s) is the spatial probability of landslide occurrence, also known as the landslide susceptibility index; P(t) is the temporal probability of a specific triggering event. When a specific triggering event was considered to have occurred, P(t) was equal to 1; P(e) is the probability of landslide failure in response to the triggering event.

In this study, 201 landslides in the area that had occurred from 1997 to 2017 are selected as samples. Among them, there were 159 rainfall-induced landslides and 41 landslides induced by a combination of rainfall, reservoir levels or human engineering activities. Therefore, rainfall was the main trigger for the landslides in the area. The landslide hazard assessment was conducted by considering rainfall as a landslide-inducing event. The daily rainfall records from 1997 to 2017 in the area were collected in the landslide inventory. The occurrence of landslide events is not only related to the rainfall on the current day of landslide occurrence but also influenced by the previous cumulative rainfall [36]. Liu (2014) had already demonstrated that the landslide event in Wanzhou was mostly correlated with the antecedent 5-day total rainfall [37]. Therefore, when comparing the probability of rainfall events with the probability of landslide occurrence, we use it to represent rainfall intensity.

3.3. Time-Series InSAR

The time-series InSAR is a more prevalent approach for measuring surface deformation and monitoring due to the capture of a broad spectrum of deformation with high precision [38–40]. The permanent scatterer interferometry (PS-InSAR) technique was proposed by Ferretti and colleagues [41]. This method often chooses one image as the primary one and the others as secondary ones from a series of multiple images over the same area. Stable and permanent scattering points were achieved by registering and interfering with both the primary and secondary images because ground objects can maintain strong, stable electromagnetic scattering properties over a long time. The phase of surface deformation is determined by analyzing the phase information of the permanent scattering points and removing the noise phase. Small Baseline Subset-Interferometric Synthetic Aperture Radar (SBAS-InSAR) is a time-series analysis method based on the nature of distributed scatterers proposed by Berardino et al. [42] It can reduce the de-correlation induced by long spatial or temporal baselines and mitigate the effects of atmospheric effects and terrain errors. The time-series ground deformation is achieved through the least squares method and the singular value decomposition method (SVD) to guarantee millimeter-level accuracy.

Both PS-InSAR and SBAS-InSAR methods have demonstrated excellent behavior in obtaining landslide surface deformation. However, due to the complex terrain, dense gullies, large fluctuations in elevation and lush vegetation, both traditional analytical methods have their limitations in high mountainous areas [43]. PS technology can generate ground control points with high coherence and it is mainly applied for obtaining linear deformation. Although SBAS technology has some subjectivity in selecting ground control points, it performs well in analyzing distributed targets. In order to develop the advantages of PS-InSAR and SBAS-InSAR, the two time-series InSAR methods were both used to obtain surface deformation in the study area. The SBAS-InSAR technology was used in the data processing. In order to enhance the accuracy of SBAS-InSAR processing, the high-coherence ground control points were obtained through PS-InSAR technology in the orbit refinement, which can eliminate residual phase information.

The Sentinel-1A satellite image data, Digital Elevation Model (DEM) and precise orbit data of Sentinel satellites were used in the analysis of time-series InSAR. The Sentinel-1 mission consists of two near-polar orbiting satellites with C-band synthetic aperture radar sensors, launched by the European Space Agency in 2014. Each satellite has a revisit time of 12 days. A total of 57 ascending orbit images of Sentinel-1A from September 2019 to October 2021 were collected for the time-series InSAR analysis to calculate the deformation results of the LOS direction (VSLOPE) in the study area.

3.4. Correlation Analysis between Rainfall and Surface Deformation

In the study area, rainfall was considered the main factor leading to landslide deformation and failure [44]. For rainfall-induced landslides, there is a certain correlation between rainfall and landslide surface deformation. The correlation analysis between two variables can be conducted by an analytical method. Pearson's correlation coefficient method is a quantitative statistical approach capable of measuring the correlation between two variables in spatial data, providing sufficient interpretability [45–47]. In this study, Pearson's coefficient is used to explore the correlation between rainfall and surface deformation in the area. By statistically analyzing multiple sets of cumulative rainfall and surface deformation over the entire time period, the variables (X_i, Y_i) $(i = 1, 2, 3, \ldots, n)$ were used to represent the two variables. The correlation coefficient can be determined using the following equation:

$$r = \frac{\sum_{i=1}^{n}\left(X_i - \overline{X}\right)\left(Y_i - \overline{Y}\right)}{\sqrt{\sum_{i=1}^{n}\left(X_i - \overline{X}\right)^2}} \quad (2)$$

The correlation coefficient r obtained from the above formula ranges from -1 to 1. The closer $|r|$ is to 1, the higher the linear correlation between X and Y. If $r = 1$, it indicates a completely positive linear correlation between x and y. If $r = -1$, it indicates a completely negative linear correlation between x and y. The spatial relationship between rainfall and surface deformation in the area was represented by the correlation coefficient. A higher correlation coefficient 'r' at a specific area implies a stronger correlation between rainfall and surface deformation, which indicates that rainfall is the primary factor causing surface deformation. Conversely, a smaller 'r' value indicates a weaker correlation between surface deformation and rainfall, implying that rainfall is not the primary factor causing surface deformation.

The period of Sentinel satellite images covers from 27 September 2019 to 10 October 2021. Considering the timing and revisit period of the Sentinel satellite, the period of approximately two years was divided into 20 time units. Each time unit is 36 days, and it is three cycles of Sentinel image, as shown in Figure 4. Based on collected rainfall data and the InSAR deformation results, correlation analysis between the accumulated rainfall and surface deformation in the area within these 20 time units was conducted.

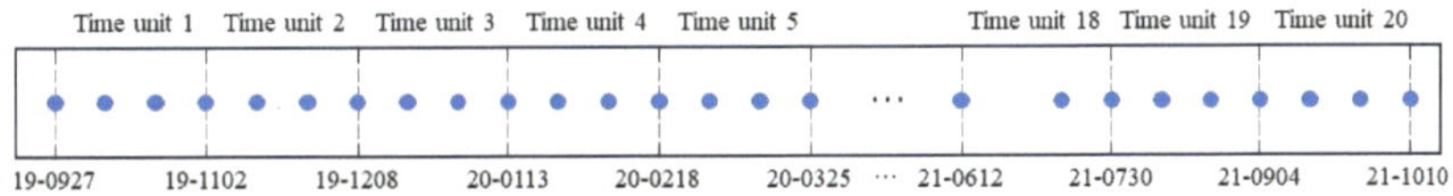

Figure 4. Distribution of time units and images.

3.5. Landslide Hazard Evaluation Considering Surface Deformation Factor

Landslide hazard assessment may become very complex due to different landslide triggers, such as rainfall, earthquakes or human activities. For example, landslide hazards may increase due to some areas undergoing surface deformation induced by human engineering activities or rainfall. Therefore, considering landslide deformation areas induced by rainfall during a specific time period becomes more precise and comprehensive in the landslide hazard assessment.

The paper explores how to combine surface deformation data and rainfall data to address such problems. Typically, in the time-series InSAR data, the positive deformation values indicate that the object is moving toward the satellite, while the negative deformation values indicate movement in the opposite direction. Researchers used positive deformation values to represent surface uplift and negative deformation values to represent subsidence. Extensive research and monitoring data demonstrated that rainfall typically promotes landslide deformation. In the source area of landslides, it generally manifests as depression and subsidence, which in InSAR deformation data appears as a movement away from the satellite, yielding negative values. Therefore, during the creeping stage of landslides, rainfall promotes downward and subsidence deformation of landslides, which in InSAR deformation data, due to its negative values, is manifested numerically as increased rainfall leading to increased subsidence, resulting in decreased InSAR deformation values, thus exhibiting a negative correlation numerically. Conversely, data that correlate positively with InSAR deformation data represent surface uplift, but in reality, there is no physical connection between surface uplift and rainfall. Therefore, InSAR deformation data that correlate positively with rainfall cannot reflect landslide deformation, indicating any correlation with landslide deformation. Based on this practical significance, this paper makes the following numerical treatment:

$$D_f = V_{nor} + |r'| \tag{3}$$

where D_f is the deformation factor, V_{nor} is the normalized value of the surface deformation rate and r' is the correlation coefficient after removing the positive value. According to the previous analysis, when the correlation coefficient r is negative and its value is closer to -1, it shows that the positive correlation between the increase of rainfall and the increase of surface deformation is greater, while the positive value of the correlation coefficient show that the two are negatively correlated, which is of no practical significance. Therefore, we remove the positive value in the correlation coefficient as the result of the correlation coefficient with practical significance and take the absolute value to calculate and construct our deformation factor. The value of the deformation factor represents the danger degree of the surface deformation in an area where rainfall is considered. The higher the value of the deformation factor, the more dangerous the deformation area is during rainfall conditions.

The deformation factor can be derived from the above equation, and be graded into four classes according to the actual situation of the study area. The landslide hazard matrix was established (Table 2). Among them, H1–H4 represents the preliminary landslide hazard class (from low to high), and Df1–Df4 represents the deformation factor class (from low to high). And the 1–4 in the table indicate represents the final hazard class (from low to high).

Table 2. Landslide hazard evaluation matrix.

	Df4	Df3	Df2	Df1
H4	4	4	3	2
H3	4	3	2	1
H2	4	3	2	1
H1	4	3	1	1

4. Results

4.1. Preliminary Landslide Hazard Assessment

4.1.1. Landslide Susceptibility Assessment

Based on the landslide inventory, landslide susceptibility assessment considering 11 evaluation factors was obtained by the SVM method, as shown in Figure 5. The spatial probability for predicting landslide occurrence ranges from 0 to 0.992. The areas where the spatial probability value is greater than 0.8 cover 9.05% of the whole area. Most of these areas are distributed on both sides of the Yangtze River and the gullies. The areas where the spatial probability value is less than 0.2 cover 61.60% of the whole area. Most of them are far away from the Yangtze River. Through the predicted landslide susceptibility map, the distance to the river and the distance to the gully are two major factors in landslide occurrence.

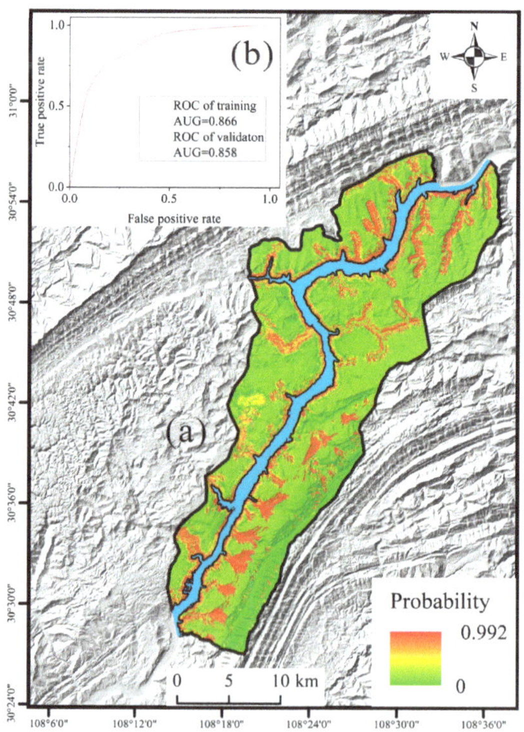

Figure 5. (**a**) Landslide susceptibility map produced by SVM; (**b**) the ROCs of SVM.

The Receiver Operating Characteristic (ROC) curve and the Area Under the ROC Curve (AUC) are used to evaluate the performance of the training model. The ROC curves of training and validation in this study are shown in Figure 5b. Their corresponding AUC values are 0.866 and 0.858, which are greater than 0.8. The statistical result of the susceptibility is shown in Figure 6. The mean and standard deviation (SD) of the susceptibility results are 0.3093 and 0.2693, respectively. The susceptibility probability is mainly distributed in 0.1–0.2 and 0.8–0.9, and the discrimination is good. These result, including standard deviation, proves that the trained model has a good performance and also verifies the reliability of landslide susceptibility assessment.

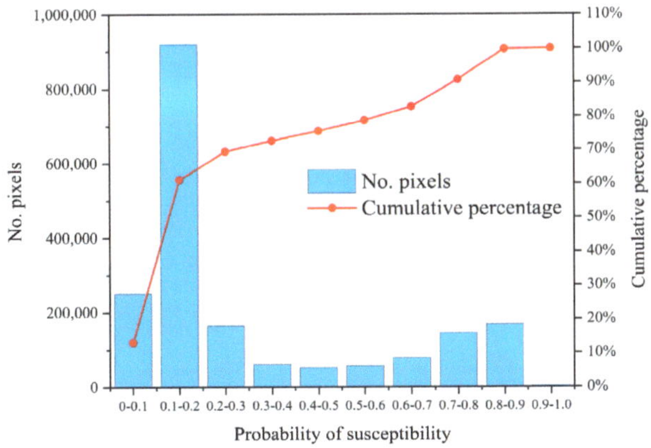

Figure 6. Statistic chart of landslide susceptibility probability.

4.1.2. Landslide Hazard Assessment Based on Rainfall

According to the definition of landslide hazard (Equation (1)), the probability of landslide occurrence under certain specific rainfall conditions was selected as the temporal probability of landslide occurrence. The recorded rainfall data from 1997–2017 was statistically analyzed. The 5-day antecedent total rainfall was divided into four levels: 0 mm, 0–50 mm, 50–100 mm and >100 mm (Table 3). During these 20 years, rainfall events of 0–50 mm occurred 5608 times, while rainfall events of 50–100 mm and >100 mm occurred 545 and 85 times, respectively. When the 5-day antecedent total rainfall was less than 50 mm, 108 landslide events occurred in the area, which accounted for 55.96% of the total landslide events. A total of 59 landslide events occurred for the 5-day antecedent total rainfall of 50–100 mm, which accounted for 30.57% of the total landslides. There were 26 landslide events for the 5-day antecedent rainfall of larger than 100 mm. It accounted for 13.47% of the total landslides. The temporal probabilities of landslide occurrence under the four different rainfall conditions are 0, 0.0193, 0.1038 and 0.3059.

According to Equation (1), the landslide hazard under different rainfall conditions was calculated based on the landslide susceptibility analysis and the temporal probability of landslide occurrence. Based on field investigation experience, the landslide hazard index was divided into four categories: H1 (0~0.0085), H2 (0.0085~0.0607), H3 (0.0607~0.0883) and H4 (>0.0883). The larger the landslide hazard index, the higher the landslide hazard zone. The preliminary landslide hazard map was obtained and shown in Figure 7.

Table 3. Statistics on rainfall events and landslide occurrence.

	5-day Antecedent Total Rainfall			
	0	**0–50**	**50–100**	**>100**
Rainfall event	1062	5608	545	85
Percentage	14.55%	76.82%	7.47%	1.16%
Landslide event	0	108	59	26
Percentage	0	55.96%	30.57%	13.47%
Time probability	0	0.019258	0.10826	0.30588

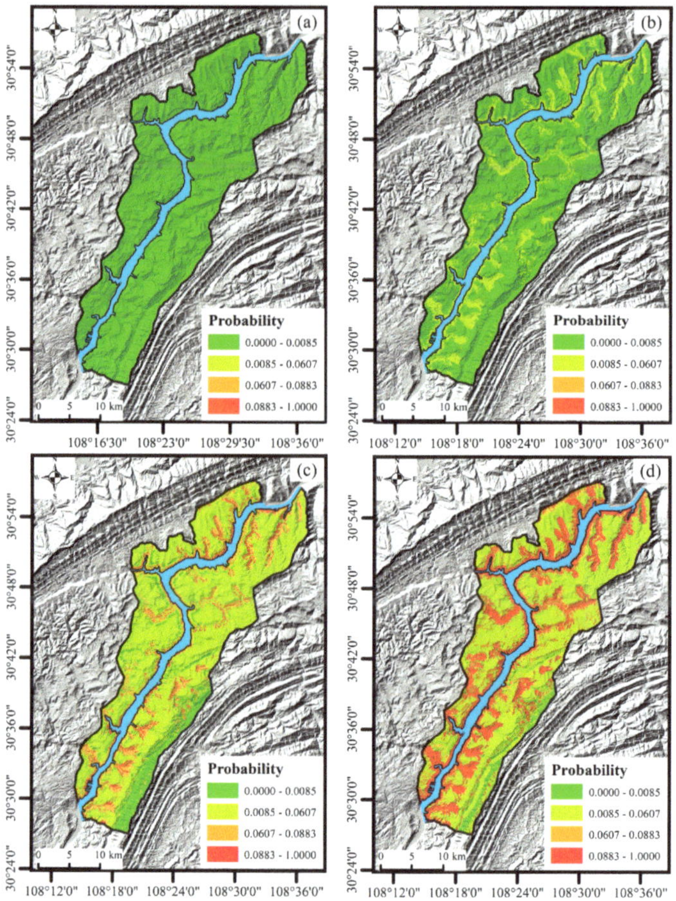

Figure 7. Preliminary landslide hazard under different rainfall conditions. (**a**) Rainfall identity of 0; (**b**) rainfall identity of 0–50; (**c**) rainfall identity of 50–100; (**d**) rainfall identity of larger than 100 mm.

4.2. Landside Hazard Assessment Considering Surface Deformation

4.2.1. Surface Deformation and Correlation Analysis

The ground control points were extracted from PS-InSAR and applied in the first inversion of SBAS-InSAR. In order to obtain more comprehensive surface deformation in the area, interferometric image pairs with better interferometric effects and more coherent areas were carefully selected for subsequent processing. The final annual surface deformation velocity is shown in Figure 8. The LOS velocities range from −97 mm/year to 70 mm/year. The areas where deformation velocity ranges from −10 mm/year to 10 mm/year account for 63.62% of the study area. For the south and northeast areas with inconsistent deformation points, the interpolation method was applied to obtain a comprehensive surface deformation rate in these areas.

In order to analyze the correlation between surface deformation and rainfall, statistical analysis of surface deformation and rainfall within each time unit was conducted. For surface deformation, the difference in surface deformation was obtained between the two images before and after the time unit. For areas without deformation monitoring points, the cumulative surface deformation for that time unit was calculated using ordinary kriging interpolation. Rainfall data were interpolated in the study area based on 12 rainfall monitoring points to obtain the cumulative rainfall distribution in each time unit.

Pearson's correlation coefficient formula (Equation (2)) was used for raster calculations on each pixel. The distribution of correlation coefficients between surface deformation and rainfall was obtained in the study area, and it is shown in Figure 9. In the figure, positive values indicate a positive correlation between rainfall and surface deformation, while negative values indicate a negative correlation. The correlation coefficient ranges from −0.2 to 0.2, which accounts for 68.94% of pixels, indicating a relatively low correlation between surface deformation and rainfall in most areas.

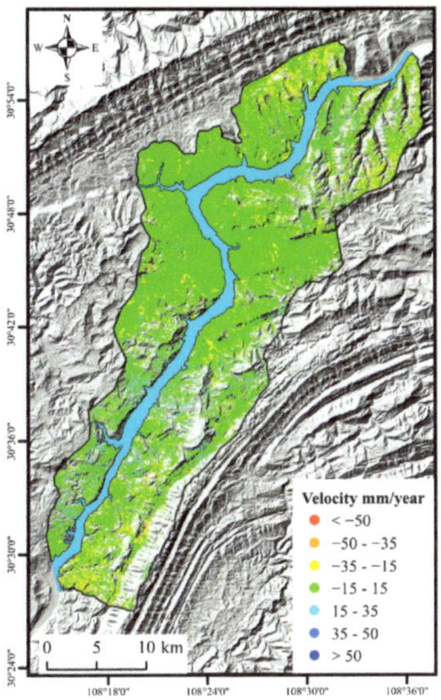

Figure 8. Annual mean rate of surface deformation.

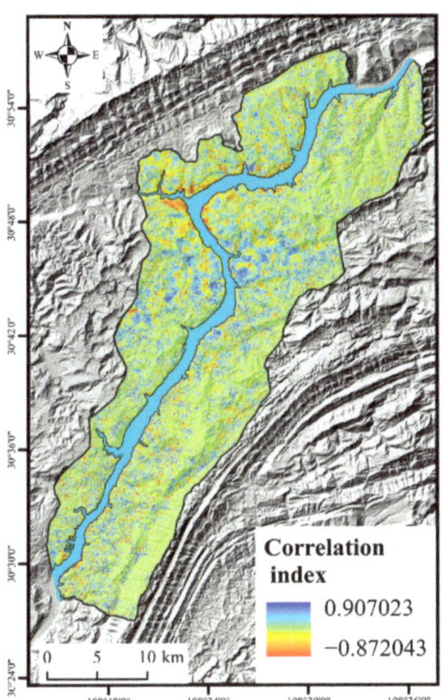

Figure 9. Distribution of correlation coefficient between surface deformation and rainfall.

4.2.2. Landslide Hazard Assessment Considering Surface Deformation

According to the landslide hazard assessment method described above (Equation (3)), the annual surface deformation rate and the processed correlation coefficient are normalized. They are equivalently summed to obtain the final deformation factor map (Figure 10). The deformation factors were divided into four levels: very low-Df1 (0~0.1), low-Df2 (0.1~0.3), moderate-Df3 (0.3~0.6) and high-Df4 (>0.6). Based on the proposed landslide hazard assessment matrix, the final landslide hazard map was obtained considering surface deformation and rainfall, which is shown in Figure 11.

Figure 12 presented statistical results on the preliminary and final landslide hazard. There is a close relationship between rainfall intensity and landslide hazard degree. With increasing rainfall, the areas with low hazards decrease while those with high hazards increase. In the preliminary landslide hazard assessment, 31.05% of pixels were at a high hazard level when the 5-day antecedent total rainfall exceeded 100 mm, while only 4.11% of pixels were at a very low hazard level. However, no area was at a high hazard level when the 5-day antecedent total rainfall was less than 50 mm.

The same characteristic was exhibited in the final landslide hazard assessment considering the deformation factor. When the 5-day antecedent total rainfall was less than 50 mm, only 6.41% of the areas were at a moderate risk level or above. When the 5-day antecedent total rainfall exceeded 100 mm, the proportion of areas at a moderate hazard level or above was 15.98%. At the same rainfall intensity, the proportion of areas at a high or moderate hazard level in the final landslide hazard assessment was smaller than in the preliminary hazard assessment. In the preliminary landslide hazard result, the proportion of pixels with high landslide hazard levels was 0%, 0%, 6.61% and 31.05% under four different rainfall conditions. In the final landslide hazard result, the proportion of pixels with high hazard levels was 0.44%, 0.44%, 1.13% and 2.78%. For rainfall intensity of larger than 100 mm, the proportion with a high hazard level in the final hazard result was significantly lower than that in the preliminary landslide hazard. After considering the surface deformation, the overall hazard in the area was relatively low.

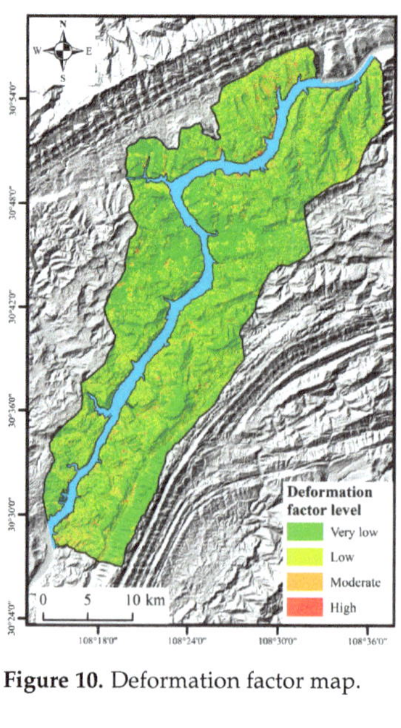

Figure 10. Deformation factor map.

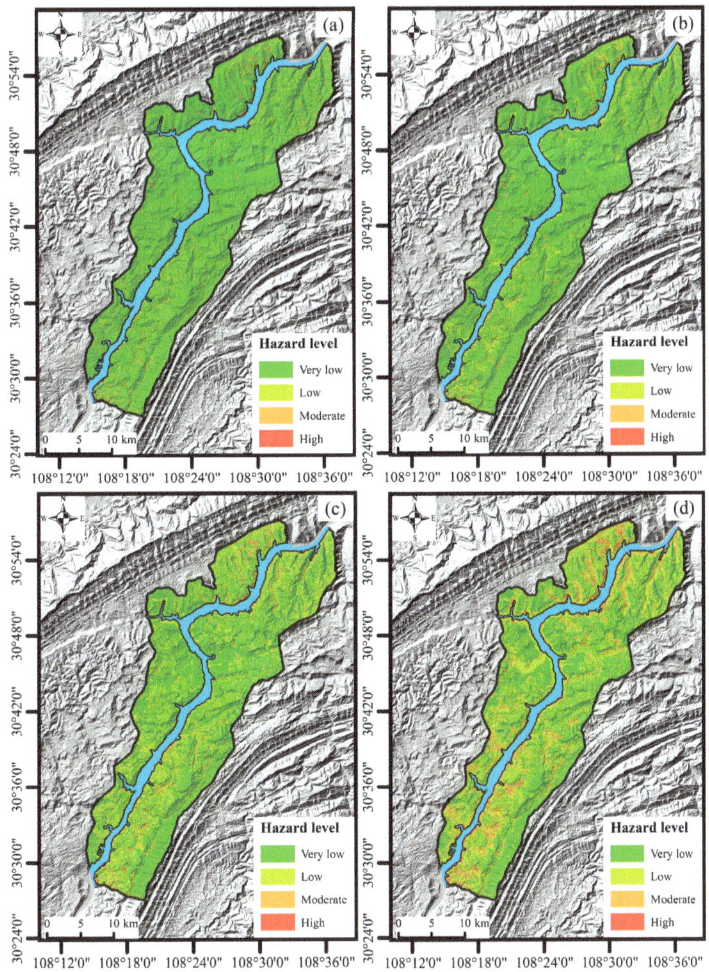

Figure 11. Final landslide hazard map for different rainfall conditions. (**a**) Rainfall intensity of 0; (**b**) rainfall intensity of 0–50; (**c**) rainfall intensity of 50–100; (**d**) rainfall intensity of larger than 100 mm.

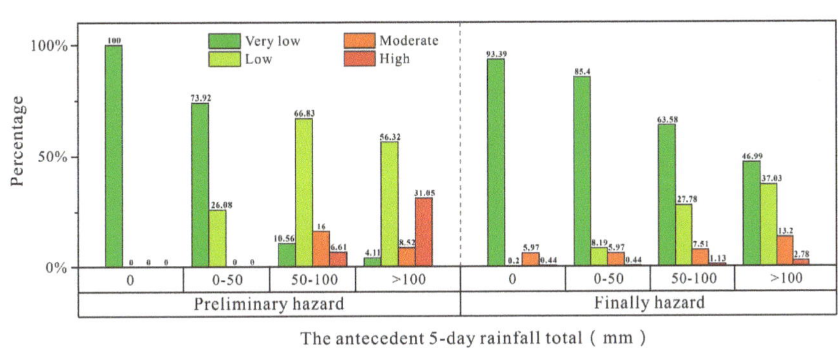

Figure 12. Landslide hazard results statistics for different rainfall conditions.

5. Discussions

Based on time-series InSAR technology, the surface deformation information in the area was obtained during the specific time period. These deformations not only reflect slope deformations caused by rainfall but also those caused by other triggering factors such as reservoir water fluctuation. When directly applying the surface deformation to landslide hazard assessment, these 'non-landslide' deformations may result in a mis-assessment, which leads to 'false positives' and 'false negatives' errors [28]. For example, it may overestimate the hazard of non-landslide deformation areas or underestimate the hazard of slow-moving landslides with less obvious deformation features, both of which can impact landslide hazard assessment results.

For small rainfall intensity, the overall landslide hazard in the preliminary hazard assessment was low, and the proportion of pixels with medium and high landslide hazard levels were both 0. However, there were still some areas with medium or high hazard levels in the final landslide hazard assessment. The number of landslides was counted at different landslide hazard levels, shown in Figure 13. In the preliminary landslide hazard assessment (Figure 13a), the landslide hazard levels undergo drastic changes with the increase in rainfall intensity. When the 5-day antecedent rainfall total was 0, the hazard level of all landslides was very low. When the 5-day antecedent total rainfall would exceed 100mm, the majority of landslides had a high hazard level. In the final landslide hazard assessment (Figure 13b), the number of landslides with the increase in rainfall intensity shows a gradual growth trend at low, moderate and high hazard levels.

Figure 14 illustrates the proportion of landslide pixels to the hazard pixels in the study region at different landslide hazard levels. In the preliminary landslide hazard level, the overall proportion is relatively low, and with cases of extreme fluctuations, the hazard assessment result was not stable enough. In the final landslide hazard level, the proportion shows the characteristics of stable low-hazard proportions and high-hazard proportions, except that the proportion of high is reduced when the 5-day antecedent rainfall total was 100 mm. This implies that the final landslide hazard assessment result was relatively stable, and the accuracy of assessing high landslide hazard areas was also higher. For small rainfall intensity, the final landslide hazard result considered those locations with significant deformation to have a certain level of hazard due to external factors. For larger rainfall intensity, the overall landslide hazard level in the final landslide hazard assessment was a noticeable decrease. However, some visibly deformed high-hazard areas were not overlooked. This indicates that incorporating the deformation factor can effectively adjust the landslide hazard based on the existing deformation.

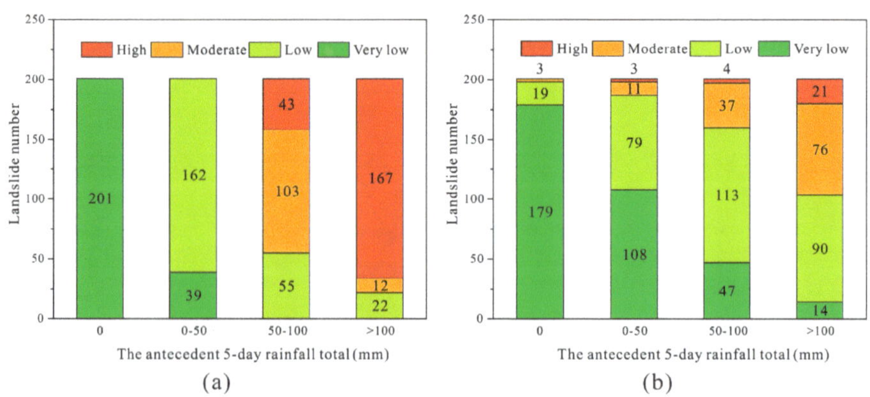

Figure 13. Number of landslides at different landslide hazard levels. (**a**) Preliminary landslide hazard result; (**b**) final landslide hazard result.

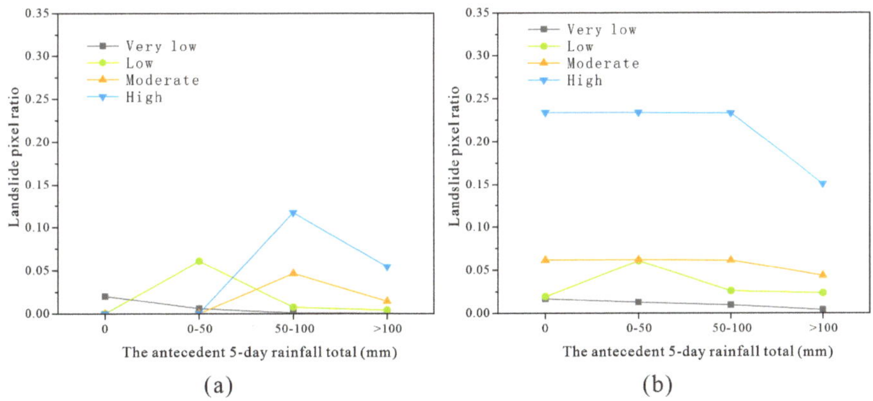

Figure 14. Proportion of landslide pixels to hazard pixels at different landslide hazard levels. (**a**) preliminary landslide hazard result; (**b**) final landslide hazard result.

In order to further validate the effectiveness of this method, a step-by-step comparative analysis combining preliminary landslide hazard, surface deformation, correlation coefficient, deformation factor and final landslide hazard was performed for the 5-day antecedent total rainfall exceeding 100 mm. Several regions with significant evaluation characteristics were identified in the process, as shown in Figure 15. In Figure 15a, the deformation at this location was caused by slope excavation near the airport, and the deformation was not directly related to rainfall through correlation analysis. The deformation factor indicated that the main deformation impact zone was near the excavated slope. Although the preliminary landslide hazard was relatively low, the landslide hazard ultimately increased near the excavation site after considering the slope excavation. In Figure 15b, there was a large number of deformation areas near the reservoir bank due to a construction project, but correlation analysis indicated that these deformations were not directly related to rainfall. There was an initial high hazard level in the preliminary landslide hazard assessment. However, the moderate landslide hazard result was obtained due to these deformations caused by human engineering activities. If the hazard of the area was evaluated directly on the basis of the surface deformation without a correlation analysis, the area might be considered to have the highest level of hazard. In Figure 15c, the overall surface deformation in the landslide area was relatively small. However, the correlation analysis revealed a close relationship between the surface deformation at certain locations and rainfall. The final result aligned with the current deformation state of the landslide's hazard. If the hazard of the area was evaluated only by the surface deformation, those locations where the surface deformation varies significantly with rainfall might be neglected, thus obtaining a lower hazard, which is very unreasonable. In Figure 15d, significant surface deformation was observed in the landslide. Deformation related to rainfall is mainly concentrated in specific locations at the trailing edge and the leading edge of the landslide. This indicated that other larger deformations may be caused by human activities such as land cultivation, which resulted in a more precise distribution of landslide hazards.

By utilizing the correlation analysis between rainfall and surface deformation to construct the deformation factor, effective surface deformation information could be identified, especially in cases where landslides were triggered by rainfall or human engineering activities. The method for landslide hazard assessment could establish the direct correction with the preliminary landslide hazard and avoid errors caused by InSAR anomalies. The landslide hazard assessment result was closer to that of the deformation areas of slopes and landslides. The areas with minor deformation but significantly influenced by rainfall were not neglected. Compared to the general method of hazard assessment in conjunction with surface deformation, the areas with large surface deformation or high landslide preliminary hazard but unrelated to rainfall-induced landslide deformation were not overestimated. Therefore, this method would enhance the efficiency of landslide hazard assessment.

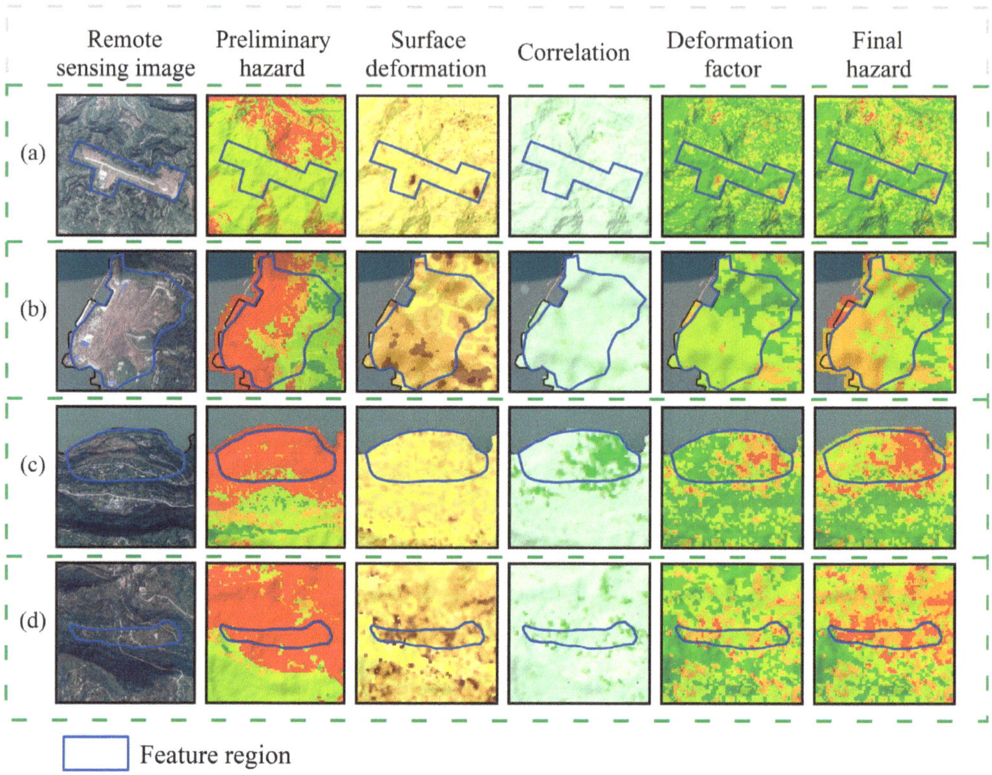

Figure 15. Comparison of preliminary hazard, surface deformation and final landslide hazard. The (**a–d**) represent four different typical regions.

6. Conclusions

In order to obtain more accurate landslide hazard assessment results, the correlation analysis of rainfall and surface deformation was applied to the landslide hazard assessment based on time series InSAR and landslide spatial-temporal probability. The research results indicate that (i) the regional landslide hazard level increases with the intensity of rainfall in preliminary hazard assessment. When the antecedent 5-day rainfall total exceeds 100 mm, 31.05% of the area is classified as a high landslide hazard subarea, while only 4.11% of the area is classified as a very low landslide hazard subarea. (ii) The final landslide hazard assessment result constructed by preliminary hazard and the deformation factor effectively adjusts and corrects the existing hazard levels in some subarea. When the antecedent 5-day rainfall total is less than 50 mm, the hazard level for 6.41% of the areas is moderate or higher. When the antecedent 5-day rainfall total exceeds 100 mm, the hazard level for 15.98% of the areas is moderate or higher, with a high-risk proportion of 2.78%. Overall, the distribution of landslide hazards is more reasonable.

As discussed above, it is possible to accurately determine the landslide hazard level, effectively eliminating potential bias in landslide hazard assessment by introducing the deformation factor obtained through the correlation analysis between the surface deformation and rainfall. Compared with general hazard assessment, this method can effectively characterize the current state of deformation and potential hazard state of landslides and slopes, and this method can reduce the hazard misclassification due to landslide information caused by different triggers of surface deformation, thus determining the landslide hazard class more accurately. The landslide hazard assessment method proposed in this study significantly enhances the reliability of hazard assessments, especially in areas where the time-series InSAR method is available and applicable. In the future, it is necessary to delve into the various influencing factors of landslides and the periodic characteristics of these factors to achieve more precise dynamic hazard assessments.

Author Contributions: X.S. processed the dataset, carried out the analysis, and wrote the original draft of this manuscript. W.Z. helped process and analyze some of the data. X.X., S.Y. and Y.S. provided full guidance on experimental processing and manuscript writing. D.L. revised and improved this manuscript. All authors contributed to the study's conception and design. All authors have read and agreed to the published version of the manuscript.

Funding: This research was funded by the project "Geological Disasters Quantitative Hazard Prediction and Dynamic Risk Evaluation of on Random Slopes" (grant no. 2023YFC3007203), which is financed by National Key Research and Development Program of China.

Data Availability Statement: The Sentinel-1A images and precise orbit data were provided by the European Space Agency. Data such as rainfall and landslides are provided by the project that supported this article. If readers want to have the data, they can request it by e-mail from the authors.

Acknowledgments: We thank Jingxue Chen for her help in mapping. We thank Bingbing Ni for providing valuable suggestions in writing this article.

Conflicts of Interest: The authors declare that they have no known competing financial interests or personal relationships that could have appeared to influence the work reported in this paper.

References

1. Petley, D.N.; Hearn, G.J.; Hart, A.; Rosser, N.J.; Dunning, S.A.; Oven, K.; Mitchell, W.A. Trends in landslide occurrence in Nepal. *Nat. Hazards* **2007**, *43*, 23–44. [CrossRef]
2. Fan, X.; Scaringi, G.; Korup, O.; West, A.J.; Van Westen, C.J.; Tanyas, H.; Hovius, N.; Hales, T.C.; Jibson, R.W.; Allstadt, K.E.; et al. Earthquake-induced chains of geologic hazards: Patterns, mechanisms, and impacts. *Rev. Geophys.* **2019**, *57*, 421–503. [CrossRef]
3. Segoni, S.; Piciullo, L.; Gariano, S.L. A review of the recent literature on rainfall thresholds for landslide occurrence. *Landslides* **2018**, *15*, 1483–1501. [CrossRef]
4. Fell, R.; Corominas, J.; Bonnard, C.; Cascini, L.; Leroi, E.; Savage, W. Guidelines for landslide susceptibility, hazard and risk zoning for land use planning. *Eng. Geol.* **2008**, *102*, 85–98. [CrossRef]
5. Constantin, M.; Bednarik, M.; Jurchescu, M.C.; Vlaicu, M. Landslide susceptibility assessment using the bivariate statistical analysis and the index of entropy in the Sibiciu Basin (Romania). *Environ. Earth Sci.* **2011**, *63*, 397–406. [CrossRef]
6. Esposito, C.; Mastrantoni, G.; Marmoni, G.M.; Antonielli, B.; Caprari, P.; Pica, A.; Schilirò, L.; Mazzanti, P.; Bozzano, F. From theory to practice: Optimisation of available information for landslide hazard assessment in Rome relying on official, fragmented data sources. *Landslides* **2023**, *20*, 2055–2073. [CrossRef]
7. Reichenbach, P.; Rossi, M.; Malamud, B.; Mihir, M.; Guzzetti, F. A review of statistically based landslide susceptibility models. *Earth-Sci. Rev.* **2018**, *180*, 60–91. [CrossRef]
8. Carrión-Mero, P.; Montalván-Burbano, N.; Morante-Carballo, F.; Quesada-Román, A.; Apolo-Masache, B. Worldwide research trends in landslide science. *Int. J. Environ. Res. Public Health* **2021**, *18*, 9445. [CrossRef] [PubMed]
9. Torizin, J.; Wang, L.; Fuchs, M.; Tong, B.; Balzer, D.; Wan, L.; Kuhn, D.; Li, A.; Chen, L. Statistical landslide susceptibility assessment in a dynamic environment: A case study for Lanzhou City, Gansu Province, NW China. *J. Mt. Sci.* **2018**, *15*, 1299–1318. [CrossRef]
10. Nefeslioglu, H.A.; Gorum, T. The use of landslide hazard maps to determine mitigation priorities in a dam reservoir and its protection area. *Land Use Policy* **2020**, *91*, 104363. [CrossRef]
11. Preuth, T.; Glade, T.; Demoulin, A. Stability analysis of a human-influenced landslides in eastern Belgium. *Geomorphology* **2010**, *120*, 38–47. [CrossRef]
12. Guzzetti, F.; Reichenbach, P.; Cardinali, M.; Galli, M.; Ardizzone, F. Probabilistic landslide hazard assessment at the basin scale. *Geomorphology* **2005**, *72*, 272–299. [CrossRef]
13. Kovrov, O.; Kolesnyk, V.; Buchavyi, Y. Development of the landslide risk classification for natural and man-made slopes based on soil watering and deformation extent. *Min. Miner. Depos.* **2020**, *14*, 105–112. [CrossRef]
14. Kontoes, C.; Loupasakis, C.; Papoutsis, I.; Alatza, S.; Poyiadji, E.; Ganas, A.; Psychogyiou, C.; Kaskara, M.; Antoniadi, S.; Spanou, N. Landslide susceptibility mapping of central and Western Greece, combining NGI and WoE methods, with remote sensing and ground truth data. *Land* **2021**, *10*, 402. [CrossRef]
15. Ciampalini, A.; Raspini, F.; Lagomarsino, D.; Catani, F.; Casagli, N. Landslide susceptibility map refinement using PSInSAR data. *Remote Sens. Environ.* **2016**, *184*, 302–315. [CrossRef]

16. Zhang, Y.; Chen, Y.; Ming, D.; Zhu, Y.; Ling, X.; Zhang, X.; Lian, X. Landslide hazard analysis based on SBAS-InSAR and MCE-CNN model: A case study of Kongtong, Pingliang. *Geocarto Int.* **2022**, *38*, 1–22. [CrossRef]
17. Ferretti, A.; Prati, C.; Rocca, F. Permanent scatterers in SAR interferometry. *IEEE Trans. Geosci. Remote Sens.* **2001**, *39*, 8–20. [CrossRef]
18. Akbarimehr, M.; Motagh, M.; Haghshenas-Haghighi, M. Slope stability assessment of the Sarcheshmeh landslide, Northeast Iran, investigated using InSAR and GPS observations. *Remote Sens.* **2013**, *5*, 3681–3700. [CrossRef]
19. Intrieri, E.; Raspini, F.; Fumagalli, A.; Lu, P.; Del Conte, S.; Farina, P.; Allievi, J.; Ferretti, A.; Casagli, N. The Maoxian landslide as seen from space: Detecting precursors of failure with Sentinel-1 data. *Landslides* **2018**, *15*, 123–133. [CrossRef]
20. Zhang, Y.; Meng, X.; Jordan, C.; Novellino, A.; Dijkstra, T.; Chen, G. Investigating slow-moving landslides in the Zhouqu region of China using InSAR time series. *Landslides* **2018**, *15*, 1299–1315. [CrossRef]
21. Yang, S.; Li, D.; Liu, Y.; Xu, Z.; Sun, Y.; She, X. Landslide Identification in Human-Modified Alpine and Canyon Area of the Niulan River Basin Based on SBAS-InSAR and Optical Images. *Remote Sens.* **2023**, *15*, 1998. [CrossRef]
22. Urgilez Vinueza, A.; Handwerger, A.L.; Bakker, M.; Bogaard, T. A new method to detect changes in displacement rates of slow-moving landslides using InSAR time series. *Landslides* **2022**, *19*, 2233–2247. [CrossRef]
23. Novellino, A.; Cesarano, M.; Cappelletti, P.; Di Martire, D.; Di Napoli, M.; Ramondini, M.; Sowter, A.; Calcaterra, D. Slow-moving landslide risk assessment combining machine learning and InSAR techniques. *Catena* **2021**, *203*, 105317. [CrossRef]
24. Zheng, X.; He, G.; Wang, S.; Wang, Y.; Wang, G.; Yang, Z.; Yu, J.; Wang, N. Comparison of machine learning methods for potential active landslide hazards identification with multi-source data. *ISPRS Int. J. Geo Inf.* **2021**, *10*, 253. [CrossRef]
25. Zhang, G.; Wang, S.; Chen, Z.; Liu, Y.; Xu, Z.; Zhao, R. Landslide susceptibility evaluation integrating weight of evidence model and InSAR results, west of Hubei Province, China. *Egypt. J. Remote Sens. Space Sci.* **2023**, *26*, 95–106. [CrossRef]
26. Shan, Y.; Xu, Z.; Zhou, S.; Lu, H.; Yu, W.; Li, Z.; Cao, X.; Li, P.; Li, W. Landslide Hazard Assessment Combined with InSAR Deformation: A Case Study in the Zagunao River Basin, Sichuan Province, Southwestern China. *Remote Sens.* **2024**, *16*, 99. [CrossRef]
27. Zhou, C.; Cao, Y.; Hu, X.; Yin, K.; Wang, Y.; Catani, F. Enhanced dynamic landslide hazard mapping using MT-InSAR method in the Three Gorges Reservoir Area. *Landslides* **2022**, *19*, 1585–1597. [CrossRef]
28. Dai, C.; Li, W.; Lu, H.; Zhang, S. Landslide Hazard Assessment Method Considering the Deformation Factor: A Case Study of Zhouqu, Gansu Province, Northwest China. *Remote Sens.* **2023**, *15*, 596. [CrossRef]
29. Zhang, X.; Chen, L.; Zhou, C. Deformation Monitoring and Trend Analysis of Reservoir Bank Landslides by Combining Time-Series InSAR and Hurst Index. *Remote Sens.* **2023**, *15*, 619. [CrossRef]
30. Miao, F.; Wu, Y.; Xie, Y.; Li, Y. Prediction of Landslide Displacement with Step-like Behavior Based on Multialgorithm Optimization and a Support Vector Regression Model. *Landslides* **2018**, *15*, 475–488. [CrossRef]
31. Guzzetti, F.; Reichenbach, P.; Ardizzone, F.; Cardinali, M.; Galli, M. Estimating the quality of landslide susceptibility models. *Geomorphology* **2006**, *81*, 166–184. [CrossRef]
32. Marjanović, M.; Kovačević, M.; Bajat, B.; Voženílek, V. Landslide susceptibility assessment using SVM machine learning algorithm. *Eng. Geol.* **2011**, *123*, 225–234. [CrossRef]
33. Dou, J.; Yunus, A.P.; Bui, D.T.; Merghadi, A.; Sahana, M.; Zhu, Z.; Chen, C.W.; Han, Z.; Pham, B.T. Improved landslide assessment using support vector machine with bagging, boosting, and stacking ensemble machine learning framework in a mountainous watershed, Japan. *Landslides* **2020**, *17*, 641–658. [CrossRef]
34. Bordoni, M.; Vivaldi, V.; Lucchelli, L.; Ciabatta, L.; Brocca, L.; Galve, J.P.; Meisina, C. Development of a data-driven model for spatial and temporal shallow landslide probability of occurrence at catchment scale. *Landslides* **2021**, *18*, 1209–1229. [CrossRef]
35. Corominas, J.; van Westen, C.; Frattini, P.; Cascini, L.; Malet, J.P.; Fotopoulou, S.; Catani, F.; Van Den Eeckhaut, M.; Mavrouli, O.; Agliardi, F.; et al. Recommendations for the quantitative analysis of landslide risk. *Bull. Eng. Geol. Env.* **2014**, *73*, 209–263. [CrossRef]
36. Lee, J.H.; Kim, H.; Park, H.J.; Heo, J.H. Temporal prediction modeling for rainfall-induced shallow landslide hazards using extreme value distribution. *Landslides* **2021**, *18*, 321–338. [CrossRef]
37. Liu, C. *Landslide Risk Assessment at Wanzhou Urban Area in the Three Gorges Reservoir*; China University of Geosciences Press: Wuhan, China, 2014.
38. Del Soldato, M.; Riquelme, A.; Bianchini, S.; Tomàs, R.; Di Martire, D.; De Vita, P.; Moretti, S.; Calcaterra, D. Multisource Data Integration to Investigate One Century of Evolution for the Agnone Landslide (Molise, Southern Italy). *Landslides* **2018**, *15*, 2113–2128. [CrossRef]
39. Wasowski, J.; Pisano, L. Long-Term InSAR, Borehole Inclinometer, and Rainfall Records Provide Insight into the Mechanism and Activity Patterns of an Extremely Slow Urbanized Landslide. *Landslides* **2020**, *17*, 445–457. [CrossRef]
40. Lu, P.; Catani, F.; Tofani, V.; Casagli, N. Quantitative Hazard and Risk Assessment for Slow-Moving Landslides from Persistent Scatterer Interferometry. *Landslides* **2014**, *11*, 685–696. [CrossRef]
41. Ferretti, A.; Prati, C.; Rocca, F. Nonlinear subsidence rate estimation using permanent scatterers in differential SAR interferometry. *IEEE Trans. Geosci. Remote Sens.* **2000**, *38*, 2202–2212. [CrossRef]
42. Berardino, P.; Fornaro, G.; Lanari, R.; Sansosti, E. A new algorithm for surface deformation monitoring based on small baseline differential SAR interferograms. *IEEE Trans. Geosci. Remote Sens.* **2002**, *40*, 2375–2383. [CrossRef]
43. Zhang, J.; Gong, Y.; Huang, W.; Wang, X.; Ke, Z.; Liu, Y.; Huo, A.; Adnan, A.; Abuarab, M.E.-S. Identification of Potential Landslide Hazards Using Time-Series InSAR in Xiji County, Ningxia. *Water* **2023**, *15*, 300. [CrossRef]

44. Wang, J.; Xiao, L.; Zhang, J.; Zhu, Y. Deformation characteristics and failure mechanisms of a rainfall-induced complex landslide in Wanzhou County, Three Gorges Reservoir, China. *Landslides* **2020**, *17*, 419–431. [CrossRef]
45. Buytaert, W.; Celleri, R.; Willems, P.; De Bievre, B.; Wyseure, G. Spatial and temporal rainfall variability in mountainous areas: A case study from the south Ecuadorian Andes. *J. Hydrol.* **2006**, *329*, 413–421. [CrossRef]
46. Singh, N.; Maddheshiya, S.K.; Jha, M.; Tignath, S.; Singh, B.N. Hydrogeomorphic assessment of badlands in part of the Mandakini River watershed, Chitrakoot, India. *Arab. J. Geosci.* **2020**, *13*, 1066. [CrossRef]
47. Dong, J.; Niu, R.; Li, B.; Xu, H.; Wang, S. Potential landslides identification based on temporal and spatial filtering of SBAS-InSAR results. *Geomat. Nat. Hazards Risk* **2023**, *14*, 52–75. [CrossRef]

remote sensing MDPI

Article

Formative Period Tracing and Driving Factors Analysis of the Lashagou Landslide Group in Jishishan County, China

Qianyou Fan [1], Shuangcheng Zhang [1,2,*], Yufen Niu [3], Jinzhao Si [1], Xuhao Li [1], Wenhui Wu [1], Xiaolong Zeng [1] and Jianwen Jiang [1]

1 College of Geological Engineering and Geomatics, Chang'an University, Xi'an 710054, China; 2020026013@chd.edu.cn (Q.F.); 2020126029@chd.edu.cn (J.S.); 2022126034@chd.edu.cn (X.L.); 2021226025@chd.edu.cn (W.W.); 2022126035@chd.edu.cn (X.Z.); jjw@chd.edu.cn (J.J.)
2 Observation and Research Station of Ground Fissure and Land Subsidence, Ministry of Natural Resources, Xi'an 710054, China
3 School of Mining and Geomatics Engineering, Hebei University of Engineering, Handan 056038, China; niuyufen@hebeu.edu.cn
* Correspondence: shuangcheng369@chd.edu.cn

Abstract: The continuous downward movement exhibited by the Lashagou landslide group in recent years poses a significant threat to the safety of both vehicles and pedestrians traversing the highway G310. By integrating geomorphological interpretation using multi-temporal optical images, interferometric synthetic aperture radar (InSAR) measurements, and continuous global navigation satellite system (GNSS) observations, this paper traced the formation period of the Lashagou landslide group, and explored its kinematic behavior under external drivers such as rainfall and snowmelt. The results indicate that the formation period can be specifically categorized into three periods: before, during, and after the construction of highway G310. The construction of highway G310 is the direct cause and prerequisite for the formation of the Lashagou landslide group, whereas summer precipitation and spring snowmelt are the external driving factors contributing to its continuous downward movement. Additionally, both the long-term seasonal downslope movement and transient acceleration events are strongly controlled by rainfall, and there is a time lag of approximately 1–2 days between the transient acceleration and heavy rainfall events. This study highlights the benefits of leveraging multi-source remote sensing data to investigate slow-moving landslides, which is advantageous for the implementation of effective control and engineering intervention to mitigate potential landslide disasters.

Keywords: shallow loess landslide; slope excavation; multi-source remote sensing; evolutionary history; triggering factors

Citation: Fan, Q.; Zhang, S.; Niu, Y.; Si, J.; Li, X.; Wu, W.; Zeng, X.; Jiang, J. Formative Period Tracing and Driving Factors Analysis of the Lashagou Landslide Group in Jishishan County, China. *Remote Sens.* **2024**, *16*, 1739. https://doi.org/10.3390/rs16101739

Academic Editors: Federico Raspini, Chao Zhou and Kunlong Yin

Received: 10 April 2024
Revised: 7 May 2024
Accepted: 7 May 2024
Published: 14 May 2024

1. Introduction

In recent years, with the progressive implementation of the Western Development Drive Strategy, China has constructed several high-level mountain roads. However, the frequent excavation of hillsides during construction disrupts the stress balance of the original slope, subsequently triggering a series of landslides [1]. Conducting research on the long-term kinematic characteristics and causative factors of landslides is crucial for implementing focused monitoring, early warning, and engineering interventions, ultimately reducing the disaster risk to a socially acceptable level.

Optical remote sensing enables the reproduction of the kinematic behavior of fast-moving landslides through the visual interpretation of surface geomorphological features [2,3]. For instance, Lacroix et al. [2] investigated the reactivation of the Harmalière landslide using Sentinel-2 optical data. The results revealed a 7-month period of low activity ($\leq$1 m), followed by a sudden acceleration of 3.2 $\pm$ 1.2 m in 3 days, before the failure of a mass of about 2 to 3.6 $\times$ 10^6 m^3. However, the method faces challenges in capturing the kinematic characteristics and stability of slow-moving landslides (ranging from mm/year to cm/year) due to the limitations of meter-level spatial resolution and temporal resolution related to cloud cover [4,5]. In contrast, InSAR technology facilitates all-day and all-weather observations with high accuracy, measuring from centimeters to millimeters [6–9]. With this technology, it becomes possible to capture the surface kinematic characteristics of landslides at a spatial scale, ranging from several meters to tens of meters [10]. In addition, massive synthetic aperture radar (SAR) data archived over the past 30 years have made it possible to retrieve the long-term spatiotemporal kinematic characteristics of landslides [11,12]. For example, Cignetti et al. [13] investigated the 26-year spatiotemporal evolution of Sauze d'Oulx slopes using four different datasets, ERS 1/2, Radarsat-1/2, COSMO SkyMed, and Sentinel-1, collected between 1992 and 2018. Xu et al. [14] retrieved the 12-year kinematic characteristics of Hooskanaden landslide from 2007 to 2019 by combining digital elevation model (DEM), ALOS-1/2, and Sentinel-1 datasets.

Due to the SAR satellite's revisited period, which ranges from several days to tens of days, the InSAR technology faces challenges in capturing the instantaneous response of landslide dynamics to external environmental factors like rainfall, mountain fires, and so forth [15–17]. GNSS technology, with its ability to continuously track Earth's surface displacements in real time at a high sampling rate (seconds or less), is a valuable tool for studying the kinematic behavior of landslides and analyzing the impact of external triggers [18,19]. Therefore, combining continuous GNSS measurements with high temporal resolution and spaceborne InSAR measurements with high spatial resolution can potentially reveal more comprehensive insights into the kinematic characteristics and deformation mechanisms of landslides [20–22]. For instance, Hu et al. [23] successfully delineated the intricate spatiotemporal variations in the kinematic characteristics of the Crescent Lake landslide by integrating data from semi-permanent GNSS stations with space-borne InSAR measurements. Moreover, they established a threshold for the amount of pre-accumulated rainfall needed to trigger the displacement of the landslide.

The Lashagou landslide group is situated in the transitional zone between the Qinghai–Tibet Plateau and the Loess Plateau (Figure 1a). Its primary cause of formation is considered to be the hillslope excavation activities conducted during the construction of the highway G310 from 2011 to 2013 [10,24]. However, no studies have yet demonstrated the impact of the construction of the highway G310 on the formation and evolution of the Lashagou landslide group. The latest research indicates that the Lashagou landslide group remained in a state of continuous downward movement from 2020 to 2021 [24], posing a significant threat to the safety of both vehicles and pedestrians on highway G310. To exacerbate the situation, a magnitude 6.2 earthquake transpired on 18 December 2023, at a distance of approximately 18 km from the Lashagou landslide group (Figure 1a). This unfortunate occurrence resulted in the tragic loss of 151 lives and induced a minimum of 3767 co-seismic landslides [25]. The research conducted by Huang et al. [26] and Liu et al. [27] indicates that the Lashagou landslide group is situated in an area highly susceptible to post-earthquake geological disasters. Therefore, the Lashagou landslide group poses a significant risk of failure during the approaching spring snowmelt season and rainy season.

Taking the Lashagou landslide group as a case study, this research examines the collaborative use of various remote sensing technologies to investigate the kinematic characteristics and driving factors influencing landslides. We firstly employed geomorphological interpretation using multi-temporal optical images and small baseline subset InSAR (SBAS-InSAR) technology to trace the formation period of the Lashagou landslide group and analyze the impact of the construction of highway G310 on its spatiotemporal evolution patterns. Meanwhile, the characteristics and mechanisms of landslide downward movement were analyzed based on InSAR time-series displacement over a period of 9 years. Finally, the kinematic behavior of the Lashagou landslide group, including seasonal variations, transient acceleration events, and episodic failure events, was investigated by GNSS measurements with high temporal sampling frequency. This study highlights the benefits of leveraging multi-source remote sensing data to investigate slow-moving landslides, which is advantageous for the implementation of effective control and engineering intervention to mitigate potential landslide disasters.

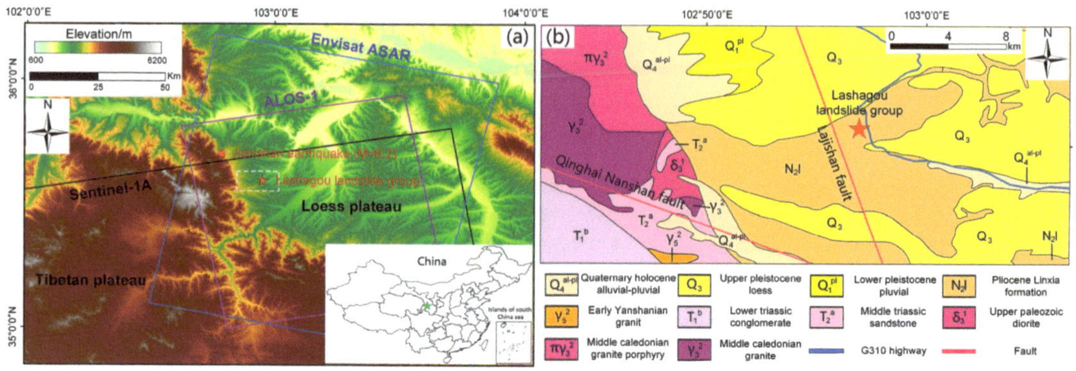

Figure 1. (**a**) Location of the study area and spatial coverage of the SAR data used in this study. The green star in the inset represents the geographical location of the Lashagou landslide group in China. (**b**) Regional engineering geological map of the study area (the dotted white area in (**a**)). The red star denotes the location of the Lashagou landslide group.

2. Study Area

Jishishan County is located in the transition zone between the Qinghai–Tibet and the Loess plateaus (Figure 1a), with mountainous terrain accounting for approximately 90% of the total area [28]. The active tectonic movement in the northeastern Qinghai–Tibet Plateau and the weak strata (high porosity and low cohesion of loess material, which makes it easily erodible and prone to collapse under pressure) of the Loess Plateau have made Jishishan County one of the regions with the most serious geological disasters in China [29]. The Lashagou landslide group, located in the southern region of Jishishan County, consists of eight small- and medium-sized landslides (Figure 2a). The average gradient where the landslide group is located measures approximately 19.5°, while the gradient at the foot ranges between 50° and 80° (e.g., steep free surface in Figure 2b,c). In addition, the landslide group is primarily distributed at an elevation ranging from 1100 m to 1200 m. Of these, the L1 landslide holds the title of being the largest, covering an area of approximately 49,600 m^2, while the L8 landslide is identified as the smallest, encompassing a mere 2900 m^2.

The surrounding area of the landslide group is widely distributed in Quaternary aeolian loess (Q_3), alluvial loess ($Q_4{}^{al-pl}$), and Neogene sand-mudstone (N_2I) from the Linxia Formation [10]. The area also features exposed strata from the Sinian, Carboniferous, Permian, and Cretaceous (Figure 1b). According to the borehole data, the sliding surface is situated at the interface of loess and silty clay (Figure 3). The sliding body primarily comprises Pliocene silty loess (Figure 3b), characterized by loose porosity and well-developed vertical joints, resulting in strong permeability. The underlying strata are composed of Middle Pleistocene silty clay (Figure 3b), which is mostly impermeable or weakly permeable mudstone. The differential permeability characteristics between the loess and clay strata create favorable conditions for the formation and subsequent development of the Lashagou landslide group [10,30].

According to records from the Jishishan County People's Government (http://www.jss.gov.cn/ (accessed on 8 March 2023)), the Jishishan region is characterized by a typical continental monsoon climate, featuring an average annual precipitation of 660.2 mm and an average annual evaporation of 880 mm. The majority of rainfall is concentrated from May to August, encompassing around 60–78% of the annual precipitation, often accompanied by heavy rain events. Located within the seasonal permafrost zone, the region experiences a winter extreme minimum temperature of −32.2 °C, with a maximum freezing depth of

1.2 m. Significantly, the initial freezing of the landslide surface loess typically occurs around mid-November each year, followed by a melting period from mid-February to early April, resulting in a high frequency of spring landslide disasters due to freeze–thaw erosion.

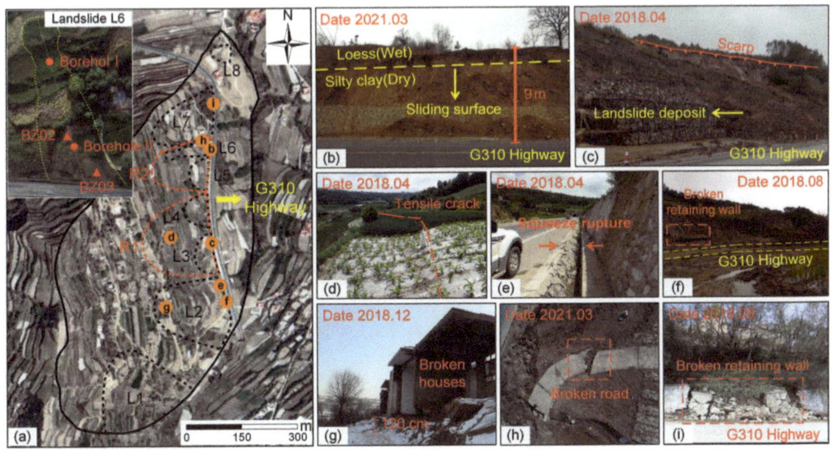

Figure 2. (**a**) Optical imagery of the Lashagou landslide group acquired from Google Earth. The red triangles and circles in the inset image are the locations of GNSS stations and boreholes, respectively. (**b–i**) Photos from the locations indicated by solid orange circles in (**a**), showing the damages caused by the landslides in 2018.

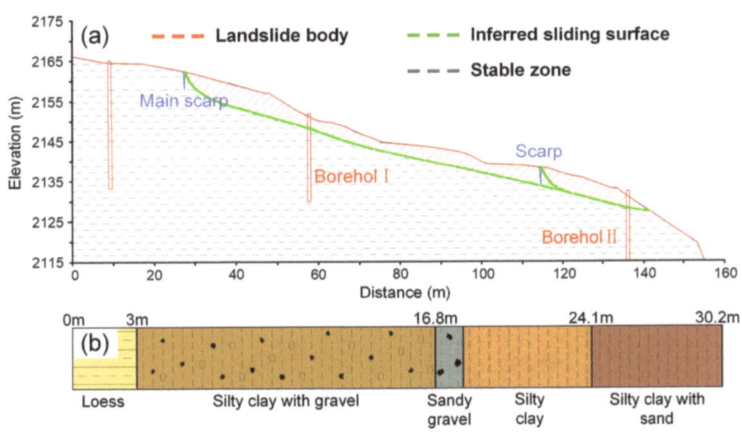

Figure 3. (**a**) Engineering geological profile of the L6 landslide, whose location is illustrated in the inset image in Figure 2a. (**b**) Stratigraphy of borehole II.

As the primary route connecting Linxia City and Qinghai Province, the construction of highway G310 (Linxia section) began in January 2011 and opened to traffic in November 2013 (Figure 4). During this period, the slope where Lashagou Village was situated (represented by the solid black line in Figure 2a) was excavated to an average depth of 3–8 m and a maximum depth of 12 m [10]. Hillslope excavation led to a steep free surface of 50–80° on the left side of the highway (Figure 2a,b), which disrupted the initial stress equilibrium of the slope and ultimately led to the Lashagou landslide group. According to records from the Emergency Management Department of the Gansu Province, landslides L3 and L4 first occurred in early April 2018 (Figure 2c). These landslides expanded the damage to the upper slope during the subsequent rainy season (Figure 2d), ultimately leading to the formation of retrogressive landslide R1. Simultaneously, landslide L2 experienced accelerated displacement and continued to exert pressure on the retaining wall at its rupture surface, resulting in the uplift and crushing of the side ditch along the highway G310 (Figure 2e). On 2 and 3 August 2018, Jishishan County encountered an unusual rainstorm, with precipitation reaching as high as 194 mm [31]. This heavy rainfall directly triggered the landslide L2, causing damage to several houses and roads in Lashagou Village (Figure 2g). Additionally, the retaining wall at the foot of the slope was damaged (Figure 2f), leading to the closure of the highway for maintenance. Subsequently, landslide L1 suffered localized failure, but there were no associated casualties or property losses. Furthermore, several active landslides, i.e., L5, L6 (Figure 2h), and L7 (Figure 2i), were identified, whereas landslide L8 was classified as a failed landslide.

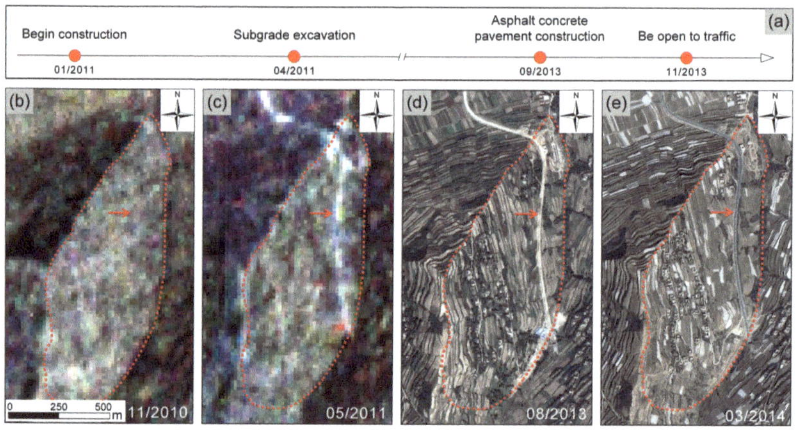

Figure 4. Optical satellite images recording the construction process of highway G310. (**a**) Important events during construction and their corresponding dates. (**b**,**c**) Landsat 7 images from November, 2010 and May, 2011. (**d**,**e**) Google Earth images from August 2013 and March 2014. The red dotted line indicates the extent of the slope where the Lashagou landslide group is located, and the red arrow points to highway G310.

3. Datasets and Methodology

3.1. Datasets

To reveal the long-term spatiotemporal evolution of the Lashagou landslide group, we analyzed three sets of SAR data with different imaging geometries acquired between 2006 and 2021. These datasets consisted of 26 descending Envisat images, 12 ascending ALOS-1 images, and 138 ascending Sentinel-1A images. Figures 1a and 5a show the spatial coverage and time span of the different SAR datasets, respectively, and Table 1 lists the fundamental parameters associated with the SAR datasets utilized in this study. In addition, four Google Earth historical images from September 2013, October 2015, November 2016, and October 2019, and an unmanned aerial vehicle (UAV) orthophoto image from August 2020 were used to reconstruct the spatiotemporal evolution history of the Lashagou landslide group.

The shuttle radar topography mission (SRTM) and ALOS world 3D (AW3D) DEMs with a resolution of 30 m were used to simulate and remove the topographic phase. Additionally, we employed UAV oblique photogrammetry to generate a high-precision DEM with a resolution of 0.05 m. Two GNSS stations (BZ02 and BZ03) installed on the landslide L6 (Figure 2a) provided 3D surface displacement with millimeter-level precision, which was used to validate the InSAR results and assist in time-series analysis. Rainfall and snowfall from 2006 onwards were provided by the Jishishan Meteorological Station, which is located approximately 14 km from the Lashagou landslide group. A soil hygrometer installed on landslide L6 measures the soil water content at a depth of 10 cm once every hour. In addition, two drilling operations (Figure 2a) were conducted on landslide L6 during the autumn of 2018 to test the fundamental physical properties and strength characteristics of the slide body.

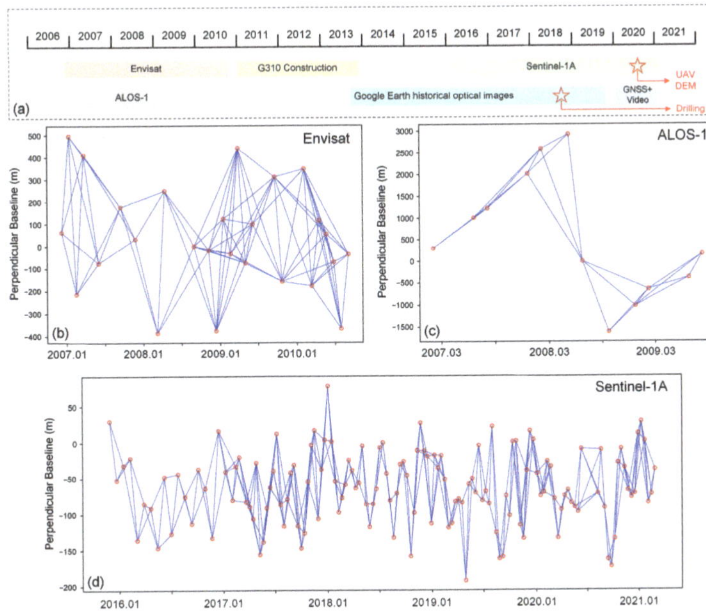

Figure 5. (**a**) Temporal coverage of SAR datasets, optical images, and GNSS observations used in this study. (**b–d**) Spatiotemporal baseline combinations of descending Envisat ASAR, ascending ALOS-1, and ascending Sentinel-1A datasets.

Table 1. Basic parameters of the three SAR datasets.

Sensor	Envisat	ALOS-1	Sentinel-1A
Band	C	L	C
Orbit direction	Descending	Ascending	Ascending
Polarization	VV	HH	VV
Heading (°)	−168.13	−10.26	−9.76
Incidence angle (°)	22.91	38.73	42.12
Pixel spacing (m)	7.8 × 4.0	4.7 × 3.2	2.3 × 13.9
Date range	December 2006 to September 2010	March 2007 to September 2009	November 2015 to February 2021
Number of images	26	12	138

3.2. Interferometric Processing and SBAS-InSAR Analysis

Because there are extensive farmlands and grasslands in the study area, we chose the SBAS-InSAR method to identify the maximum number of high-coherence pixels possible. This method is characterized by a redundant interferogram network that effectively reduces orbit errors and incoherent noise. Differential interference processing was performed on all SAR datasets using the GAMMA (version: 20171206) software. The generated interferogram network is shown in Figure 5. During this process, the SRTM DEM (used for the Envisat and ALOS-1 datasets) and AW3D DEM (used for the Sentinel-1A dataset) were used to simulate and remove the topographic phase. We did not perform any spectral filtering on the SAR images to preserve image resolution and maximize the density of the observation points [6].

SBAS-InSAR processing of multi-master interferograms was conducted using StaMPS software [6], which primarily involved high-coherence pixel selection, spatially uncorrelated noise correction, phase unwrapping, and spatiotemporal filtering. The amplitude dispersion index (≤0.65) and pixel phase noise characteristics were combined to select the high-coherence pixels. The errors caused by the orbital ramp were modeled by using a quadratic polynomial function, and the errors resulting from topographic undulation were removed through the operation of the DEM-assisted ortho-rectification of SAR images. Generic Atmospheric Correction Online Service (GACOS) products were downloaded and adopted to correct the tropospheric delay. We applied the quadratic polynomial to estimate and remove phase ramps that may be caused by residual tropospheric or ionospheric delays and orbital errors. The wrapped phase was unwrapped using a 3D phase-unwrapping method, and spatially correlated noise was separated using an iterative filter to estimate the phase component of the deformation. Notably, during filtering, a small filtering window was selected to avoid excessive information loss. Finally, the phase component of the deformation was converted into the line-of-sight (LOS) velocity and time-series displacement using the singular value decomposition (SVD) method.

3.3. GNSS Data Processing

In January 2020, two consecutive GNSS stations (BZ02 and BZ03), equipped with Septentrio M2 receivers, were installed on the leading edge of landslide L6 (Figure 2a). As BZ03 is installed on a highly active sub-block, it is anticipated to capture a more significant deformation signal. The landslide monitoring system is constructed by integrating the GAMIT/GLOBK (version: 10.71) software, TrackRT module, and CEEMD algorithm, allowing for a near-real-time baseline solution and real-time single ephemeris solution. This system can automatically perform the entire process from real-time data acquisition to the release of monitoring results. A real-time single ephemeris solution was used in this study. For more details, please refer to Zhong et al. [32].

4. Results

4.1. LOS Displacements from December 2006 to February 2021

Figure 6a–c show the mean displacement velocities along the LOS direction of the Lashagou landslide group from December 2006 to September 2010 (Envisat), March 2007 to September 2009 (ALOS-1), and November 2015 to February 2021 (Sentinel-1A), respectively, which were used to trace the formation period of the Lashagou landslide group and the influence of the construction of the highway G310 (January 2011 to November 2013) on the spatiotemporal evolution patterns.

Prior to the construction of the highway (2006–2010), we used both the Envisat and ALOS-1 datasets to cross-validate the reliability of the InSAR displacements. The mean and standard deviation of the difference in displacement velocities between the Envisat and ALOS-1 maps for corresponding points (closest and within 50 m proximity) were determined to be 3.69 and 3.50 mm/y, respectively, indicating that they have a high degree of agreement. The observed disparity can be primarily attributed to the distinct imaging geometries of the two independent SAR datasets (Table 1). Harmonizing these datasets into the same benchmark for comparison is challenging, particularly when the geometry of the landslide movement, including its sliding direction and inclination, remains unknown [33]. In addition, the incomplete agreement between the observation periods of the Envisat and ALOS-1 datasets may also explain the above discrepancy because the displacement of landslides is typically nonlinear (e.g., Figure 7b,d,f). After the completion of the highway construction (2015–2021), we examined the InSAR measurements of Sentinel-1A using high-precision GNSS observations. As shown in Figure 7d, the time-series displacement derived from Sentinel-1A datasets using the SBAS-InSAR method exhibited a strong agreement with the GNSS (BZ02) observations, and both have highly similar displacement trends and magnitudes, with the mean and standard deviation of the differences being 5.99 and 4.42 mm/y, respectively. This slight disparity is acceptable for landslide displacement monitoring and may be because the spatial locations of the GNSS sites and the corresponding InSAR measurement points do not exactly correspond.

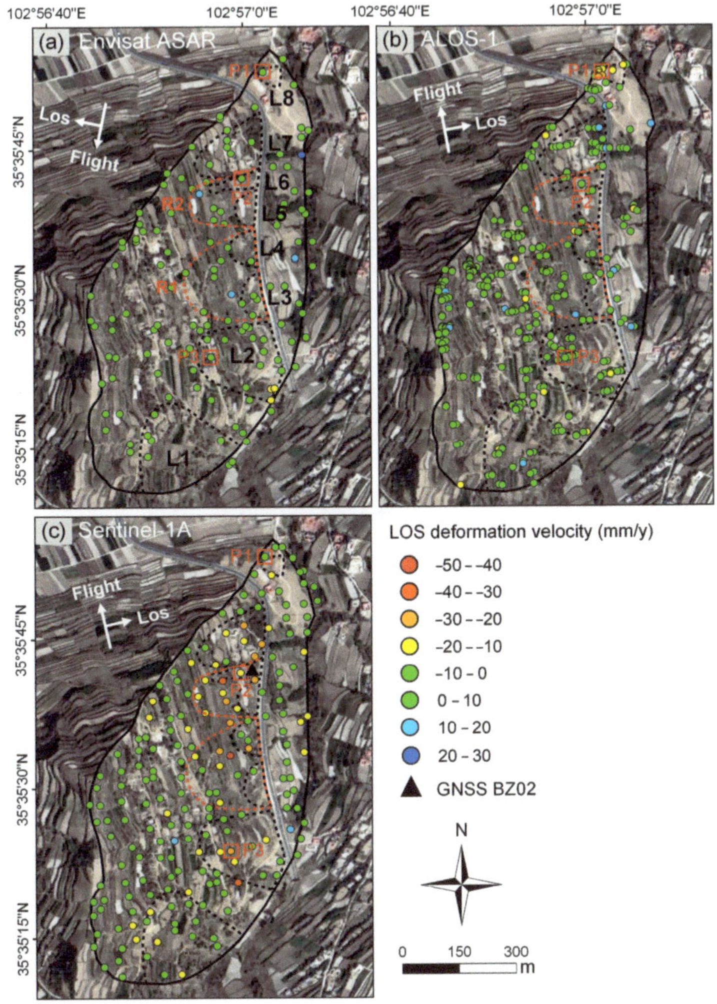

Figure 6. LOS displacement velocity maps derived from (**a**) descending Envisat images (December 2006 to September 2010); (**b**) ascending ALOS-1 images (March 2007 to September 2009); and (**c**) ascending Sentinel-1A images (November 2015 to February 2021). P1–P3 are used for time-series analysis.

As shown in Figure 6a,b, the Lashagou landslide group was in a relatively stable condition prior to the construction of highway G310, and approximately 70% of the monitored points in the Envisat displacement map have deformation magnitudes in the range of ±5 mm/y, whereas approximately 85% of the monitored points in the ALOS-1 displacement map have deformation magnitudes in the range of ±8 mm/y. Because of the large incidence angle in the ALOS-1 dataset (Table 1), the projected landslide displacement component in the LOS direction is greater. Consequently, the overall displacement magnitude measured using the ALOS-1 dataset is slightly higher than that obtained using the Envisat dataset. In addition, active landslide L8 was identified at the foot of the northeastern part of the slope with a maximum displacement velocity of −21.7 mm/y. As shown in Figure 6c, the Lashagou landslide group exhibited continuous downward movement after the construction of the highway, and most of the landslides experienced displacements exceeding 20 mm/y, either across the whole or in specific localized regions. The most severely deformed region is situated in the central part of retrogressive landslide R1, which has a maximum displacement velocity of −47.4 mm/y. In addition, under the constant traction of active landslides L5 and L6, the overlying slopes began to deform owing to the reduction in the support force, gradually forming retrogressive landslide R2.

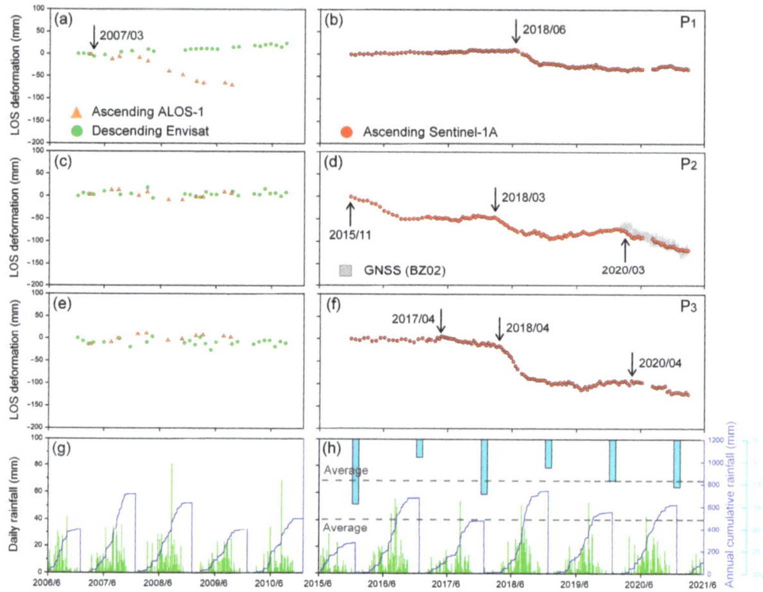

Figure 7. Time series displacement for Points P1–P3 (marked in Figure 6) from December 2006 to February 2021. (**a**,**b**) P1; (**c**,**d**) P2; and (**e**,**f**) P3. The green circles, orange triangles, and red circles represent the time-series displacement obtained from the Envisat, ALOS-1, and Sentinel-1A datasets, respectively, whereas the gray rectangles represent the GNSS (marked in Figure 6c) displacements along the LOS direction. (**g**,**h**) Rainfall and snowfall during the monitoring period, where the green line, blue line, and light blue column indicate the daily rainfall, annual cumulative rainfall, and annual cumulative snowfall, respectively. Additionally, the gray dashed line represents the average rainfall and snowfall over the last decade.

4.2. Formative Period Tracing

We traced the formation period and evolutionary history of the Lashagou landslide group by combining the InSAR time-series displacement (Figure 7) and historical optical imagery (Figure 8). Landslide formation periods were separated into three categories.

Before the construction of highway G310 (landslide L8): As shown in Figure 7a, P1 is situated at the crown of landslide L8, and the time-series displacement obtained from both the Envisat and ALOS-1 datasets indicates that landslide L8 began to deform in March 2017 and continued until September 2010. Specifically, landslide L8 had a cumulative LOS displacement of 24 mm in the descending Envisat results from December 2006 to September 2010, whereas the ascending ALOS-1 results had a cumulative LOS displacement reaching 69 mm from March 2007 to September 2009. The ratio r_1 of the cumulative LOS displacement obtained from similar time periods between the ALOS-1 and Envisat datasets was approximately 4.6 and was primarily attributed to the different imaging geometries of these datasets. Assuming that landslide L8 exhibited translational movement, we obtained its average slope aspect and gradient using the UAV DEM as the direction and inclination of the landslide displacement. Combining the imaging geometry of the Envisat and ALOS-1 datasets (Table 1), we derived the scale factors for the projection of the displacement vector of landslide L8 into the InSAR LOS direction to be 0.16 and 0.92, respectively (with a ratio

r_2 of approximately 5.75). The deviation between r_1 and r_2 is acceptable, considering the uncertainty of the translational movement assumption and InSAR measurements.

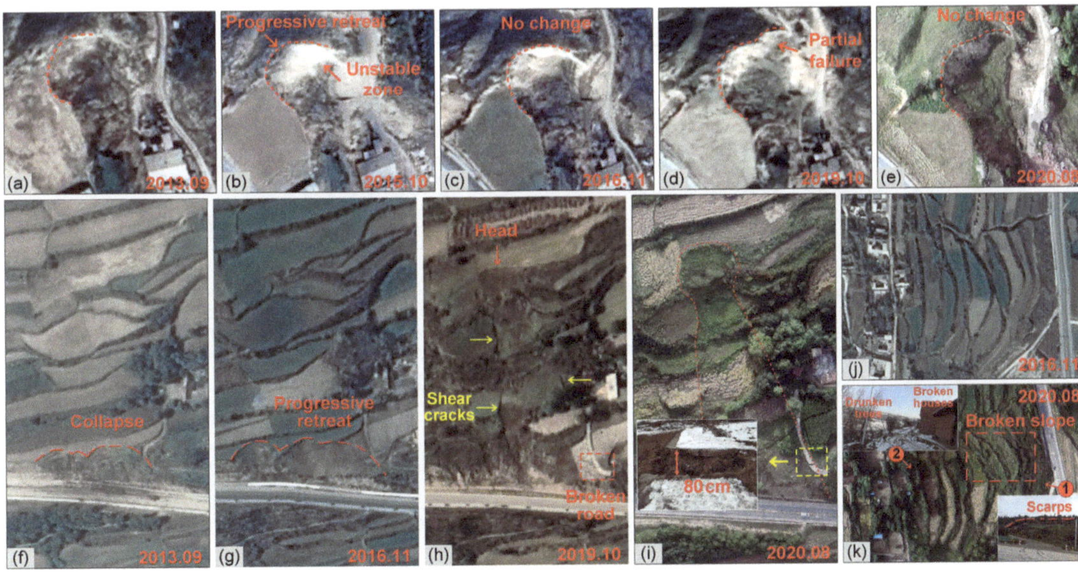

Figure 8. Historical optical images of typical landslides in the Lashagou landslide group. (**a**–**e**) Landslide L8; (**f**–**i**) Landslide L6; (**j**,**k**) Landslide L2. Optical images from 2013 to 2019 were obtained from Google Earth, whereas the image from August 2020 was a UAV-generated orthophoto.

Landslide L8 formed a through-crown tension crack prior to September 2013, as shown in Figure 8a, which allowed it to gain greater freedom of movement from the surrounding materials. Subsequently, landslide L8 continued to drag the upper slope until October 2015 when significant damage occurred (Figure 8b). Disturbed landslide L8 experienced a decrease in its gravitational potential energy and established a new state of stress equilibrium with the basal moraines. Subsequently, the landslide entered a period of stabilization from November 2015 to May 2018, was not reactivated until June 2018, and continued to deform until November 2018 (Figure 7b). Furthermore, the surface texture information obtained from historical optical imagery further confirmed that landslide L8 remained in a stable condition following the initial activity (Figure 8b,c). Additionally, localized reactivation occurred between November 2016 and October 2019 (Figure 8d). Combined with the InSAR time-series displacement (note that P1 in the Sentinel-1A map is situated outside the crown of the locally damaged sub-block, and the displacement is a result of traction from this sub-block), we can shorten the timeframe of this localized damage from June to November 2018, and landslide L8 has been stabilized since then (Figures 7b and 8e).

During the construction of highway G310 (landslides L3, L4, L5, and L6): For landslide L6, the time-series displacement at P2, situated in its center, indicates that the landslide was in a stable state until the construction of the highway (Figure 7c). During construction (before September 2013), a local collapse occurred at the foot of landslide L6 (Figure 8f), which subsequently caused progressive damage to the upper slope over time (Figure 8g). The time-series displacement derived from the Sentinel-1A dataset (Figure 7d) similarly captured the continuous downhill signal of landslide L6 after the construction of the highway (November 2015 to September 2016) and showed that the landslide was in a stabilized state between October 2016 and March 2018. Historical optical imagery showed that by October 2019 (Figure 8h), the posterolateral tensile and flank shear cracks of landslide L6 had penetrated to form a trap, indicating the complete formation of the landslide. Based on our analysis, landslide L6 underwent a severe sliding event from March 2018 to October 2019, and the InSAR time-series displacement revealed that the event occurred between April 2018 and early August 2018 (Figure 7d). Subsequently, landslide L6 transitioned to a stable state, reactivated in February 2020, and continued to deform until February 2021.

After the construction of the highway (landslides L1, L2, L7, R1, and R2): For landslide L2, the time-series displacement at P3, situated in its center, indicates that the landslide was in a stable state until the construction of the highway (Figure 7e). After construction was completed, the landslide remained stable until April 2017, when it was initially triggered and rapidly transitioned to a critical destabilization state within a year. Subsequently, in August 2018, a significant failure occurred (Figure 7f). Historical optical imagery (Figure 8j,k) documented changes in the surface morphology before and after the failure of landslide L8, which resulted in the destruction and collapse of houses at the trailing edge (Figure 2g), fragmentation of the surface, and failure of the retaining wall at the front edge (Figure 2f). The destructive event resulted in the dislocation of landslide L8 by 1–2 m without depleting its gravitational potential energy; thus, it was reactivated in April 2020 and continued its downward movement until September of the same year (Figure 7f).

5. Discussion

Following the analysis presented in Section 4.2, it was determined that the landslides within the Lashagou landslide group, with the exception of L8, were formed either during the construction of highway G310 (L3, L4, L5, and L6) or after its completion (L1, L2, L7, R1, and R2). Excavation conducted during the construction of the highway is widely acknowledged as a significant factor and an essential condition for the formation of the Lashagou landslide group (e.g., [10,24]). This excavation activity altered the original state of the stress equilibrium in the slope, thereby directly inducing slope deformation, as exemplified by the occurrence of landslide L6. Moreover, the steep free surface created by excavation deprives the slip-resistant structure of the natural slope, potentially lowering the threshold for shallow landslides triggered by rainfall [34], such as the occurrence of landslide L2.

As shown in Figure 7, the majority of landslides were activated during the snowmelt period (mid-February to early April) in 2017, 2018, and 2020, and then continued to accelerate downslope during the subsequent wet season (mid-May to early October). Therefore, snowmelt and precipitation are considered the main drivers of the continued downward movement of the Lashagou landslide group, which is consistent with the findings of Zhou et al. [30] and Zhang et al. [10]. It is worth noting that 2018 appears to be the most exceptional year of the 2006–2021 period, with landslides L3 and L4 first occurring in early April, followed by landslides L2 and L1 in their succession of total or localized failure in August. As shown in Figure 7h, the cumulative snowfall during the winter of 2017 (resulting in snowmelt in the spring of 2018) and the cumulative rainfall during the summer of 2018 were higher than the averages recorded in the last 10 years. As a result of the shallow burial depth of the sliding surface (Figure 3), a significant influx of snowmelt and precipitation infiltrated and saturated the basal shear zone of the landslide. This caused an elevation in the pore water pressure and a reduction in the soil's frictional resistance and shear strength [4,35,36], ultimately leading to the failure of the landslide. Interestingly, we observed that freeze–thaw landslides L3 and L4 occurred at the end of the snowmelt season, coinciding with a decrease in surface soil moisture (Figure 9c). This suggested that the seasonal frozen soil within the landslides was completely thawed, and the groundwater inside the landslides was rapidly discharged from the lower part of the free surface as mudflow (See Figure 10c in Zhang et al. [10]), which was elevated due to the "freezing stagnation effect" [37]. The flow and release of groundwater may cause weak structural surfaces to penetrate each other, ultimately leading to catastrophic landslides [10].

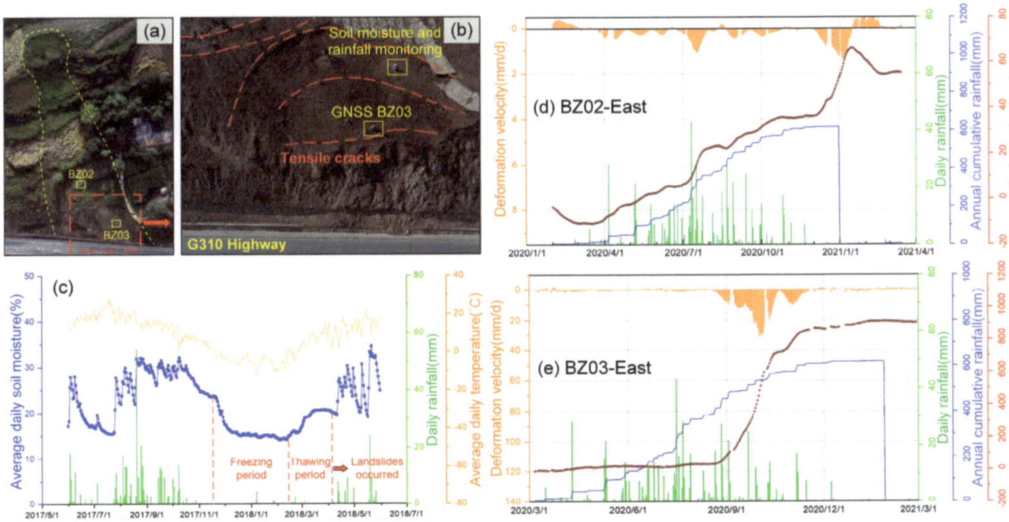

Figure 9. (**a**) Orthographic aerial photograph of landslide L6. (**b**) Locations of GNSS station (BZ03), rain gauge, and soil moisture sensor on landslide L6. (**c**) Daily rainfall (green bars), average daily temperature (orange line), and average daily soil moisture (blue dotted line) in the study area from June 2017 to June 2018. (**d**,**e**) The cumulative displacement along the E–W direction (red dotted line) and its corresponding velocity (orange bars) for BZ02 and BZ03 and compared with daily rainfall (green bars) and cumulative rainfall (blue line), respectively.

The 9-year InSAR time-series deformation provides valuable insights into the long-term displacement history of the Lashagou landslide group and facilitates the exploration of the causes and drivers contributing to its formation. However, the lower temporal resolution of InSAR measurements (typically no less than 6 d) makes it difficult to capture the rapid response of downslope movements triggered by snowmelt or rainfall. Therefore, we investigated the kinematic behavior of the Lashagou landslide group, including seasonal variations, transient acceleration events, and episodic failure events, by combining GNSS measurements with higher temporal sampling frequency. Figure 9a,b illustrate the precise positions of BZ02, BZ03, the soil moisture meter, and rain gauge on landslide L6. BZ03 was situated on a highly active sub-block in the upper part of the free surface at the foot of the slope. Consequently, BZ03 exhibited a displacement magnitude approximately 18 times greater than that of BZ02. The red scatter curves shown in Figure 9d,e represent the cumulative displacements of BZ02 and BZ03 along the east–west direction, as landslide L6 predominantly moved horizontally along that direction [10], whereas the orange bar graphs indicate the velocity changes at the corresponding times.

Both BZ02 and BZ03 began downslope movement around 25 March 2020, when snowmelt infiltrated the shear zone at the foot of the slope (Figure 2b), causing the downward sliding force to exceed the frictional resistance. During the subsequent rainy season, a significant correlation was observed between landslide displacement and cumulative rainfall (Figure 9d). This correlation indicates that both the long-term seasonal downslope movement and transient accelerated movement were predominantly influenced by rainfall. In addition, during the monitoring period, there were five heavy rainfall events, each with precipitation exceeding 25 mm/d. Among these events, 90% were associated with the transient acceleration of landslides, exhibiting an average velocity increase of 1.56 times. Moreover, the transient acceleration event of the landslide occurred on the second and first days immediately after the heavy rainfall in the dry and rainy seasons, respectively, indicating that the movement of shallow loess landslides is not only highly sensitive to rainfall intensity but is also influenced by the preceding rainfall. By calculating Pearson's correlation coefficient, our previous study [36] also demonstrated that the deformation of landslides exhibited a delay of approximately 2 days following rainfall, with a peak correlation coefficient of 0.627. Compared with most deep-seated landslides (e.g., [14,23]), the time lag of 1–2 d is short, and this is potentially related to the time required for the increase in pore water pressure to be transmitted to the basal sliding surface. Specifically, during the rainy season, the upper layer of the landslide became nearly saturated (Figure 9c), resulting in a shorter duration for the downward transfer of pore-water pressure compared to the dry season.

In addition, Figure 9e suggests that the active sub-block at the foot of landslide L6 may have been fully triggered and rapidly accelerated to a critical failure state after a cumulative rainfall of approximately 400 mm, which in turn may trigger the overall collapse of landslide L6. We also observed that the impact of rainfall on landslide movement was persistent, as landslides continued to experience slow downslope movement for a considerable period even after the conclusion of the rainy season (Figure 9d,e). Notably, both BZ02 and BZ03 exhibit abnormal and rapid downslope movements during the freezing period (mid-November to mid-February of the following year). This phenomenon may have been caused by the winter seasonally frozen layer obstructing the groundwater discharge channel connecting the interior and exterior of the slope. Consequently, groundwater concentrates inside the slopes, leading to the generation of significant hydrostatic pressure [37]. At this moment, the hydrostatic pressure within the landslide was considerably higher than the atmospheric pressure outside, resulting in a substantial pressure gradient that propelled the leading edge of the landslide to persistently move outward [33].

6. Conclusions

By integrating geomorphological interpretation using multi-temporal optical images, InSAR measurements with high spatial resolution, and continuous GNSS observations with high temporal resolution, we have traced the spatial and temporal evolution of the Lashagou landslide group over the past 15 years. Additionally, we have explored its kinematic behavior under external drivers such as rainfall and snowmelt, including seasonal variations, transient acceleration events, and episodic failure events. This study highlights the benefits of leveraging multi-source remote sensing data to investigate shallow deforming landslides, which is advantageous for the implementation of effective control and engineering interventions to mitigate potential landslide disasters. The main conclusions are as follows:

1. The formation of the Lashagou landslide group has been specifically categorized into three periods: L8 was formed before the construction of highway G310; L3, L4, L5, and L6 were formed during construction; and L1, L2, L7, R1, and R2 were formed within five years of the completion of the highway.
2. Hillslope excavation during the construction of the highway was the direct cause and prerequisite for the formation of the landslide group, whereas summer precipitation and spring snowmelt were the primary driving factors contributing to its continuous downward movement.
3. The occurrence of freeze–thaw landslides in spring may be related to the release of internal groundwater rather than the infiltration of meltwater.
4. Both the long-term seasonal downslope movement and transient acceleration events of the Lashagou landslide group were strongly controlled by rainfall, and there was a time lag of approximately 1–2 days between the transient acceleration and heavy rainfall events. More importantly, the movement of shallow loess landslides is not only highly sensitive to rainfall intensity but is also influenced by the preceding rainfall.

Author Contributions: Conceptualization, Q.F., S.Z. and Y.N.; methodology, Q.F., S.Z., J.S. and X.Z.; software, Q.F. and J.J.; validation, Q.F., S.Z., W.W. and X.L.; writing—original draft preparation, Q.F. and S.Z.; writing—review and editing, S.Z. and Y.N.; supervision, S.Z. and Y.N. All authors have read and agreed to the published version of the manuscript.

Funding: This work was financially supported by the National Key Research and Development Program of China (Grant No. 2020YFC1512000); The National Natural Science Foundation of China Projects (grant nos. 42074041 and 42307255); Shaanxi Province Science and Technology Innovation Team (Ref. 2021TD-51); The innovation team of ShaanXi Provincial Tri-Qin Scholars with Geoscience Big Data and Geohazard Prevention (2022) and the Open Project from Observation and Research Station of Ground Fissure and Land Subsidence in Ministry of Natural Resources (grant no. 2022-04).

Data Availability Statement: The data used to support the findings of this study are available from the corresponding author upon reasonable request.

Acknowledgments: We thank Andrew Hooper for sharing the StaMPS. We thank the European Space Agency (ESA) for providing the ENVISAT ASAR and Sentinel-1A datasets, the National Aeronautics and Space Administration (NASA) for providing the SRTM DEM, and the Japan Aerospace Exploration Agency (JAXA) for providing the ALOS-1 and AW3D DEM datasets.

Conflicts of Interest: The authors declare no conflicts of interest.

References

1. Xi, H.P.; Wei, M.P.; Xuan, C.Q. Analysis of landslide stability of a highway interchange section in Linxia. *Subgrade Eng.* **2022**, *6*, 220–225. [CrossRef]
2. Lacroix, P.; Bièvre, G.; Pathier, E.; Kniess, U.; Jongmans, D. Use of Sentinel-2 images for the detection of precursory motions before landslide failures. *Remote Sens. Environ.* **2018**, *215*, 507–516. [CrossRef]
3. Walton, G.; Christiansen, C.; Kromer, R.; Silaev, A. Evaluation of rockfall trends at a sedimentary rock cut near Manitou Springs, Colorado, using daily photogrammetric monitoring: Evaluation of rockfall trends at a sedimentary rock cut. *Landslides* **2023**, *20*, 2657–2674. [CrossRef]
4. Dille, A.; Kervyn, F.; Handwerger, A.L.; d'Oreye, N.; Derauw, D.; Bibentyo, T.M.; Samsonov, S.; Malet, J.P.; Kervyn, M.; Dewitte, O. When image correlation is needed: Unravelling the complex dynamics of a slow-moving landslide in the tropics with dense radar and optical time series. *Remote Sens. Environ.* **2021**, *258*, 112402. [CrossRef]
5. Liu, X.J.; Zhao, C.Y.; Zhang, Q.; Lu, Z.; Li, Z.H.; Yang, C.S.; Zhu, W.; Zeng, J.L.; Chen, L.Q.; Liu, C.J. Integration of Sentinel-1 and ALOS/PALSAR-2 SAR datasets for mapping active landslides along the Jinsha River corridor, China. *Eng. Geol.* **2021**, *284*, 106033. [CrossRef]
6. Hooper, A.; Segall, P.; Zebker, H. Persistent scatterer interferometric synthetic aperture radar for crustal deformation analysis, with application to Volcán Alcedo, Galápagos. *J. Geophys. Res. Solid Earth* **2007**, *112*, B07407. [CrossRef]
7. Bayer, B.; Simoni, A.; Schmidt, D.; Bertello, L. Using advanced InSAR techniques to monitor landslide deformations induced by tunneling in the Northern Apennines, Italy. *Eng. Geol.* **2017**, *226*, 20–32. [CrossRef]
8. Rosi, A.; Tofani, V.; Tanteri, L.; Tacconi, S.C.; Agostini, A.; Catani, F.; Casagli, N. The new landslide inventory of Tuscany (Italy) updated with PS-InSAR: Geomorphological features and landslide distribution. *Landslides* **2018**, *15*, 5–19. [CrossRef]
9. Roy, P.; Martha, T.R.; Khanna, K.; Jain, N.; Kumar, K.V. Time and path prediction of landslides using InSAR and flow model. *Remote Sens. Environ.* **2022**, *271*, 112899. [CrossRef]
10. Zhang, S.C.; Fan, Q.Y.; Niu, Y.F.; Qiu, S.C.; Si, J.Z.; Feng, Y.H.; Zhang, S.Q.; Song, Z.W.; Li, Z.H. Two-dimensional deformation monitoring for spatiotemporal evolution and failure mode of Lashagou landslide group, Northwest China. *Landslides* **2023**, *20*, 447–459. [CrossRef]
11. Costantini, M.; Ferretti, A.; Minati, F.; Falco, S.; Trillo, F.; Colombo, D.; Novali, F.; Malvarosa, F.; Mammone, C.; Vecchioli, F.; et al. Analysis of surface deformations over the whole Italian territory by interferometric processing of ERS, Envisat and COSMO-SkyMed radar data. *Remote Sens. Environ.* **2017**, *202*, 250–275. [CrossRef]
12. Wu, S.B.; Yang, Z.F.; Ding, X.L.; Zhang, B.C.; Zhang, L.; Lu, Z. Two decades of settlement of Hong Kong International Airport measured with multi-temporal InSAR. *Remote Sens. Environ.* **2020**, *248*, 111976. [CrossRef]
13. Cignetti, M.; Godone, D.; Notti, D.; Lanteri, L.; Giordan, D. Impacts on mountain settlements of a large slow rock-slope deformation: A multi-temporal and multi-source investigation. *Landslides* **2024**, *21*, 327–337. [CrossRef]
14. Xu, Y.K.; Kim, J.; George, D.L.; Lu, Z. Characterizing seasonally rainfall-driven movement of a translational landslide using SAR imagery and SMAP soil moisture. *Remote Sens.* **2019**, *11*, 2347. [CrossRef]
15. Wasowski, J.; Bovenga, F. Investigating landslides and unstable slopes with satellite Multi Temporal Interferometry: Current issues and future perspectives. *Eng. Geol.* **2014**, *174*, 103–138. [CrossRef]
16. Moretto, S.; Bozzano, F.; Mazzanti, P. The role of satellite InSAR for landslide forecasting: Limitations and openings. *Remote Sens.* **2021**, *13*, 3735. [CrossRef]
17. Xiao, T.; Huang, W.; Deng, Y.k.; Tian, W.M.; Sha, Y.L. Long-Term and Emergency Monitoring of Zhongbao Landslide Using Space-Borne and Ground-Based InSAR. *Remote Sens.* **2021**, *13*, 1578. [CrossRef]
18. Rodriguez, J.; Deane, E.; Macciotta, R.; Evans, T.; Gräpel, C.; Skirrow, R. Practical evaluation of single-frequency dGNSS for monitoring slow-moving landslides. *Landslides* **2021**, *18*, 3671–3684. [CrossRef]

19. Huang, G.W.; Du, S.; Wang, D. GNSS techniques for real-time monitoring of landslides: A review. *Satell. Navig.* **2023**, *4*, 5. [CrossRef]
20. Carlà, T.; Tofani, V.; Lombardi, L.; Raspini, F.; Bianchini, S.; Bertolo, D.; Thuegaz, P.; Casagli, N. Combination of GNSS, satellite InSAR, and GBInSAR remote sensing monitoring to improve the understanding of a large landslide in high alpine environment. *Geomorphology* **2019**, *335*, 62–75. [CrossRef]
21. Cenni, N.; Fiaschi, S.; Fabris, M. Integrated use of archival aerial photogrammetry, GNSS, and InSAR data for the monitoring of the Patigno landslide (Northern Apennines, Italy). *Landslides* **2021**, *18*, 2247–2263. [CrossRef]
22. Yang, H.Q.; Song, K.L.; Chen, L.C.; Qu, L.L. Hysteresis effect and seasonal step-like creep deformation of the Jiuxianping landslide in the Three Gorges Reservoir Region. *Eng. Geol.* **2023**, *317*, 107089. [CrossRef]
23. Hu, X.; Lu, Z.; Pierson, T.C.; Kramer, R.; George, D.L. Combining InSAR and GPS to determine transient movement and thickness of a seasonally active low-gradient translational landslide. *Geophys. Res. Lett.* **2018**, *45*, 1453–1462. [CrossRef]
24. Zhao, L.D.; Ma, X.P.; Xiang, Z.F.; Zhang, S.C.; Hu, C.; Zhou, Y.; Chen, G.C. Landslide deformation extraction from terrestrial laser scanning data with weighted least squares regularization iteration solution. *Remote Sens.* **2022**, *14*, 2897. [CrossRef]
25. Chen, B.; Song, C.; Chen, Y.; Li, Z.; Yu, C.; Liu, H.; Jiang, H.; Liu, Z.; Cai, X.; Nai, Y.; et al. Emergency identification and influencing factor analysis of coseismic landslides and building damages induced by the 2023 Ms 6.2 Jishishan (Gansu, China) earthquake. *Geomat. Inf. Sci. Wuhan Univ.* **2024**. [CrossRef]
26. Huang, G.; Jing, C.; Li, D.; Huang, X.; Wang, L.; Zhang, K.; Yang, H.; Xie, S.; Bai, Z.; Wang, D. Deformation analysis of Jishishan Mw 6.2 earthquake on the landslide hazard areas. *Geomat. Inf. Sci. Wuhan Univ.* **2023**. [CrossRef]
27. Liu, S.; He, B.; Wang, T.; Liu, J.; Cao, J.; Wang, H.; Zhang, S.; Li, K.; Li, R.; Zhang, Y.; et al. Development characteristics and susceptibility assessment of co-seismic geological hazards in Jishishan Ms 6.2 earthquake, Gansu Province. *J. Geomech.* **2024**. [CrossRef]
28. Cai, G.Z.; Yang, Z.H.; Wang, D.P.; Sun, Y.L.; Zhou, S.D. Cause analysis and defense countermeasures of geological hazards in Linxia city, Gansu Province. *J. Agric. Catastrophology* **2015**, *5*, 32–35. [CrossRef]
29. Yang, W.M.; Wan, F.P.; Ma, S.Q.; Qu, J.K.; Zhang, C.S.; Tang, H.B. Hazard assessment and formation mechanism of debris flow outbursts in a small watershed of the Linxia Basin. *Front. Earth Sci.* **2023**, *10*, 994593. [CrossRef]
30. Zhou, X.; Zhang, S.C.; Zhang, Q.; Liu, Q.; Ma, Z.M.; Wang, T.; Tian, J.; Li, X.R. Research of deformation and soil moisture in loess landslide simultaneous retrieved with ground-based GNSS. *Remote Sens.* **2022**, *14*, 5687. [CrossRef]
31. Qiang, D.X.; Ma, H.Z.; Zhu, Z.P.; Xun, Y.M. Spatial distribution and analysis of debris flow in Jishishan county of Gansu province. *Bull. Surv. Mapp.* **2022**, *7*, 107–111. [CrossRef]
32. Zhong, C.H.; Li, X.R.; Zhang, S.C.; Wang, X.Q. GAMIT/TrackRT used in landslide real-time GNSS monitoring. *Sci. Surv. Mapp.* **2022**, *47*, 57–65. [CrossRef]
33. Shi, X.G.; Zhang, L.; Zhou, C.; Li, M.H.; Liao, M.S. Retrieval of time series three-dimensional landslide surface displacements from multi-angular SAR observations. *Landslides* **2018**, *15*, 1015–1027. [CrossRef]
34. Pradhan, S.; Toll, D.G.; Rosser, N.J.; Brain, M.J. An investigation of the combined effect of rainfall and road cut on landsliding. *Eng. Geol.* **2022**, *307*, 106787. [CrossRef]
35. Iverson, R.M.; George, D.L.; Allstadt, K.E.; Reid, M.E.; Collins, B.D.; Vallance, J.W.; Schilling, S.P.; Godt, J.W.; Cannon, C.M.; Magirl, C.S.; et al. Landslide mobility and hazards: Implications of the 2014 Oso disaster. *Earth Planet. Sci. Lett.* **2015**, *412*, 197–208. [CrossRef]
36. Hu, X.; Wang, T.; Pierson, T.C.; Lu, Z.; Kim, J.; Cecere, T.H. Detecting seasonal landslide movement within the Cascade landslide complex (Washington) using time-series SAR imagery. *Remote Sens. Environ.* **2016**, *187*, 49–61. [CrossRef]
37. Kong, J.X.; Zhuang, J.Q.; Peng, J.B.; Ma, P.H.; Zhan, J.W.; Mu, J.Q.; Wang, J.; Zhang, D.; Zheng, J.; Fu, Y.T.; et al. Failure mechanism and movement process of three loess landslides due to freeze-thaw cycle in the Fangtai village, Yongjing County, Chinese Loess Plateau. *Eng. Geol.* **2023**, *315*, 107030. [CrossRef]

remote sensing

Article

Deformation Characteristics and Activation Dynamics of the Xiaomojiu Landslide in the Upper Jinsha River Basin Revealed by Multi-Track InSAR Analysis

Xu Ma [1], Junhuan Peng [1,*], Yuhan Su [1], Mengyao Shi [2], Yueze Zheng [1,3], Xu Li [1] and Xinwei Jiang [1]

1 School of Land Science and Technology, China University of Geosciences, Beijing 100083, China; maxu@email.cugb.edu.cn (X.M.); s_yuhan@163.com (Y.S.); zheng_cugb@163.com (Y.Z.); lixu2021@163.com (X.L.); gzhzy1996@163.com (X.J.)
2 Faculty of Land Resources Engineering, Kunming University of Science and Technology, Kunming 650093, China; shimengyao@kust.edu.cn
3 Beijing Institute of Surveying and Mapping, Beijing 100038, China
* Correspondence: pengjunhuan@163.com

Abstract: The upper Jinsha River, located in a high-mountain gorge with complex geological features, is highly prone to large-scale landslides, which could result in the formation of dammed lakes. Analyzing the movement characteristics of the typical Xiaomojiu landslide in this area contributes to a better understanding of the dynamics of landslides in the region, which is of great significance for landslide risk prediction and analysis. True displacement data on the surface of landslides are crucial for understanding the morphological changes in landslides, providing fundamental parameters for dynamic analysis and risk assessment. This study proposes a method for calculating the actual deformation of landslide bodies based on multi-track Interferometric Synthetic Aperture Radar (InSAR) deformation data. It iteratively solves for the optimal true deformation vector of the landslide on a per-pixel basis under a least-squares constraint based on the assumption of consistent displacement direction among adjacent points on the landslide surface. Using multi-track Sentinel data from 2017 to 2023, the line of sight (LOS) accumulative de-formation of the Xiaomojiu landslide was obtained, with a maximum LOS deformation of −126 mm/year. The true surface displacement of the Xiaomojiu landslide after activation was calculated using LOS deformation. The development of two rotational sub-slipping zones on the landslide body is inferred based on the distribution of actual displacements along the central profile line. Analysis of temporal changes in water body area data revealed that the Xiaomojiu landslide was activated after a barrier lake event and continuously moved due to the influence of higher water levels' in the river channel. In conclusion, the proposed method can be applied to calculate the true surface displacement of landslides with complex mechanisms for analyzing the movement status of landslide bodies. Furthermore, the spatiotemporal analysis of the Xiaomojiu landslide characteristics can support analyzing the mechanisms of similar landslides in the Jinsha River Basin.

Keywords: Sentinel-1; Xiaomojiu landslide; 3D displacements; SBAS-InSAR; Multi-Track InSAR

Citation: Ma, X.; Peng, J.; Su, Y.; Shi, M.; Zheng, Y.; Li, X.; Jiang, X. Deformation Characteristics and Activation Dynamics of the Xiaomojiu Landslide in the Upper Jinsha River Basin Revealed by Multi-Track InSAR Analysis. *Remote Sens.* **2024**, *16*, 1940. https://doi.org/10.3390/rs16111940

Academic Editor: Alex Hay-Man Ng

Received: 16 April 2024
Revised: 20 May 2024
Accepted: 21 May 2024
Published: 28 May 2024

1. Introduction

Landslides, a common geological hazard worldwide, cause billions of dollars in losses and thousands of casualties annually [1,2]. The frequency of landslides has been increasing annually due to human engineering activities and the increase in extreme rainfall caused by global climate change [3]. In the Qinghai–Tibet Plateau region of China, where geological conditions are complex, the significant uplift of the plateau crust and the rapid incision of rivers have led to widespread instability on both sides of river valleys [4,5]. The Jinsha River valley's steep terrain and narrow channels make it prone to landslides that can easily block the river, forming barrier lakes [6]. In 2018, the Baige landslide, a high-altitude, long-range event, occurred in the area, blocking the main channel of the upper reaches of the Jinsha River and creating a barrier lake [7,8]. Consequently, the breach of the barrier lake dam led to a flood disaster, resulting in a direct economic loss of USD 960 million [9,10]. Moreover, the abrupt fluctuations in the river channel's water level following the dam break stimulated the reactivation of the Xiaomojiu ancient landslide situated 5 km upstream [11]. Therefore, continuous monitoring of the Xiaomojiu landslide is crucial to prevent similar disasters [12].

In landslide-prone areas, various remote sensing technologies are currently being utilized for comprehensive surface deformation monitoring [13–16]. Among these technologies, InSAR is widely used due to its broad coverage and high accuracy [17,18]. For instance, InSAR technology has been utilized to capture the pre-failure deformation of the Baige landslide, confirming the adherence of the landslide deformation history to the three-stage creep theory [19]. Particularly during the initial phase of landslide deformation, InSAR technology can track the surface displacement of the landslide mass, which, in turn, facilitates the deduction of geometric characteristics of the landslide surface [20]. These results help analyze the structure of landslides and serve as an essential technical tool for assisting landslide motion analysis.

InSAR technology has proven to be a valuable tool for monitoring surface deformation; however, it is not without limitations. In particular, the polar orbit side-looking mode of InSAR satellites means that a single synthetic aperture radar SAR platform typically only captures one-dimensional deformation along the radar line of sight (LOS) [21]. This limitation is compounded by topographic shadowing, vegetation coverage, and surface deformation gradients, which can significantly impact monitoring results. The challenge arises when the LOS displacement, which is the projection of the surface displacement vector onto the LOS, is perpendicular to the surface displacement [22], making it challenging to detect surface deformation accurately. To overcome the limitations of InSAR technology and capture a broader range of surface deformation scenarios, obtaining measurements in three different directions or having equivalent prior information is essential [23]. By restoring the true surface deformation through multiple measurements, researchers can enhance the accuracy and reliability of InSAR monitoring results.

In estimating true three-dimensional deformation from one-dimensional observations, the rank deficiency caused by insufficient observations can be improved by imposing additional constraints or including unrelated observations. Therefore, landslide deformation inversion based on InSAR technology can be categorized into two main approaches. The first method involves increasing constraints by incorporating true observations, which can be further categorized into homogeneous and heterogeneous subclasses. In the homogeneous subclass, it is defined as utilizing Spaceborne SAR data to obtain horizontal deformation in the north-south direction through alternative processing methods [23] (e.g., Pixel Offset Tracking (POT) [24], Multi-Aperture Interferometry (MAI) [25], and various combinations of satellite observations). However, the accuracy of POT technology is influenced by data spatial resolution. In contrast, MAI technology is constrained by coherence, resulting in lower accuracy than LOS deformation obtained through time-series InSAR. In the heterogeneous subclass utilizing non-spaceborne SAR data, such as Global Navigation Satellite Systems (GNSS) [26], unmanned aerial vehicles [27], and ground-based SAR [28,29]. Nonetheless, obtaining diverse multi-source observational data is often challenging due to factors like the spatial distribution of landslides and data acquisition costs.

The second method, which involves adding constraints through prior information to reduce the required true observations [30], has become a standard technique for landslide deformation inversion. For example, in east-west landslides, the north-south horizontal deformation is often neglected [20], or models like the Surface Parallel Flow Model (SPFM) are used [31–33]. SPFM is frequently applied to translational landslides [34] but is unsuitable for rotational landslides with arcuate sliding surfaces [35,36]. However, assumptions made in advance regarding the sliding direction of the landslide mass may significantly differ from the true complex movements of the landslide. Therefore, in some studies [37], based on the local parallel flow assumption, landslides are assumed to be composed of multiple landslide sub-blocks, where each sub-block shares a uniform sliding surface motion direction. The partitioning of landslide sub-blocks plays a significant role in the reliability of this approach's deformation inversion.

Therefore, this study does not initially divide landslide blocks but infers the real 3D (three-dimensional) displacements on the landslide surface by assuming that adjacent points on the surface are in the same direction. We apply this approach to the Xiaomojiu landslide in the Jinsha River Basin. The true motion displacements and directions on the landslide surface were inferred using multi-track Sentinel data. Compared to conventional methods, this approach does not require prior knowledge of the landslide block boundaries and better reproduces the true motion characteristics of the landslide body. The method calculates the actual deformation of landslides, which can be used to analyze the spatiotemporal attributes of other landslides in the area. Additionally, calculating the actual landslide deformation can help deduce the sliding surface and thickness of the landslide body during the slow deformation stages [31,38,39], providing data support for assessing the extent of damage caused by landslide failure in subsequent simulations. Based on the actual deformation results and water area data, this study discusses the spatiotemporal variations and driving factors of the Xiaomojiu landslide.

2. Study Area

The Jinsha River Basin, situated in the eastern part of the Qinghai-Tibet Plateau, is characterized by intense crustal movement and frequent seismic activity [40]. This basin exhibits complex geological and geomorphological conditions, forming numerous large-scale, deep-seated creeping landslides [41]. One such significant landslide, the Xiaomojiu landslide, is situated in the V-shaped valley of the upper Jinsha River. The main steep slope of this landslide has a peak elevation of approximately 3655 m, while the toe of the landslide sits at around 2895 m, resulting in a relative height difference of 760 m between the two points. The overall slope inclination ranges between 30~35° (Figure 1). Due to lateral erosion from the Jinsha River, the landslide is typical of a high-steep valley-shaped fragmented loose body [42]. The geological lithology in the area mainly comprises metamorphic ultramafic rocks (including chlorite schist and serpentinite), metamorphic clastic rocks (such as quartz-sericite schist, biotite-muscovite quartz schist, hornfels, and slate), as well as marble [43]. The structural context of the region is primarily influenced by the Jinsha River fault zone, with the main fault trending NW-SE located approximately 1 km from the Xiaomojiu landslide. The presence of this fault significantly impacts the development of the landslide [11,44]. Historical records document destructive events associated with this landslide [45]. Previous exploration studies have revealed that the primary components of the landslide body consist of gravel and clay, with no bedrock detected even at a drilling depth of 60 m, highlighting the thick accumulation characteristics of the landslide body [42].

The Baige landslide event had a significant impact on the water level in the region. Before the Baige landslide, the average elevation of the riverbed in that section was 2870 m. However, after the secondary dam breach, the elevation at the bottom of the spillway rose to 2903 m [46]. This event resulted in an overall increase in the water level upstream of the Baige landslide, significantly affecting the surrounding geological environment. As of 21 April 2023, optical images show that the remaining dam debris still blocks the Jinsha River channel (Figure 1d). The dam breach event directly triggered the activation of the Xiaomojiu landslide [11], leading it into the uniform deformation stage [47]. In the long term, the continued higher water level may have a lasting adverse effect on the stability of the Xiaomojiu landslide mass.

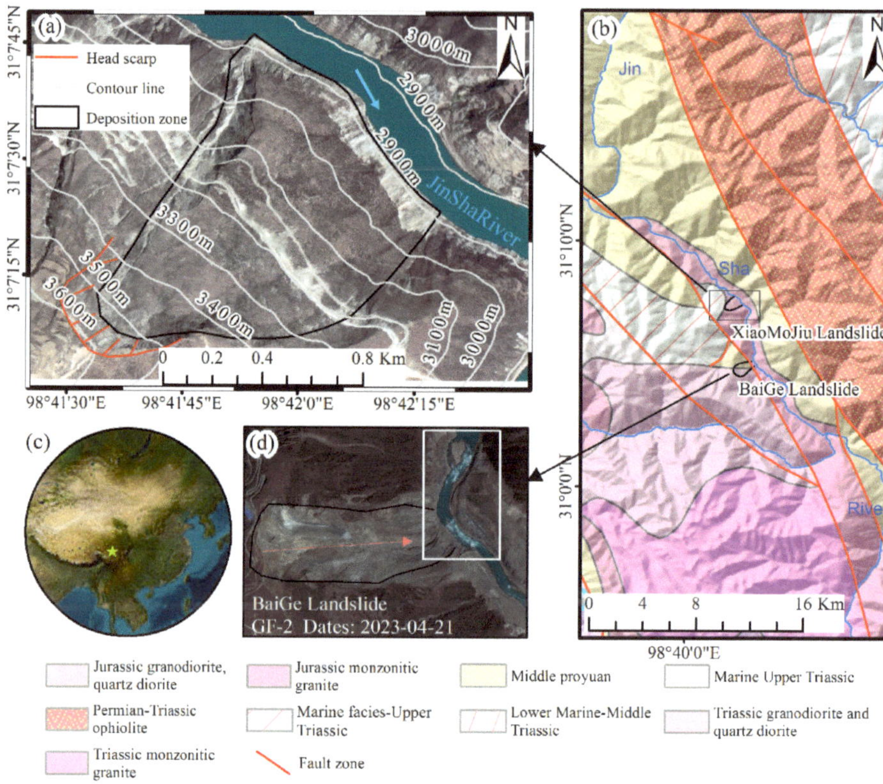

Figure 1. (**a**) Optical image of the Xiaomojiu landslide, (**b**) geological map of the study area at 1:2,500,000 scale, (**c**) map showing the location of the Xiaomojiu landslide, (**d**) schematic image showing the remaining debris blocking the river channel after the breach of the Baige landslide barrier lake.

3. Data and Methods

3.1. Data

This study to obtain the long-term deformation of the Xiaomojiu landslide, collected Sentinel-1A/B images from October 2017 to September 2023, including 139 ascending orbit images from path 99 (A99), 153 descending orbit images from path 33 (D33), and 168 descending orbit images from path 106 (D106). The detailed information on the Sentinel-1 SAR data is shown in Table 1. We sequentially processed the data from each direction to obtain time-series deformation data. The ALOS World 3D-30m (AW3D30) Digital Elevation Model (DEM) was obtained from the JAXA Earth Observation Research Center [48], and precise orbit determination (POD) products were acquired from the Copernicus Sentinels POD Data Hub. Additionally, monthly total precipitation data from 2017 to 2022, with a spatial resolution of 1 km, were obtained from the Institute of Tibetan Plateau Research, Chinese Academy of Sciences [49]. Optical image data from GF-2 and GF-7 were obtained from the China Aero Geophysical Survey and Remote Sensing Center for Land and Resources (AGRS).

Table 1. Detailed information on Sentinel-1 data.

Period	Orbit	Incidence Angle (°)	Heading Angle (°)	Path	Frame	Number of Scenes
8 October 2017~ 19 September 2023	Ascending	35.76	347.22	99	1280	139
15 October 2017~ 20 October 2023	Descending	43.92	192.77	33	487	153
8 October 2017~ 7 September 2023	Descending	31.92	192.80	106	486	168

3.2. Methods

This section focuses on the operations performed to obtain the true displacement of the landslide surface. We can split the proposed approach into two processing steps: First, SBAS-InSAR obtains three LOS cumulative deformations. Second, adjacent orientation consistency is used to invert the landslide to true displacement. The workflow diagram illustrating the InSAR data processing, and the method proposed in this paper is detailed in Figure 2.

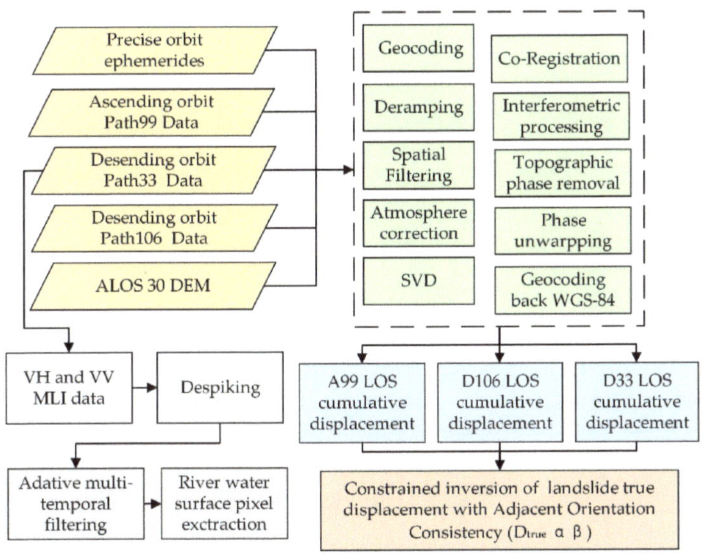

Figure 2. Flowchart for the SAR data processing and true deformation inversion.

3.2.1. InSAR Processing

This study employs the Small Baseline Subset InSAR (SBAS-InSAR) [50] processing method to obtain a long-term displacement time series of surface movement in the landslide area along three orbital LOS directions. Firstly, the ascending and descending images covering the study area are individually geocoded and registered. Subsequently, the registered data are geocoded, and corresponding mask files are generated to remove geometric distortions and shadow regions. A multi-look ratio of 4 (range) and 1 (azimuth) was employed on SAR images to enhance the signal-to-noise ratio and image consistency while balancing the density of landslide monitoring points. This ratio was chosen to match the spatial resolution of the imagery, with DEM resolution interpolated to the same level. Next, differential interferograms are generated from the registered images, with the temporal and spatial baselines of interferograms illustrated in Figure 3. A Goldstein filter with a window size of 32 × 32 pixels is applied to the interferogram to remove the topographic and flat-earth phase components to enhance the signal-to-noise ratio. The phase unwrapping is then performed using the Minimum Cost Flow algorithm (MCF) [51]. At this stage, we can examine the primary components of the unwrapped phase interferograms in detail in Equation (1). Linear regression is applied to remove the terrain residual phase components φ_{res} from the unwrapped phase. Using empirical linear elevation models and spatiotemporal filtering eliminates atmospheric phase components φ_{atm} [52]. Finally, the deformation phase components φ_{disp} are estimated using Singular Value Decomposition (SVD), cumulative deformation is calculated, and the results are re-encoded into the WGS84 coordinate system for further assessment of landslide 3D movement.

$$\varphi_{unw} = \varphi_{res} + \varphi_{atm} + \varphi_{disp} + \varphi_{noise} \tag{1}$$

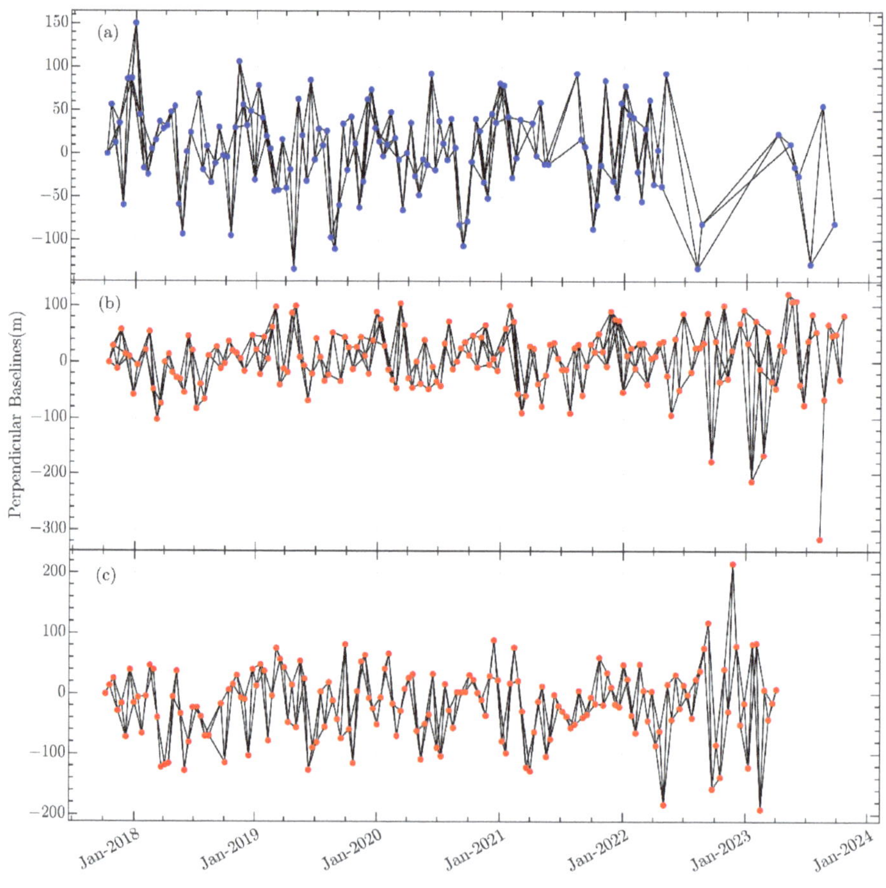

Figure 3. This study used the spatial and temporal baselines of the Sentinel-1 data from Ascending orbit Path 99 (**a**), Descending orbit Path 33 (**b**), and Descending orbit Path 106 (**c**). Blue dots represent ascending track data, and red dots represent descending track data.

3.2.2. Landslide True Ground Displacement and LOS Conversion Coefficients

The LOS deformation obtained from time-series InSAR represents the projection of true surface displacement in that direction, as depicted in Figure 4. Unit vectors of LOS direction in the north, earth, and up (NEU) directions $[\sin\theta\sin\varphi \quad -\sin\theta\cos\varphi \quad \cos\theta]^T$ represent the satellite's incidence angle and denote the satellite's azimuth angle. The unit vectors of true ground deformation in the NEU directions can be represented as $[\cos\alpha\cos\beta \quad \cos\alpha\sin\beta \quad \sin\alpha]^T$, where α is the angle between the true displacement vector and the horizontal plane $\alpha \in [-\pi/2, \pi/2]$ and $\alpha < 0$ when the vertical component of the true deformation vector is downward. The β represent the azimuth angles of the true displacement vector [53].

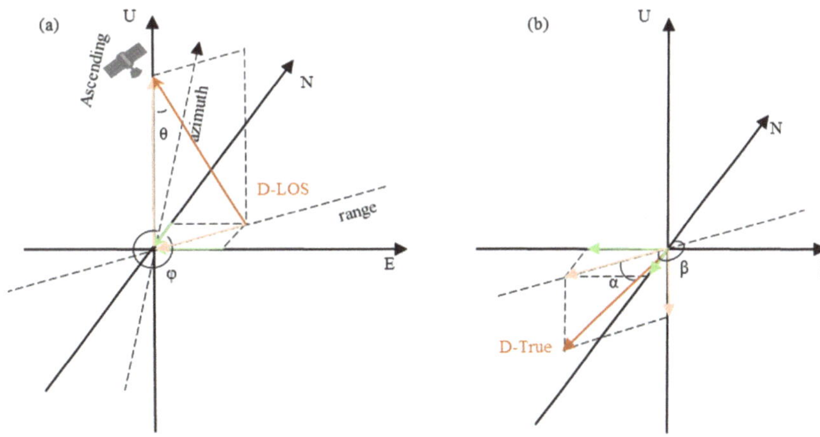

Figure 4. (**a**) NEU-directional projection delineates the satellite LOS motion vector. (**b**) NEU-directional projection of the authentic ground deformation vector.

The geometric transformation relationship between the observed LOS displacement L_{los} and the true surface deformation D_{true} can be expressed by the Equation (2).

$$L_{los} + \Delta l = C \cdot D_{true} \tag{2}$$

where L_{los} takes negative values when far from the satellite and positive values when close to the satellite; Δl represents the observation error, and C represents the projection coefficient of the LOS observation direction onto the unit vector of the true deformation direction [53]. Its calculation formula is shown in the Equation (3).

$$\begin{aligned} C &= \sin\theta \sin\varphi \cdot \cos\alpha \cos\beta - \sin\theta \cos\varphi \cdot \cos\alpha \sin\beta + \cos\theta \cdot \sin\alpha \\ &= \sin\theta \cdot \cos\alpha \cdot \sin(\varphi - \beta) + \cos\theta \cdot \sin\alpha \end{aligned} \tag{3}$$

The true surface displacement vector consists of D_{true}, α, and β, representing the magnitude and direction of the true displacement vector. When the assumption of surface parallel flow is employed, $\alpha = -\angle Slope$, $\beta = \angle Aspect$. In this case, the conversion coefficients are called C-index [54–56], commonly used to evaluate the sensitivity of detecting landslide deformations in the LOS direction or to assist in landslide identification [2,55]. The C-index is equivalent to the projection coefficient of surface deformation in the LOS direction for translational landslides, where the movement direction aligns with the slope gradient.

To preliminarily analyze the movement characteristics of the Xiaomojiu landslide, we computed the C-index results for the Xiaomojiu landslide area on three orbits based on the slope and aspect derived from the AW3D DEM (see Figure 5), following Equation (3). The C-index distribution reveals that, under the assumption of surface parallel flow, the sensitivity to deformations associated with movement along the slope gradient is in the order of A99 > D33 > D106. Specifically, deformations projected onto A99 are predominantly negative, while those on the descending orbits D33 and D106 are primarily positive. Due to differences in the incidence angles, the absolute values of the C-index on the D106 orbit are smaller than those on the D33 orbit. Under the assumption of surface parallel flow, the distribution of positive and negative LOS deformations theoretically aligns with the C-index distribution when the landslide body moves along the slope gradient.

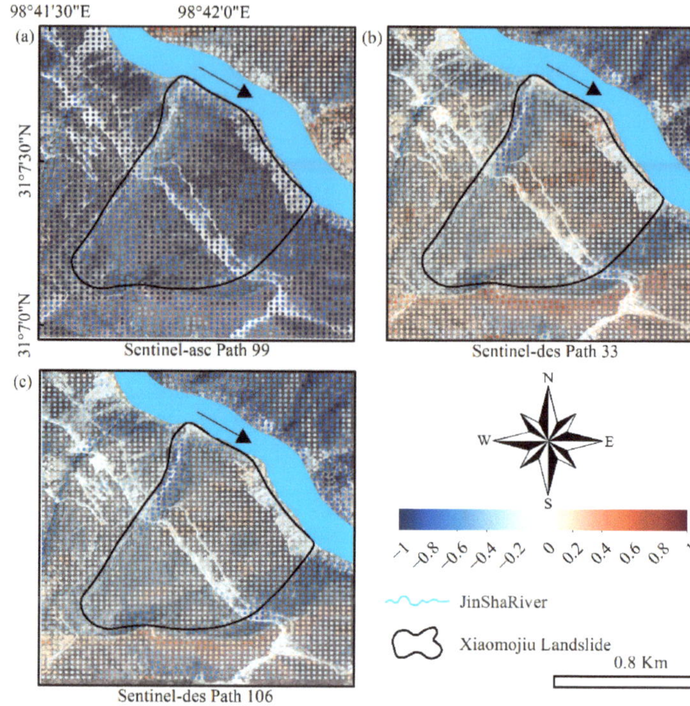

Figure 5. C-index distribution for the Xiaomojiu landslide on three orbits: (**a**) orbit A99, (**b**) orbit D33, and (**c**) orbit D106. The black arrow represents the direction of the river flow. Positive values indicate proximity to the satellite, while negative values indicate distance from the satellite. When $|C| = 1$, the LOS direction is parallel to the direction of the steepest slope, and when $C = 0$, the LOS direction is perpendicular to the direction of the steepest slope.

In cases of non-translational landslide movement, the C-index may fail to accurately reflect the relationship between the actual deformation of the landslide and the LOS deformation. For a point on the landslide body, the observations in different LOS directions represent the projection of the true deformation vector onto those LOS directions. The detectability and sensitivity of landslide deformation measurements are determined by factors such as the ground deformation gradient, InSAR imaging geometry, and coherence [57]. When the landslide deformation is not affected by geometric distortions and is detectable, the geometric relationship between the LOS direction determined by InSAR imaging geometry and the true deformation vector is crucial for estimating the true deformation. Therefore, this study computed the projection coefficients of true deformation in arbitrary directions onto the three LOS directions (see Figure 6). The azimuth look direction (ALD) is the clockwise angle between the LOS projection on the horizontal plane and the north direction [58]. Given the possibility of landslide expansion and toe bulging, the deformation of the landslide body may include a vertical component. Thus, two scenarios were considered: corresponding to Figure 6a–c, while corresponding to Figure 6d–f. In observable regions on the ground, when the LOS deformation vector is nearly perpendicular to the actual deformation vector, the projection of actual deformation onto the LOS direction approaches zero, indicating low sensitivity of the LOS direction to the actual deformation at that location. Figure 6 shows that data from different orbits can complement each other, except for horizontal deformations in the north–south direction.

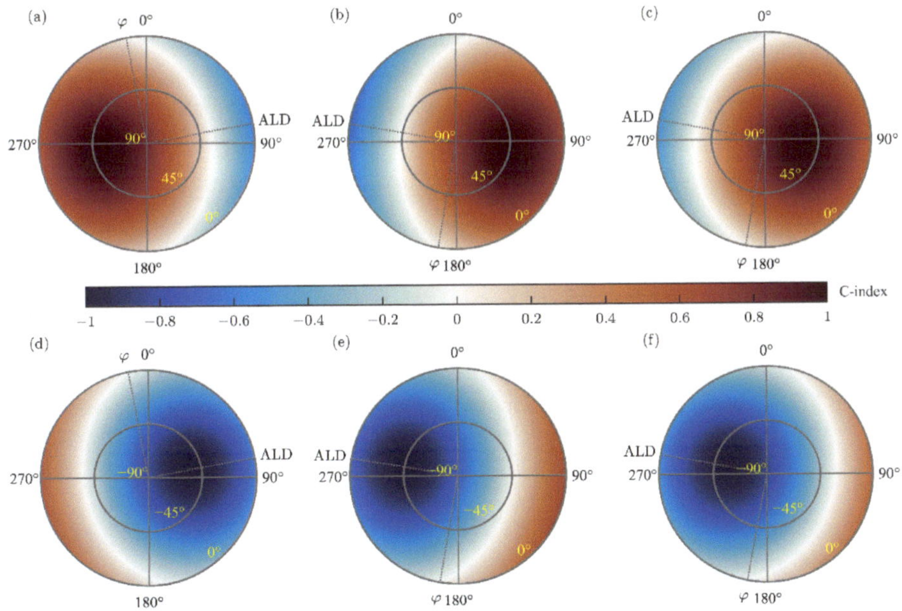

Figure 6. Different combinations of ground displacement directions correspond to projection coefficient C-index values, (**a**,**d**)ascending path 99; (**b**,**e**) descending path 33; (**c**,**f**) descending path 106.

3.2.3. Adjacent Orientation Consistency Constrained Inversion of True Displacement

In addition to determining the direction of true displacement in advance using the SPFM to solve for the true deformation vector, other methods, such as prior information, can be employed to partition the landslide body into regions. This approach assumes that the sliding direction within individual landslide blocks is consistent, enabling the solution of the true deformation direction within each landslide sub-block by combining observations from different directions [37,59]. Alternatively, fixed motion directions can be assumed to solve for the true deformation displacement, as demonstrated by models such as the Aspect Parallel Flow Model (APFM) and the Steepest Terrain Following Model (STFM) [60–62]. Considering the similarity in the motion direction of adjacent points in space, this study assumes that the true displacement direction of neighboring points in the landslide space is consistent, facilitating the solution of the true deformation of the landslide body.

Assuming there are m different orbits or types of data observations for n spatially adjacent points within the same sliding unit on the landslide body, it can be assumed that they share a consistent displacement direction $\begin{bmatrix}\alpha & \beta\end{bmatrix}^T$. However, given that the true displacement $\begin{bmatrix}x_1 & x_2 & \cdots & x_n\end{bmatrix}^T$ of the n points may vary, a total of $m{\cdot}n$ observation equations can be constructed. For the jth point in the ith orbit observation, the projection of the true deformation in its observation direction can be represented by the following nonlinear function:

$$\begin{array}{c}\widetilde{L}_k = [\sin\theta_i\cdot\cos\alpha\cdot\sin(\varphi_i-\beta)+\cos\theta_i\cdot\sin\alpha]\cdot x_j \\ i\in[1,m], j\in[1,n]\end{array} \tag{4}$$

where $\widetilde{L}_k$ represents the corresponding function for this observation and $k = (i-1)n + j$ denotes the observation sequence number in the observation equation set. Letting

$f_k(\alpha, \beta, x_j) = [\sin\theta_i \cdot \cos\alpha \cdot \sin(\varphi_i - \beta) + \cos\theta_i \cdot \sin\alpha] \cdot x_j$, a system of the following nonlinear equations can be constructed for all points in the point set.

$$\begin{aligned} \underset{mn\times 1}{\mathbf{F}}(\mathbf{X}) &= \begin{bmatrix} f_1 & f_2 & \cdots & f_k & \cdots & f_{mn} \end{bmatrix}^T \\ \underset{n+2}{\mathbf{X}} &= \begin{bmatrix} \alpha & \beta & x_1 & x_2 & \cdots & x_n \end{bmatrix}^T \end{aligned} \tag{5}$$

Based on the equation set above, the observation function can be constructed, and the specific form of the error model is as follows:

$$\begin{aligned} \underset{mn\times 1}{\mathbf{L}} &= \underset{mn\times 1}{\mathbf{F}}(\mathbf{X}) - \underset{mn\times 1}{\boldsymbol{\Delta}} \\ D_{\mathbf{L}} &= \sigma_0^2 \mathbf{P}^{-1} \end{aligned} \tag{6}$$

where $\underset{mn\times 1}{\mathbf{L}}$ represents the matrix of observed values, $\underset{mn\times 1}{\mathbf{F}}(\mathbf{X})$ is the corresponding matrix of nonlinear functions, and $\underset{mn\times 1}{\boldsymbol{\Delta}}$ is the error matrix. The angles represent the direction of the true deformation vector α and β are bounded $\alpha \in [\frac{-\pi}{2}, \frac{\pi}{2}]$, $\beta \in [0, 2\pi]$. The true deformation vector can be estimated using nonlinear least squares constrained optimization [63].

$$\hat{\mathbf{X}} = \text{argmin}||G(\bullet)||^2 \tag{7}$$

where the function G represents the loss function, which can be chosen based on the specific circumstances, when employing nonlinear least squares optimization, it is as follows:

$$\begin{aligned} \mathbf{V} &= \mathbf{F}(\hat{\mathbf{X}}) - \mathbf{L} \\ G &= \mathbf{V}^T \mathbf{P} \mathbf{V} \end{aligned} \tag{8}$$

where $\mathbf{V}$ is the error vector, and $\hat{\mathbf{X}}$ represents the estimated values of the true deformation. Based on the above method, the true surface deformation can be calculated according to the LOS deformations obtained from different orbits.

4. Results: Xiaomojiu Landslides Displacement between October 2017 to September 2023

4.1. InSAR LOS Deformation Results

Based on the Sentinel-1A Ascending and Descending Orbit Data spanning from October 2017 to September 2023, annual average deformation velocities for three orbits were computed, revealing distinctive patterns illustrated in Figure 7a,b,d. Negative values denote deformation moving away from the satellite, while positive values signify deformation towards the satellite. The deformation velocity for orbit A99 is predominantly negative, with deformation mainly concentrated in the middle part of the landslide, reaching a maximum deformation of −126 mm/year. In contrast, the results for orbit D33 show predominantly positive deformation in the middle and lower parts of the landslide, with a maximum deformation of 62 mm/year, while the upper part exhibits negative deformation, reaching a maximum of −53 mm/year. For orbit D106, except for the toe of the landslide where the deformation is positive with a maximum of 26 mm/year, the rest of the areas show negative values, with a maximum deformation of −86 mm/year. The differing distribution of positive and negative deformation rates between the two descending orbits can primarily be attributed to variations in their incident angles. Upon analysis of the three velocity maps, it is evident that the magnitude of deformation rates in the ascending orbit significantly surpasses those in the descending orbits. The variation in deformation rates between the two descending orbit datasets is influenced mainly by the disparity in their incident angles. The discrepancy in the allocation of positive and negative values in the descending orbit data, as depicted in Figure 5, indicates a discrepancy between the deformation direction of the landslide and the slope's direction.

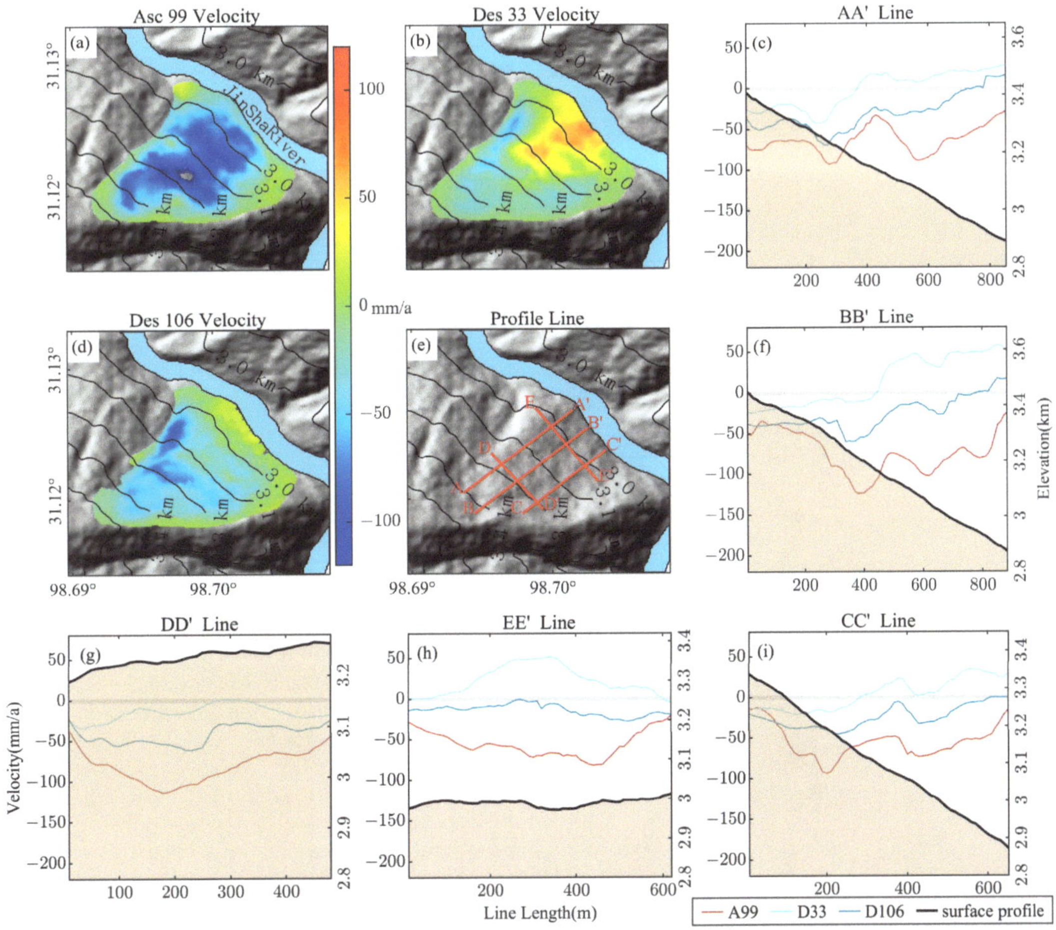

Figure 7. LOS Annual Average Displacement Velocity and Velocity Profile Maps. (**a**,**b**,**d**) Depict the velocity maps for ascending path 99, descending path 33, and descending path 106, respectively. In (**e**), AA′, BB′, and CC′ represent three profiles along the downslope direction, while DD′ and EE′ denote two cross profiles. (**c**,**f**–**i**) Correspond to the velocity distribution maps of the three orbits for profiles AA′, BB′, CC′, DD′, and EE′, respectively.

To further analyze the discrepancies, five cross-sectional profiles were set within the landslide area: three profiles along the downslope direction, AA′, BB′, and CC′, and two transverse profiles, DD′ and EE′. From Figure 7g,h, it can be observed that along the central profile line DD′, the projection of the true motion velocity of the landslide body on the D33 orbit is minimal, followed by D106, with all three orbit results being negative. These results indicate that at DD′, the predominant true deformation of the landslide occurs vertically and is approximately perpendicular to the LOS direction of the D33 orbit. However, in the middle-lower part of the landslide along profile line EE′, the direction of true deformation is approximately perpendicular to the LOS direction of the D106 orbit, with positive values observed on the D33 orbit and negative values projected on the D106 orbit. At EE′, the horizontal deformation of the landslide predominates over the vertical deformation. These conditions arise because the local incidence angle of the D33 orbit is greater than that of D106, resulting in differences in the measured true deformation between the two orbits. Moreover, the distribution of positive and negative projection coefficients in the D106 orbit in Figure 5 does not match the distribution of positive and negative deformation velocities corresponding to this orbit in Figure 7d. These discrepancies further illustrate that the direction of true displacement does not align with the slope direction, indicating that the surface parallel flow model cannot accurately reproduce the true deformation results of the Xiaomojiu landslide.

In the existing drilling data for the Xiaomojiu landslide [42], the maximum depth of the three boreholes is 60 m, none of which reached the bedrock. Based on the drilling results and other geological and topographical data, the landslide mass is a typical high-steep valley-shaped fragmented and loose structure. From the profiles along the steepest downhill direction shown in Figure 7c,f,i, it can be observed that the landslide deformation exhibits distinct segments. Two main deformation zones within the landslide are located in the upper-middle and lower-middle parts of the slope.

4.2. Temporal Deformation of the Xiaomojiu Landslide

In the simulation study of the Xiaomojiu landslide [42], it was suggested that the Xiaomojiu landslide might block the river and create the risk of forming a barrier lake after its collapse. To further analyze the development characteristics of the landslide, we selected the time-series deformation results of four points at different heights in the center of the landslide for temporal analysis.

From Figure 8a,c,e, it can be observed that before the occurrence of the Baige landslide damming event, the overall deformation of the landslide was relatively small. However, after the disaster, the Xiaomojiu landslide began to accelerate. Notably, the maximum temporal deformation was observed at point P2 in the upper-middle part of the landslide. Specifically, for point P2, the cumulative deformation from October 2017 to July 2023 was −56 cm for the A99 orbit, −10 cm for the D33 orbit, and −33 cm for the D106 orbit. The differences in deformation magnitudes reflect variations in the projection coefficients of true deformation in different observation directions. Analysis of the cumulative deformation results for the four points indicates that the D33 orbit is less sensitive to deformations in the middle part of the landslide.

In contrast, the D106 orbit is less sensitive to deformations in the lower part. As depicted in Figure 8b, after the Baige landslide damming event, the deformation at points P4 and P3 in the lower-middle part of the landslide initially increased before decreasing. This phenomenon may be attributed to the immersion of the landslide toe in the dammed lake water, resulting in initial deformation at the toe followed by displacement in the middle and lower parts. The transfer of displacement from lower to higher areas exhibits significant lag.

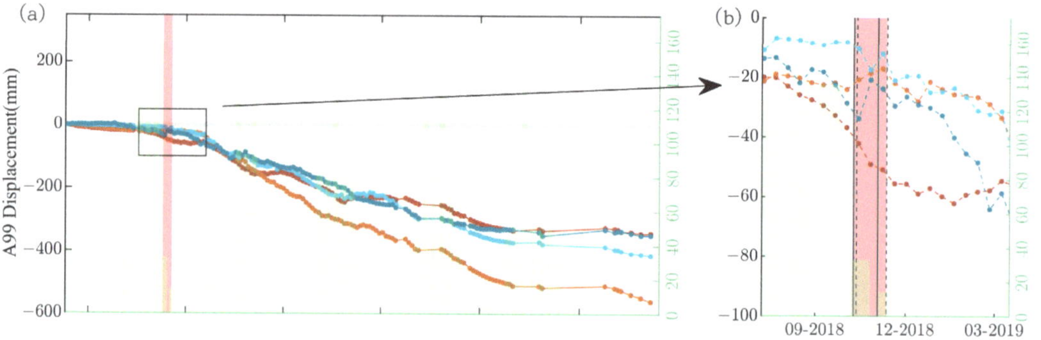

Figure 8. *Cont.*

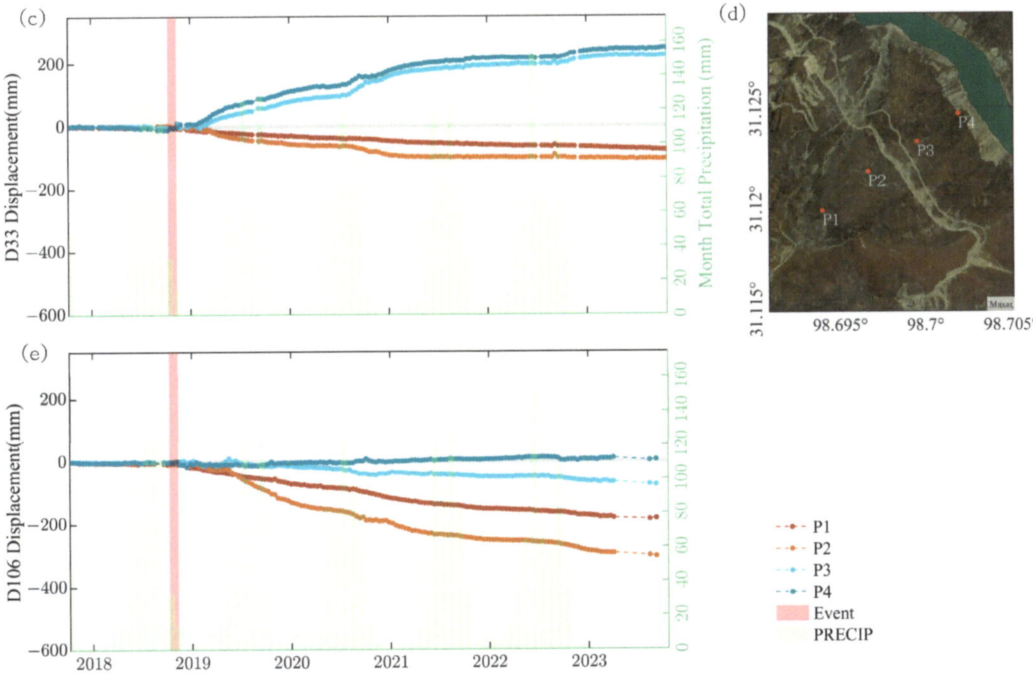

Figure 8. (**a**) P1 to P4 cumulative ascending orbit path 99 temporal cumulative strain map, (**c**) P1 to P4 cumulative descending orbit path 33 temporal cumulative deformation map, (**d**) distribution map of four points P1 to P4, (**e**) P1 to P4 cumulative ascending orbit path 106 temporal cumulative strain map. (**b**) This sub-figure is a magnified view of (**a**), where solid black lines represent the dates of the two failures of the Baige landslide, and black dashed lines represent the dates of the debris flow events from the barrier lake outburst. In all sub-figures, the light red EVENT represents the barrier lake events formed by the two failures of the Baige landslide. PRECIP represents the monthly total precipitation in the region from 2017 to 2022 [49].

5. Analysis and Discussion

5.1. The 3D Deformation Characteristics of the Xiaomojiu Landslide

The actual spatial deformation characteristics of the Xiaomojiu landslide were computed using the methods mentioned in this paper. Considering the relatively long data gap in the ascending orbit data after May 2021, cumulative deformations were calculated from October 2018 to May 2021 using data from three orbits. These results were then compared with those obtained from other methods, such as the SPFM [32], SPFM coupled with the least-squares method (SPF-LSM) [60], and the APFM [62]. The Adjacent Orientation Consistency Model (AOCM) mentioned in this paper does not utilize slope and aspect as the displacement direction for surface deformation. Instead, it solves for the unknown true deformation direction and decomposes the true deformation into the NEU directions, as illustrated in Figure 9.

From the results in Figure 9, it can be observed that the east-west and vertical deformations obtained by various methods exhibit relatively consistent spatial distributions, but there are differences in numerical values. The north-south deformation obtained using AOCM differs significantly from the results obtained by the previous three methods, indicating the presence of some degree of lateral horizontal movement in the middle part of the landslide. These results suggest a disparity between the movement direction of landslide sub-blocks on the landslide body and the overall steepest movement direction. Overall, there is a significant spatial distribution difference in the vertical deformation of the landslide body, with eastward movement predominating horizontally. In conclusion, optimizing the true deformation displacement using multi-track LOS data can provide more accurate vertical deformation of the landslide body, thereby offering more precise data support for analyzing the profile characteristics of the landslide. To further analyze the landslide structure, we extracted the actual deformation results corresponding to the BB′ profile, including the actual deformation values on the BB′ profile calculated by AOCM and the corresponding angles with the horizontal plane (Figure 10).

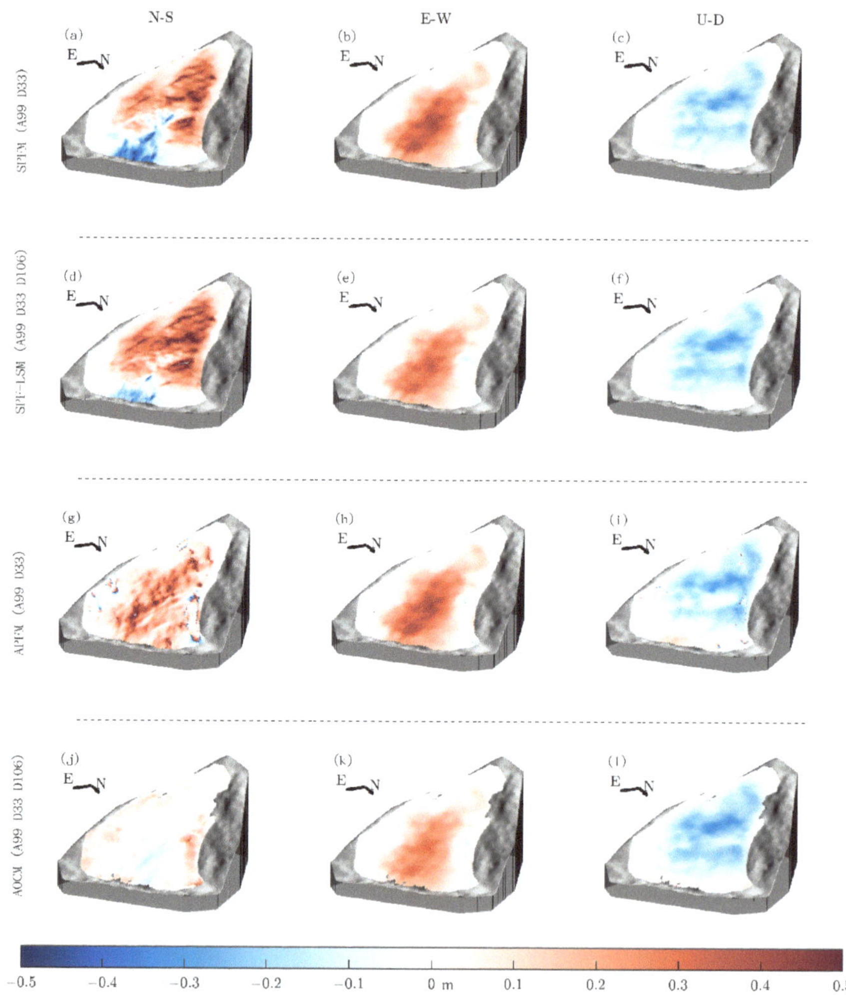

Figure 9. Different methods yield 3D deformation maps of the Xiaomojiu landslide. The first row displays the NEU direction deformation maps computed using the SPFM method: (**a–c**). The second row showcases the NEU direction deformation maps computed using the SPF-LSM method: (**d–f**). The third row exhibits the NEU direction deformation maps obtained from the APFM method: (**g–i**). Finally, the fourth row presents the NEU direction deformation maps calculated using the AOCM method: (**j–l**).

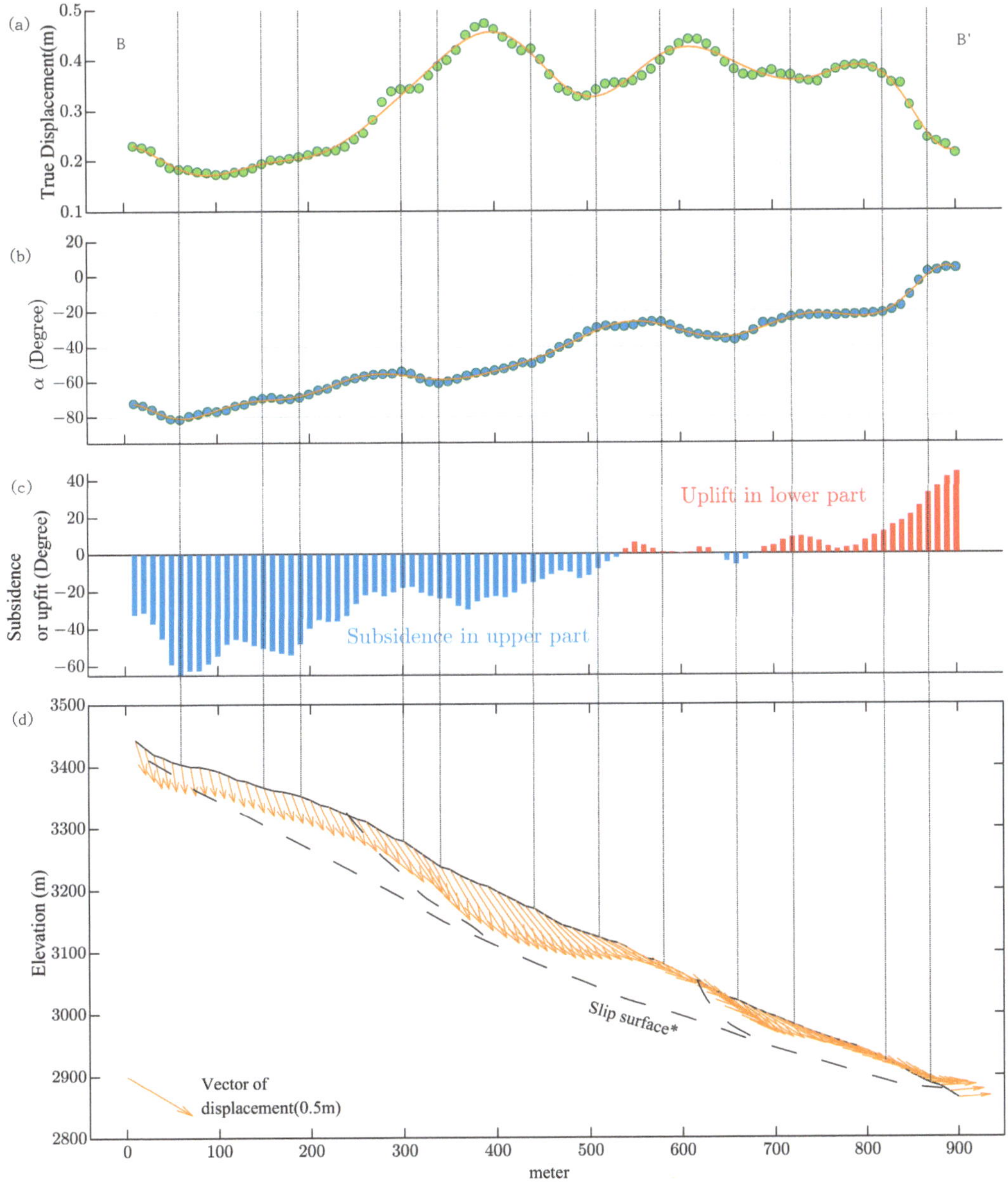

Figure 10. (**a**) Accumulated true deformation along profile BB′ (from October 2018 to May 2021), (**b**) angle between the true deformation direction along profile BB′ and the horizontal plane, (**c**) deviation between the corresponding inclination angle along profile BB′ and the negative slope angle, (**d**) the distribution of displacement vectors along profile BB′ is illustrated, with vertical dashed lines employed to delineate variations in angles. The dashed line (slip surface*) indicates the inferred sliding surface derived from prior research and computational outcomes.

The displacement on the landslide profile reflects the main movement characteristics of the landslide body. Figure 10a shows that the landslide body exhibits three deformation peaks concentrated in the middle-lower part of the landslide. The maximum deformation peak is located in the middle part of the landslide, with a cumulative deformation of 0.47 m. In contrast, the upper part primarily experiences vertical movement, with a cumulative deformation of approximately 0.2 m. Figure 10b indicates that the middle-lower part is subjected to compression from the upper part, resulting in the most significant degree of deformation. Figure 10b,c depict changes in the direction of surface movement of the landslide body. The average angle α of the upper part is approximately 70°, with deformation primarily manifested as subsidence, resulting in compression in the middle part. The angle α gradually approaches horizontal from the middle to the toe, while deformation gradually decreases, indicating a change in stress state and increased frictional resistance. These angular changes suggest the presence of an arc-shaped sliding zone within the landslide body. Additionally, at least two main landslide sub-blocks correspond to two arc-shaped sliding surfaces in the middle-lower part of the landslide, resulting in the superposition of landslide subzones and a pronounced toe extrusion phenomenon.

For further analysis of the landslide area, the real deformation direction, magnitude, and spatial information of each point obtained using the methods mentioned in this study were utilized as features. The K-means method [64] was employed for cluster analysis, and based on the elbow method, the landslide was divided into 12 sub-blocks, as shown in Figure 11b. For the five landslide sub-block regions located along the central axis (C12, C9, C8, C5, C2), each region's pixel points correspond to two angles, alpha and beta, representing the actual deformation directions. The magnitudes of these angles exhibit a relatively concentrated distribution. This result indicates that within each sub-block region, the actual displacement directions of each pixel are closely aligned (Figure 12a,b). When the deformation within the sub-block was relatively small, such as in the case of the C11 and C3 sub-blocks, the estimated real displacement direction within the sub-block was more dispersed, and the standard deviation of the actual deformation obtained was also more prominent.

Additionally, the overall direction of the slope in the C4 area was close to the north direction, consistent with the actual terrain distribution. The horizontal directions in the remaining areas were relatively similar except for the C3, C4, and C11. From the results of the average temporal cumulative deformation, it can be observed that among the C2, C5, and C8 sub-blocks arranged from low to high elevation, after the occurrence of the damming event, the bottom C2 sub-block began accelerating movement first, followed by a certain lag in the C5 sub-block at a higher elevation, with the C8 sub-block showing even longer lag time. This distribution indicates that the damming event progressively influenced the landslide activation from low to high elevations in a chain reaction along the altitude.

Considering the calculated 3D features of the landslide sub-blocks, it can be observed that the lower-middle parts of the landslide, represented by sub-blocks C8 and C5, exhibit characteristics of rotational sliding surfaces, consistent with the two rotational sliding sub-zones mentioned earlier. This phenomenon corresponds to the thick accumulation layers within the landslide body. In addition to the overall movement of the Xiaomojiu landslide, the presence of these two sliding sub-zones adds complexity to the landslide morphology and diversifies the sliding patterns. Consequently, during the subsequent continuous deformation of the landslide, the instability of the landslide body is increased.

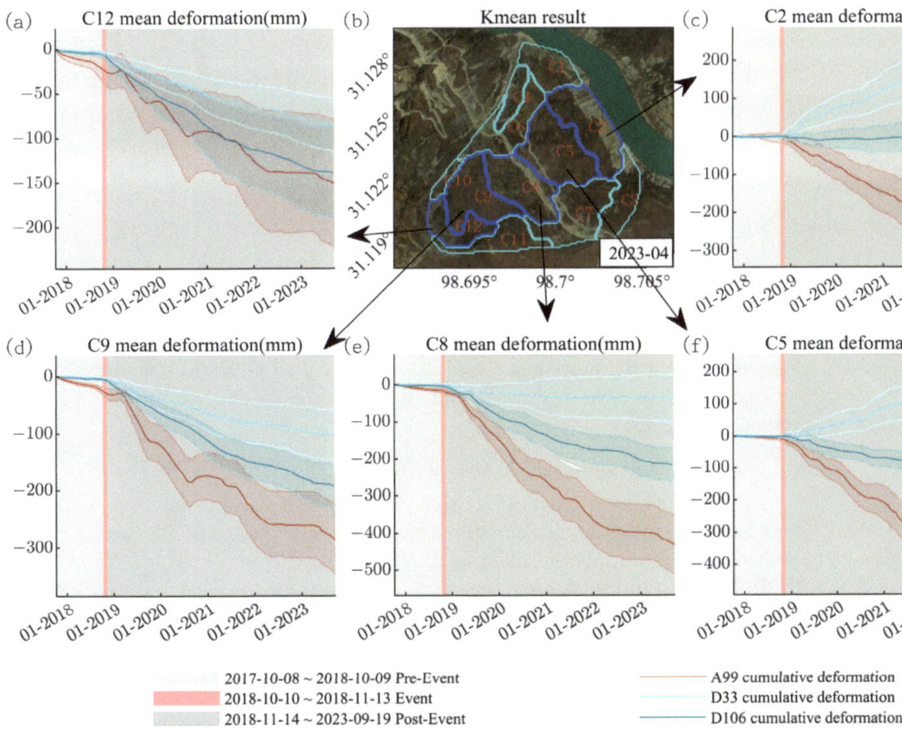

Figure 11. K-means clustering results of landslide blocks and typical multi-track average temporal results; trends of average cumulative deformation for landslide blocks across three orbits, (**a**) Block C12, (**c**) Block C2, (**d**) Block C9, (**e**) Block C8, (**f**) Block C5. The temporal data's light-colored upper and lower bounds represent the standard deviation range. (**b**) Spatial distribution map of clustering results.

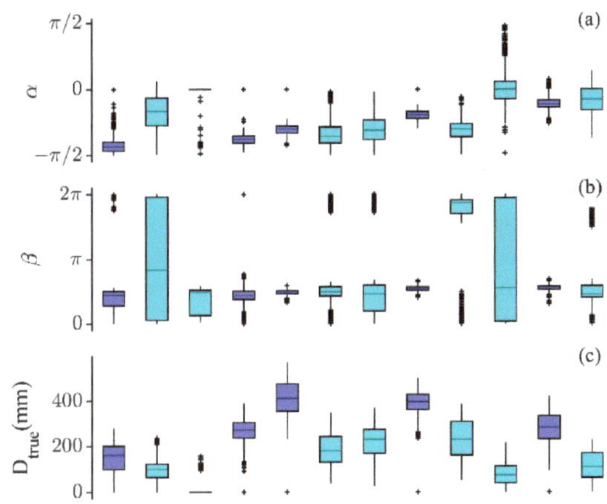

Figure 12. *Cont.*

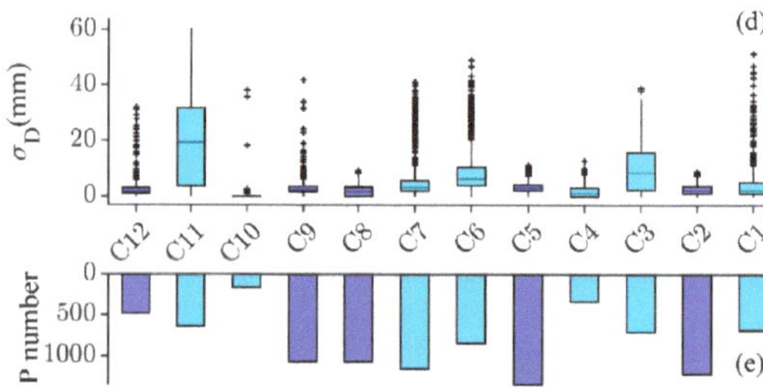

Figure 12. Box plots of the actual deformation characteristics within sub-blocks obtained from K-means clustering, (**a**) comparative distribution of angle α in different sub-block regions, (**b**) comparative distribution of angle β in different sub-block regions, (**c**) comparative distribution of angle D_{true} in different sub-block regions, (**d**) comparative distribution of standard deviation D_{true} in different sub-block regions, (**e**) comparative distribution of the number of ground points in different sub-block regions.

5.2. The Driving Factors of the Xiaomijiu Landslide Movement

To analyze the impact of changes in river water level caused by the damming lake event on the landslide movement process, intensity images of D33 descending orbit VH and VV polarization were used as the data source. The Block Matching 3D algorithm (BM3D) filtering method [65] was employed to filter the selected intensity image data, removing speckle noise. Time-series charts of water area extraction from the beginning of the Xiaomojiu landslide to the upstream river were generated to indirectly represent changes in river water level. Considering the significant surface undulations in the area and the presence of many shadow regions in the intensity images, masking was applied to avoid the influence of shadow areas on the delineation of water bodies. Additionally, river buffer zones were delineated based on elevation information, and the threshold segmentation method [66] was utilized to delineate water bodies within the buffer zones. Comparing Figure 13b,c, it is evident that the water area increased after the damming lake event. As shown in Figure 13c, the acquisition time of the D33 orbit data was UTC 23:21 on 10 October 2018, approximately 9 h after the Baige landslide had dammed the lake, resulting in a significant rise in water level.

The temporal results of the water body area (Figure 13e) reflect the following points. Firstly, following the formation of the landslide dam, the water area upstream of the Baige landslide rapidly increased, indicating a swift rise in the upstream water level due to the damming event. During the landslide dam occurrence, the water area underwent two cycles of increase followed by a decrease, corresponding to the two damming events and consistent with the cumulative deformation results in Figure 8b. Secondly, in the aftermath of the damming events, the water area upstream did not return to its pre-event level, indicating that the remaining debris from the landslide blocked the river channel, causing a certain degree of elevation in the water level upstream. Additionally, erosion and immersion of the upstream river channel might have occurred during the damming events, leading to the widening of the river channel. Thirdly, after the release of the landslide dam, the elevated water level in the upstream river channel was in an overall slow declining state. Therefore, the overall movement characteristics of the Xiaomojiu landslide are closely related to changes in the water level of the Jinsha River after being activated.

Furthermore, upon further comparison with optical data in Figure 14, it can be observed that due to immersion from the damming lake and erosion by the river water, the exposed area of the Xiaomojiu landslide toe gradually increased. Figure 14a,c,h demonstrate the comparison of river channels during the same season. After the Baige landslide event, the width of the river channel noticeably increased, consistent with the temporal results representing the pixel count of the river channel in Figure 13e.

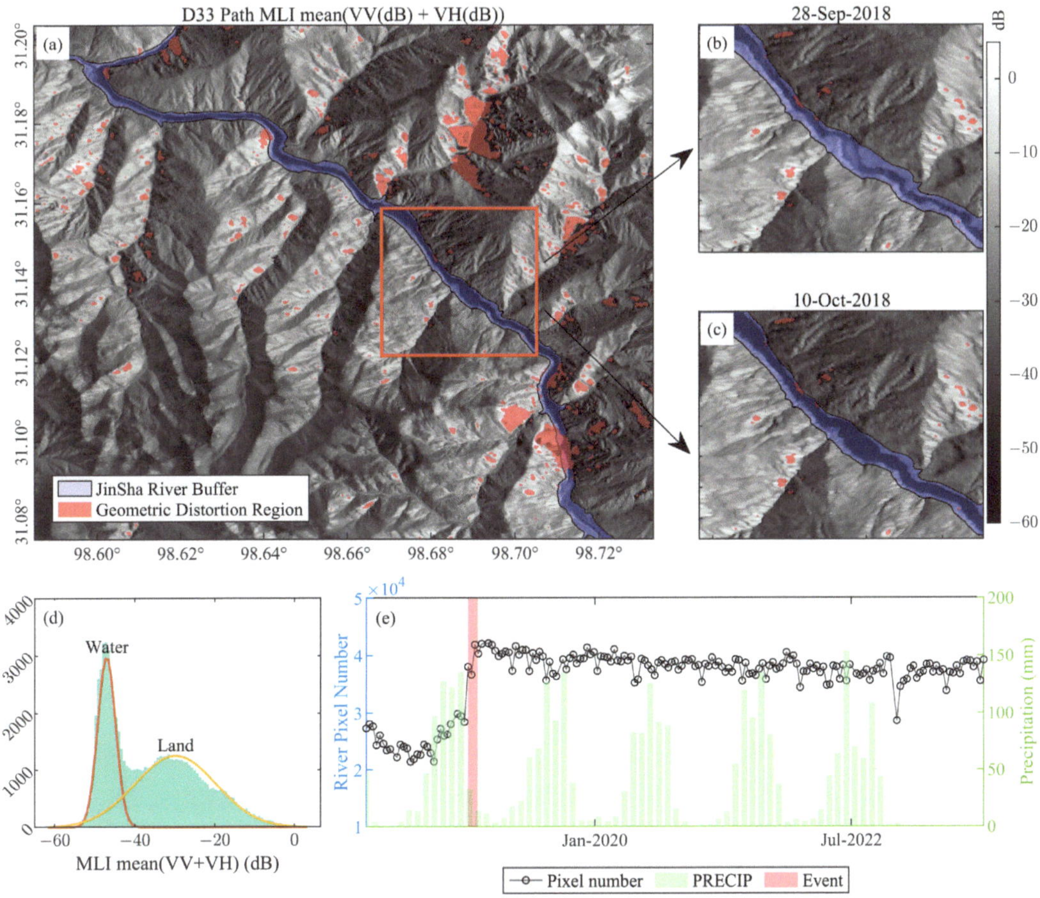

Figure 13. (**a**) Overlay image of D33 orbit VV and VH backscatter images (in dB), with the light purple area representing the river buffer zone generated based on elevation. (**b**) Partial VV+VH image before the first occurrence of the Baige landslide damming lake on 28 September 2018. (**c**) Partial VV+VH image two hours after the first occurrence of the Baige landslide damming lake on 10 October 2018. (**d**) Distribution map of pixel values within the river buffer zone, where the first peak represents water bodies, and the second peak represents land. (**e**) The time series plot of the total pixel count within the river buffer zone was calculated using the threshold segmentation method.

According to the time-series results of the water area shown in Figure 13e, the following conclusions can be drawn: Firstly, after the barrier lake event, the water area in the upstream region of the Baige landslide expanded rapidly, indicating a swift rise in water level upstream due to the impact of the barrier lake. During the barrier lake event, the quantity of water area experienced two increases followed by decreases, corresponding to the two occurrences of the barrier lake. The changes in water level are consistent with the cumulative deformation results shown in Figure 8b. Secondly, following the damming lake event, the water area in the upstream area did not return to the pre-event level, indicating that the remaining debris from the Baige landslide blocked the river channel to some extent, raising the water level in the upstream area. Additionally, channel widening may have been due to erosion and immersion during the damming lake event. Thirdly, after the damming lake was discharged, the elevated water level in the upstream channel of the Baige landslide gradually decreased overall. Therefore, after being activated, the overall movement characteristics of the Xiaomojiu landslide are closely related to the changes in the Jinsha River water level.

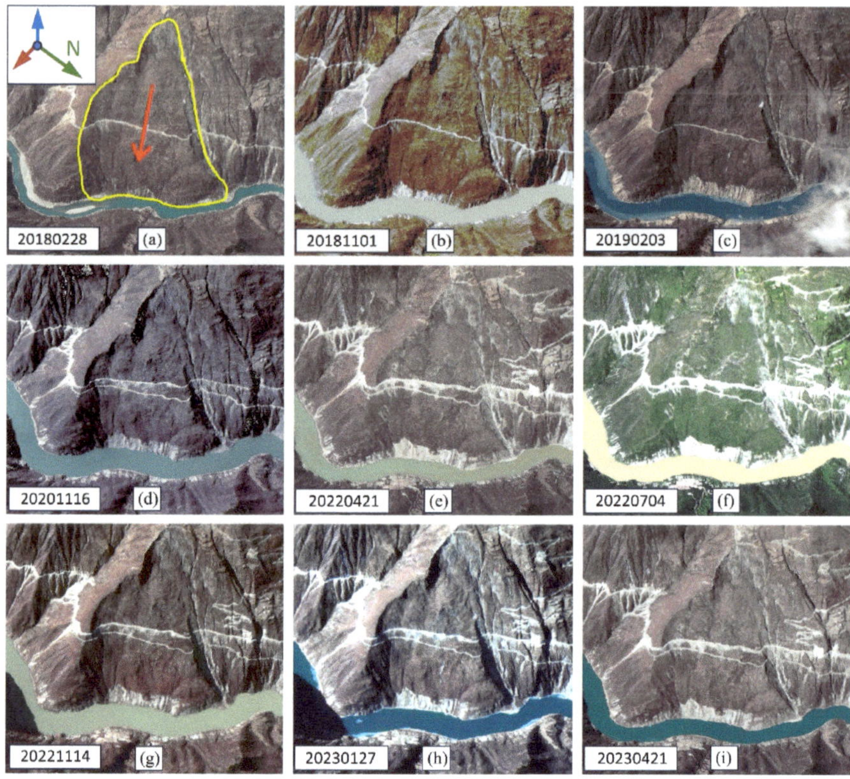

Figure 14. Shows optical remote sensing images from the GF-2 satellite on 28 February 2018 (**a**), 1 February 2019 (**c**), 16 November 2020 (**d**), 21 April 2022 (**e**), 4 July 2022 (**f**), 14 November 2022 (**g**), 27 January 2023 (**h**), and 21 April 2023 (**i**). Another optical remote sensing image was taken from a GF-7 satellite on 1 November 2018 (**b**).

Due to the influence of the residual accumulation of the Baige landslide on the water level in the upstream channel, the moisture content within the landslide has changed. Prolonged higher water levels can cause a redistribution of water within the landslide, increasing the soil moisture content and reducing its shear strength, thereby accelerating the deformation process of the landslide. Results obtained through InSAR technology reveal that the upper part of the Xiaomojiu landslide is undergoing subsidence, while the bottom part is experiencing extrusion and uplift. This deformation pattern indicates a redistribution of internal stresses within the landslide when subjected to external forces. Furthermore, the presence of two sub-blocks in the lower part of the landslide exhibiting characteristics similar to soil-rotational landslides further complicates the movement characteristics of the landslide mass. These sub-blocks may play a critical role in the overall stability of the landslide during its movement process, influencing its overall stability. In conclusion, the Xiaomojiu landslide is a deep-seated accumulation creep-type landslide formed under complex geological conditions and external influences. Reactivated and continuously deforming due to the rise in river water levels, the landslide remains in a phase of sustained deformation.

6. Conclusions

In this study, we propose a method based on the consistency constraint of neighboring directions to invert the true deformation displacement. Utilizing three-track cumulative deformation data calculated by SBAS-InSAR, we invert the true deformation of the landslide body. Compared to methods that pre-determine the direction of landslide movement, the proposed method in this study more accurately reflects the true movement characteristics of the landslide body. Based on this, we draw the following conclusions when analyzing the Xiaomojiu landslide movement.

(1) This study presents the results of SBAS-InSAR technology to obtain multi-track LOS velocity data of the Xiaomojiu landslide. From 2017 to 2023, the maximum LOS displacement rate of the Xiaomojiu landslide was found to be −126 mm/yr. Variations in the C-index values and the distribution of LOS velocity signs differ among the orbits, suggesting that the motion pattern of the Xiaomojiu landslide does not fit the translation-type landslide model. This difference indicates a significant discrepancy between the true displacement direction of the Xiaomojiu landslide body's surface and the deformation direction calculated based on the surface slope. The vertical deformation is further constrained by utilizing the LOS observations from three orbits. The proposed method in this study provides a more accurate real displacement vector of the landslide motion compared to several surface parallel flow methods.

(2) Large landslides typically exhibit complex movement characteristics. By analyzing the angle distribution along the profile line, it is evident that the upper part of the Xiaomojiu landslide is mainly dominated by downward movement. The sinking in the upper part compresses the middle part of the landslide, resulting in the most significant deformation in the middle, while the bottom part shows extrusion phenomena. Additionally, two rotational sliding zones are present in the lower part of the landslide. The landslide area is partitioned using the K-means classification method based on the 3D deformation features. The two rotational sliding zones correspond to the C8 and C5 blocks obtained from the classification, indicating the complexity of the Xiaomojiu landslide structure.

(3) In analyzing driving factors for the movement of the Xiaomojiu landslide, we extracted the water area of certain rivers in the study area using Sentinel-1 intensity information. Changes in water areas can reflect changes in water levels to some extent. The activation of the landslide body due to the debris dam event occurs gradually from low to high elevations, with a phenomenon of deformation lag. Following the event, the upstream water level remains consistently high for an extended period, gradually decreasing. This state has profound implications for the landslide body, maintaining it in continuous deformation.

In this study, the proposed method is a practical approach for accurately measuring surface deformation in landslides characterized by intricate sliding surfaces without assuming the direction of movement. However, a fundamental limitation of this method is its reliance on a minimum of three observation tracks, which constrains its applicability. Moreover, the method's utility is further compromised by the inherent constraints of InSAR technology, particularly in managing substantial displacement and rapid accelerations during landslide events. By integrating the Xiaomojiu calculation outcomes with the specific topographic and geological features of the landslide area, it is deduced that the Xiaomojiu landslide exhibits characteristics of a deep-seated accumulation-type creep landslide with a complex movement pattern. The continuous water-driven influence poses a considerable destabilization risk to the landslide area, underscoring the critical necessity for intensified deformation monitoring and comprehensive ground surveys.

Author Contributions: Conceptualization, X.M. and J.P.; methodology, X.M.; validation, X.M., Y.S. and M.S.; investigation, X.M., Y.Z. and X.L.; data curation, X.M. and X.J.; writing—original draft preparation, X.M.; writing—review and editing, X.M., M.S. and Y.S.; visualization, X.M.; supervision, J.P., Y.S. and M.S.; funding acquisition, J.P. All authors have read and agreed to the published version of the manuscript.

Funding: This research was funded by the National Natural Science Foundation of China under Grant 42074004 and in part by Shanxi Transportation Holdings Group, Company, Ltd., under Grant 19-JKKJ-3.

Data Availability Statement: The Sentinel-1 data have been made freely available by the European Space Agency and distributed and archived by the Alaska Satellite Facility, https://www.asf.alaska.edu/sentinel/, (accessed on 4 November 2023); Precise Orbit Ephemerides (POD) used in this study were provided by the ESA, https://scihub.copernicus.eu/gnss/#/home (accessed on 4 November 2023); and the AW3D DEM was freely downloaded from the website https://opentopography.org/ (accessed on 4 November 2023). Meteorological observation data are provided by National Tibetan Plateau Data Center http://data.tpdc.ac.cn, (accessed on 4 November 2023).

Acknowledgments: The authors thank the European Space Agency (ESA) for providing Sentinel-1 SAR data. The optical image data were kindly provided by AGRS.

Conflicts of Interest: The authors declare no conflicts of interest.

References

1. Wang, J.B.; Lei, T.J.; Liu, W.K.; Chen, Y.J.; Yue, J.W.; Liu, B.Y. Prediction analysis of landslide displacement trajectory based on the gradient descent method with multisource remote sensing observations. *Geomat. Nat. Hazards Risk* **2023**, *14*, 143–175. [CrossRef]
2. Xiong, Z.Q.; Zhang, M.Z.; Ma, J.; Xing, G.L.; Feng, G.C.; An, Q. InSAR-based landslide detection method with the assistance of C-index. *Landslides* **2023**, *20*, 2709–2723. [CrossRef]
3. Haque, U.; da Silva, P.F.; Devoli, G.; Pilz, J.; Zhao, B.X.; Khaloua, A.; Wilopo, W.; Andersen, P.; Lu, P.; Lee, J.; et al. The human cost of global warming: Deadly landslides and their triggers (1995–2014). *Sci. Total Environ.* **2019**, *682*, 673–684. [CrossRef] [PubMed]
4. Du, J.; Glade, T.; Woldai, T.; Chai, B.; Zeng, B. Landslide susceptibility assessment based on an incomplete landslide inventory in the Jilong Valley, Tibet, Chinese Himalayas. *Eng. Geol.* **2020**, *270*, 105572. [CrossRef]
5. Fan, X.M.; Xu, Q.; Alonso-Rodriguez, A.; Subramanian, S.S.; Li, W.L.; Zheng, G.; Dong, X.J.; Huang, R.Q. Successive landsliding and damming of the Jinsha River in eastern Tibet, China: Prime investigation, early warning, and emergency response. *Landslides* **2019**, *16*, 1003–1020. [CrossRef]
6. Liu, X.J.; Zhao, C.Y.; Zhang, Q.; Lu, Z.; Li, Z.H.; Yang, C.S.; Zhu, W.; Liu-Zeng, J.; Chen, L.Q.; Liu, C.J. Integration of Sentinel-1 and ALOS/PALSAR-2 SAR datasets for mapping active landslides along the Jinsha River corridor, China. *Eng. Geol.* **2021**, *284*, 106033. [CrossRef]
7. Yang, Z.W.; Liu, W.M.; Garcia-Castellanos, D.; Ruan, H.C.; Luo, J.P.; Zhou, Y.L.; Sang, Y.Y. Geomorphic response of outburst floods: Insight from numerical simulations and observations-The 2018 Baige outburst flood in the upper Yangtze River. *Sci. Total Environ.* **2022**, *851*, 158378. [CrossRef] [PubMed]
8. Ouyang, C.J.; An, H.C.; Zhou, S.; Wang, Z.W.; Su, P.C.; Wang, D.P.; Cheng, D.X.; She, J.X. Insights from the failure and dynamic characteristics of two sequential landslides at Baige village along the Jinsha River, China. *Landslides* **2019**, *16*, 1397–1414. [CrossRef]
9. Zhang, Z.; He, S.M.; Liu, W.; Liang, H.; Yan, S.X.; Deng, Y.; Bai, X.Q.; Chen, Z. Source characteristics and dynamics of the October 2018 Baige landslide revealed by broadband seismograms. *Landslides* **2019**, *16*, 777–785. [CrossRef]
10. FENG, W.; ZHANG, G.; BAI, H.; ZHOU, Y.; XU, Q.; ZHENG, G. A preliminary analysis of the formation mechanism and development tendency of the huge Baige landslide in Jinsha River on October 11, 2018. *J. Eng. Geol.* **2019**, *27*, 415–425. [CrossRef]
11. Gao, Y.J.; Zhao, S.Y.; Deng, J.H.; Yu, Z.Q.; Rahman, M. Flood assessment and early warning of the reoccurrence of river blockage at the Baige landslide. *J. Geogr. Sci.* **2021**, *31*, 1694–1712. [CrossRef]
12. Tian, S.F.; Chen, N.S.; Wu, H.; Yang, C.Y.; Zhong, Z.; Rahman, M. New insights into the occurrence of the Baige landslide along the Jinsha River in Tibet. *Landslides* **2020**, *17*, 1207–1216. [CrossRef]
13. Casagli, N.; Intrieri, E.; Tofani, V.; Gigli, G.; Raspini, F. Landslide detection, monitoring and prediction with remote-sensing techniques. *Nat. Rev. Earth Environ.* **2023**, *4*, 51–64. [CrossRef]
14. Torres, R.; Snoeij, P.; Geudtner, D.; Bibby, D.; Davidson, M.; Attema, E.; Potin, P.; Rommen, B.; Floury, N.; Brown, M.; et al. GMES Sentinel-1 mission. *Remote Sens. Environ.* **2012**, *120*, 9–24. [CrossRef]
15. Su, Y.; Peng, J.; Shi, M.; Guo, C.; Ma, X.; Li, X.; Wang, J.; Wang, W. An M-Estimation Method for InSAR Nonlinear Deformation Modeling and Inversion. *IEEE Trans. Geosci. Remote Sens.* **2024**, *62*, 1–12. [CrossRef]
16. Zhao, C.; Lu, Z. Remote Sensing of Landslides—A Review. *Remote Sens.* **2018**, *10*, 279. [CrossRef]
17. Miano, A.; Mele, A.; Calcaterra, D.; Martire, D.D.; Infante, D.; Prota, A.; Ramondini, M. The use of satellite data to support the structural health monitoring in areas affected by slow-moving landslides: A potential application to reinforced concrete buildings. *Struct. Health Monit.* **2021**, *20*, 3265–3287. [CrossRef]
18. Solari, L.; Del Soldato, M.; Raspini, F.; Barra, A.; Bianchini, S.; Confuorto, P.; Casagli, N.; Crosetto, M. Review of Satellite Interferometry for Landslide Detection in Italy. *Remote Sens.* **2020**, *12*, 1351. [CrossRef]
19. Xiong, Z.Q.; Feng, G.C.; Feng, Z.X.; Miao, L.; Wang, Y.D.; Yang, D.J.; Luo, S.R. Pre- and post-failure spatial-temporal deformation pattern of the Baige landslide retrieved from multiple radar and optical satellite images. *Eng. Geol.* **2020**, *279*, 105880. [CrossRef]
20. Intrieri, E.; Frodella, W.; Raspini, F.; Bardi, F.; Tofani, V. Using Satellite Interferometry to Infer Landslide Sliding Surface Depth and Geometry. *Remote Sens.* **2020**, *12*, 1462. [CrossRef]

21. Liu, X.J.; Zhao, C.Y.; Zhang, Q.; Yin, Y.P.; Lu, Z.; Samsonov, S.; Yang, C.S.; Wang, M.; Tomás, R. Three-dimensional and long-term landslide displacement estimation by fusing C- and L-band SAR observations: A case study in Gongjue County, Tibet, China. *Remote Sens. Environ.* **2021**, *267*, 112745. [CrossRef]
22. Eriksen, H.O.; Lauknes, T.R.; Larsen, Y.; Corner, G.D.; Bergh, S.G.; Dehls, J.; Kierulf, H.P. Visualizing and interpreting surface displacement patterns on unstable slopes using multi-geometry satellite SAR interferometry (2D InSAR). *Remote Sens. Environ.* **2017**, *191*, 297–312. [CrossRef]
23. Hu, J.; Li, Z.W.; Ding, X.L.; Zhu, J.J.; Zhang, L.; Sun, Q. Resolving three-dimensional surface displacements from InSAR measurements: A review. *Earth-Sci. Rev.* **2014**, *133*, 1–17. [CrossRef]
24. Zhang, C.L.; Li, Z.H.; Yu, C.; Chen, B.; Ding, M.T.; Zhu, W.; Yang, J.; Liu, Z.J.; Peng, J.B. An integrated framework for wide-area active landslide detection with InSAR observations and SAR pixel offsets. *Landslides* **2022**, *19*, 2905–2923. [CrossRef]
25. Bechor, N.B.D.; Zebker, H.A. Measuring two-dimensional movements using a single InSAR pair. *Geophys. Res. Lett.* **2006**, *33*. [CrossRef]
26. Del Soldato, M.; Confuorto, P.; Bianchini, S.; Sbarra, P.; Casagli, N. Review of Works Combining GNSS and InSAR in Europe. *Remote Sens.* **2021**, *13*, 1684. [CrossRef]
27. Meng, Q.; Li, W.; Raspini, F.; Xu, Q.; Peng, Y.; Ju, Y.; Zheng, Y.; Casagli, N. Time-series analysis of the evolution of large-scale loess landslides using InSAR and UAV photogrammetry techniques: A case study in Hongheyan, Gansu Province, Northwest China. *Landslides* **2020**, *18*, 251–265. [CrossRef]
28. Lombardi, L.; Nocentini, M.; Frodella, W.; Nolesini, T.; Bardi, F.; Intrieri, E.; Carlà, T.; Solari, L.; Dotta, G.; Ferrigno, F.; et al. The Calatabiano landslide (southern Italy): Preliminary GB-InSAR monitoring data and remote 3D mapping. *Landslides* **2016**, *14*, 685–696. [CrossRef]
29. Su, Y.; Yang, H.; Peng, J.; Liu, Y.; Zhao, B.; Shi, M. A Novel Near-Real-Time GB-InSAR Slope Deformation Monitoring Method. *Remote Sens.* **2022**, *14*, 5585. [CrossRef]
30. Samsonov, S. Three-dimensional deformation time series of glacier motion from multiple-aperture DInSAR observation. *J. Geod.* **2019**, *93*, 2651–2660. [CrossRef]
31. Baum, R.L.; Messerich, J.; Fleming, R.W. Surface deformation as a guide to kinematics and three-dimensional shape of slow-moving, clay-rich landslides, Honolulu, Hawaii. *Environ. Eng. Geosci.* **1998**, *4*, 283–306. [CrossRef]
32. Ren, K.; Yao, X.; Li, R.; Zhou, Z.; Yao, C.; Jiang, S. 3D displacement and deformation mechanism of deep-seated gravitational slope deformation revealed by InSAR: A case study in Wudongde Reservoir, Jinsha River. *Landslides* **2022**, *19*, 2159–2175. [CrossRef]
33. Samsonov, S.; Dille, A.; Dewitte, O.; Kervyn, F.; d'Oreye, N. Satellite interferometry for mapping surface deformation time series in one, two and three dimensions: A new method illustrated on a slow-moving landslide. *Eng. Geol.* **2020**, *266*, 105471. [CrossRef]
34. He, L.M.; Pei, P.K.; Zhang, X.N.; Qi, J.; Cai, J.Y.; Cao, W.; Ding, R.B.; Mao, Y.C. Sensitivity Evaluation of Time Series InSAR Monitoring Results for Landslide Detection. *Remote Sens.* **2023**, *15*, 3906. [CrossRef]
35. Xia, Z.G.; Motagh, M.; Li, T.; Peng, M.M.; Roessner, S. A methodology to characterize 4D post-failure slope instability dynamics using remote sensing measurements: A case study of the Aniangzhai landslide in Sichuan, Southwest China. *Isprs J. Photogramm. Remote Sens.* **2023**, *196*, 402–414. [CrossRef]
36. Ma, S.Y.; Qiu, H.J.; Zhu, Y.R.; Yang, D.D.; Tang, B.Z.; Wang, D.Z.; Wang, L.Y.; Cao, M.M. Topographic Changes, Surface Deformation and Movement Process before, during and after a Rotational Landslide. *Remote Sens.* **2023**, *15*, 662. [CrossRef]
37. Zhang, S.; Fan, Q.; Niu, Y.; Qiu, S.; Si, J.; Feng, Y.; Zhang, S.; Song, Z.; Li, Z. Two-dimensional deformation monitoring for spatiotemporal evolution and failure mode of Lashagou landslide group, Northwest China. *Landslides* **2022**, *20*, 447–459. [CrossRef]
38. Carter, M.; Bentley, S.P. The Geometry of Slip Surfaces beneath Landslides—Predictions from Surface Measurements. *Can. Geotech. J.* **1985**, *22*, 234–238. [CrossRef]
39. Hungr, O.; Leroueil, S.; Picarelli, L. The Varnes classification of landslide types, an update. *Landslides* **2013**, *11*, 167–194. [CrossRef]
40. Hu, G.S.; Tian, S.F.; Chen, N.S.; Liu, M.; Somos-Valenzuela, M. An effectiveness evaluation method for debris flow control engineering for cascading hydropower stations along the Jinsha River, China. *Eng. Geol.* **2020**, *266*, 105472. [CrossRef]
41. Guo, C.; Yan, Y.; Zhang, Y.; Wu, R.; Yang, Z.; Li, X.; Ren, S.; Zhang, Y.; Wu, Z.; Liu, J. Research progress and prospect of the failure mechanism of large deep-seated creeping landslides in the Tibetan Plateau, China. *Earth Sci.* **2022**, *47*, 3677–3700.
42. Zhang, C.L.; Li, Z.H.; Ding, M.T.; Zhu, W.; Chen, B.; Zhuang, J.Q.; Du, J.T.; Peng, J.B. Dynamic deformation monitoring and scenario simulation of the Xiaomojiu landslide in the Jinsha River Basin, China. *Landslides* **2023**, *20*, 2343–2358. [CrossRef]
43. Cao, W.T.; Yan, D.P.; Qiu, L.; Zhang, Y.X.; Qiu, J.W. Structural style and metamorphic conditions of the Jinshajiang metamorphic belt: Nature of the Paleo-Jinshajiang orogenic belt in the eastern Tibetan Plateau. *J. Asian Earth Sci.* **2015**, *113*, 748–765. [CrossRef]
44. Chen, Z.; Zhou, H.F.; Ye, F.; Liu, B.; Fu, W.X. The characteristics, induced factors, and formation mechanism of the 2018 Baige landslide in Jinsha River, Southwest China. *Catena* **2021**, *203*, 105337. [CrossRef]
45. Liu, X.; Yao, X.; Yao, J. Accelerated Movements of Xiaomojiu Landslide Observed with SBAS-InSAR and Three-Dimensional Measurements, Upper Jinsha River, Eastern Tibet. *Appl. Sci.* **2022**, *12*, 9758. [CrossRef]
46. Liu, D.; Cui, Y.; Wang, H.; Jin, W.; Wu, C.; Bazai, N.A.; Zhang, G.; Carling, P.A.; Chen, H. Assessment of local outburst flood risk from successive landslides: Case study of Baige landslide-dammed lake, upper Jinsha river, eastern Tibet. *J. Hydrol.* **2021**, *599*, 126294. [CrossRef]

47. Zhang, J.; Tang, H.; Li, C.; Gong, W.; Zhou, B.; Zhang, Y. Deformation stage division and early warning of landslides based on the statistical characteristics of landslide kinematic features. *Landslides* **2024**, *21*, 717–735. [CrossRef]
48. Takaku, J.; Tadono, T.; Doutsu, M.; Ohgushi, F.; Kai, H. Updates of 'Aw3d30' Alos Global Digital Surface Model with Other Open Access Datasets. *Int. Arch. Photogramm. Remote Sens. Spat. Inf. Sci.* **2020**, *XLIII-B4-2020*, 183–189. [CrossRef]
49. Shouzhang, P. 1-km monthly precipitation dataset for China (1901–2021). *Earth Syst. Sci. Data* **2020**, *4*, 1931–1946. [CrossRef]
50. Berardino, P.; Fornaro, G.; Lanari, R.; Sansosti, E. A new algorithm for surface deformation monitoring based on small baseline differential SAR interferograms. *Ieee Trans. Geosci. Remote Sens.* **2002**, *40*, 2375–2383. [CrossRef]
51. Pepe, A.; Lanari, R. On the extension of the minimum cost flow algorithm for phase unwrapping of multitemporal differential SAR interferograms. *Ieee Trans. Geosci. Remote Sens.* **2006**, *44*, 2374–2383. [CrossRef]
52. Shi, M.; Peng, J.; Chen, X.; Zheng, Y.; Yang, H.; Su, Y.; Wang, G.; Wang, W. An Improved Method for InSAR Atmospheric Phase Correction in Mountainous Areas. *IEEE J. Sel. Top. Appl. Earth Obs. Remote Sens.* **2021**, *14*, 10509–10519. [CrossRef]
53. Hu, X.; Wang, T.; Pierson, T.C.; Lu, Z.; Kim, J.; Cecere, T.H. Detecting seasonal landslide movement within the Cascade landslide complex (Washington) using time-series SAR imagery. *Remote Sens. Environ.* **2016**, *187*, 49–61. [CrossRef]
54. Notti, D.; Meisina, C.; Zucca, F.; Colombo, A. Models to predict Persistent Scatterers data distribution and their capacity to register movement along the slope. In Proceedings of the Fringe 2011 Workshop, Frascati, Italy, 19–23 September 2011; pp. 19–23.
55. Notti, D.; Herrera, G.; Bianchini, S.; Meisina, C.; García-Davalillo, J.C.; Zucca, F. A methodology for improving landslide PSI data analysis. *Int. J. Remote Sens.* **2014**, *35*, 2186–2214. [CrossRef]
56. Plank, S.; Singer, J.; Minet, C.; Thuro, K. Pre-survey suitability evaluation of the differential synthetic aperture radar interferometry method for landslide monitoring. *Int. J. Remote Sens.* **2012**, *33*, 6623–6637. [CrossRef]
57. Li, M.H.; Zhang, L.; Yang, M.S.; Liao, M.S. Complex surface displacements of the Nanyu landslide in Zhouqu, China revealed by multi-platform InSAR observations. *Eng. Geol.* **2023**, *317*, 107069. [CrossRef]
58. van Natijne, A.L.; Bogaard, T.A.; van Leijen, F.J.; Hanssen, R.F.; Lindenbergh, R.C. World-wide InSAR sensitivity index for landslide deformation tracking. *Int. J. Appl. Earth Obs. Geoinf.* **2022**, *111*, 102829. [CrossRef]
59. Song, C.; Yu, C.; Li, Z.H.; Pazzi, V.; Del Soldato, M.; Cruz, A.; Utili, S. Landslide geometry and activity in Villa de la Independencia (Bolivia) revealed by InSAR and seismic noise measurements. *Landslides* **2021**, *18*, 2721–2737. [CrossRef]
60. Sharifi, S.; Macciotta, R.; Hendry, M.; Rotheram-Clarke, D.; Huntley, D. Evaluating topography-based methods in 3D decomposition of InSAR 1D velocities obtained for translational landslides: Thompson River valley in Canada. *Landslides* **2023**, *21*, 411–427. [CrossRef]
61. Kang, Y.; Lu, Z.; Zhao, C.Y.; Qu, W. Inferring slip-surface geometry and volume of creeping landslides based on InSAR: A case study in Jinsha River basin. *Remote Sens. Environ.* **2023**, *294*, 113620. [CrossRef]
62. Zhu, Y.F.; Yao, X.; Yao, L.H.; Zhou, Z.K.; Ren, K.Y.; Li, L.J.; Yao, C.C.; Gu, Z.K. Identifying the Mechanism of Toppling Deformation by InSAR: A Case Study in Xiluodu Reservoir, Jinsha River. *Landslides* **2022**, *19*, 2311–2327. [CrossRef]
63. Grant, M.; Boyd, S.; Cvx, Y.Y. *Matlab Software for Disciplined Convex Programming*, version 1.0 beta 3; CVX Research, Inc.: Austin, TX, USA, 2006.
64. Krishna, K.; Narasimha Murty, M. Genetic K-means algorithm. *IEEE Trans. Syst. Man Cybern. Part B (Cybern.)* **1999**, *29*, 433–439. [CrossRef] [PubMed]
65. MV, S.; MN, G. A Modified BM3D Algorithm for SAR Image Despeckling. *Procedia Comput. Sci.* **2015**, *70*, 69–75. [CrossRef]
66. Huang, W.L.; DeVries, B.; Huang, C.Q.; Lang, M.W.; Jones, J.W.; Creed, I.F.; Carroll, M.L. Automated Extraction of Surface Water Extent from Sentinel-1 Data. *Remote Sens.* **2018**, *10*, 797. [CrossRef]

remote sensing MDPI

Article

A Strategy for Neighboring Pixel Collaboration in Landslide Susceptibility Prediction

Xiao Wang [1,*], Di Wang [2], Mengmeng Zhang [3], Xiaochuan Song [4], Luting Xu [1], Tiegang Sun [5], Weile Li [6], Sizhi Cheng [7] and Jianhui Dong [1]

1 School of Architecture and Civil Engineering, Chengdu University, Chengdu 610106, China; xuluting@cdu.edu.cn (L.X.); dongjianhui@cdu.edu.cn (J.D.)
2 The State Key Laboratory of Information Engineering in Surveying, Mapping and Remote Sensing, Wuhan University, Wuhan 430079, China; wd10486@whu.edu.cn
3 College of Earth Sciences, Chengdu University of Technology, Chengdu 610059, China; zhangmengmeng@stu.cdut.edu.cn
4 Sichuan 402 Surveying and Mapping Technology Corp, Chengdu 412108, China; 2019010043@cdut.edu.cn
5 China Building Materials Southwest Survey and Design Co., Ltd., Chengdu 610052, China; stg@cnbm.com.cn
6 State Key Lab Geohazard Prevent & Geoenvironm Prot, Chengdu University of Technology, Chengdu 610059, China; liweile08@mail.cdut.edu.cn
7 Sichuan Earthquake Agency, Chengdu 610041, China; csz7895@scdzj.gov.cn
* Correspondence: wangxiao@cdu.edu.cn

Abstract: Landslide susceptibility prediction usually involves the comprehensive analysis of terrain and other factors that may be distributed with spatial patterns. Without considering the spatial correlation and mutual influence between pixels, conventional prediction methods often focus only on information from individual pixels. To address this issue, the present study proposes a new strategy for neighboring pixel collaboration based on the Unified Perceptual Parsing Network (UPerNet), the Vision Transformer (ViT), and Vision Graph Neural Networks (ViG). This strategy efficiently utilizes the strengths of deep learning in feature extraction, sequence modeling, and graph data processing. By considering the information from neighboring pixels, this strategy can more accurately identify susceptible areas and reduce misidentification and omissions. The experimental results suggest that the proposed strategy can predict landslide susceptibility zoning more accurately. These predictions can identify flat areas such as rivers and distinguish between areas with high and very high landslide susceptibility. Such refined zoning outcomes are significant for landslide prevention and mitigation and can help decision-makers formulate targeted response measures.

Keywords: landslide susceptibility; neighboring Pixels; UPerNet; ViG; ViT

Citation: Wang, X.; Wang, D.; Zhang, M.; Song, X.; Xu, L.; Sun, T.; Li, W.; Cheng, S.; Dong, J. A Strategy for Neighboring Pixel Collaboration in Landslide Susceptibility Prediction. *Remote Sens.* **2024**, *16*, 2206. https://doi.org/10.3390/rs16122206

Academic Editor: Fumio Yamazaki

Received: 27 May 2024
Revised: 15 June 2024
Accepted: 16 June 2024
Published: 18 June 2024

1. Introduction

Human activities severely challenge nature's tolerance toward increasing anthropogenic changes during societal development. Decades of rapid socioeconomic development in China took place with escalating conflicts between resources, energy, and the environment, resulting in geological hazards that harm society [1]. The movement of earth materials leads to the complex formation of geological hazards, and both human activities and natural variations can lead to changes in the geological environment [2]. As such changes accumulate, consequences resulting from geological disasters can occur [3]. Since the beginning of the new century, landslides and other geological disasters have occurred frequently due to expanding human activities and global climate change, leading to surging ecological and environmental issues, including soil erosion and reduction in vegetation [4,5]. Landslides involve various ground movements such as rock falls, deep slope failure, and shallow slope instability [6]. Gravity is the primary cause of landslides on excessively steep slopes. Sliding sand, rocks, and soil can block rivers and cause further disasters, posing direct threats to people's lives and property [7]. Landslides are characterized by multiple occurrences, rapid development, and significant impacts. The development of landslides is also a complex process [8]. However, the probability of landslides occurring in an area can be predicted based on hazardous environmental conditions influenced by multiple factors. Predictions can help us better understand potential locations of instability that could lead to landslides and thus control their development in high-risk areas [9,10]. In this way, landslide prevention and mitigation can achieve better efficiency and help reduce the risk of hazards.

Landslide susceptibility analysis considers the geological environment, topography, meteorological conditions, etc., and evaluates landslides' susceptibility to and danger in a given area [11]. Predicting potential landslide areas correlates with major hazard factors and historical landslide hazards and involves determining the functional relationships between the variable of hazard factors and the dependent variable of landslide occurrence [12,13]. Landslides are complex natural phenomena whose formation process can be influenced by many hazard factors, including geology, topography, hydrology, meteorology, vegetation, soil properties, and anthropogenic activities. In addition, the interactions between multiple hazard factors complicate landslide susceptibility predictions [14,15], not only due to the large number of input variables but also because of the complex and time-consuming analysis methods involved [16]. Challenges in landslide susceptibility research methods emerge from the requirement to thoroughly understand the interrelationships and mechanisms between hazard factors to accurately predict landslides [17]. In addition, since various factors influence landslide susceptibility, it is necessary to incorporate different data sources and technologies such as remote sensing, geographic information systems (GIS), and geological surveys that can obtain and integrate data [18,19]. In data processing and analysis, the latest statistical models and machine learning methods are needed to address complex non-linear relations and pattern recognition problems among factors [20].

Since the 1990s, the advancement of algorithms, remote sensing, and global positioning systems have made 3S technology (remote sensing, global position systems, and geographical information systems) an important tool for landslide susceptibility prediction. Beyond replacing conventional methods, this technology also provides more effective models for landslide susceptibility prediction, thus filling the gaps in previous studies [21–23]. However, from the perspective of geographic information science, problems still exist in current research on landslide susceptibility prediction. Due to the particularities of landslide hazards, determining their locations may require expert knowledge or on-site investigations. Moreover, the samples in landslide interpretation databases are limited in size [24], greatly restricting data-driven learning models. Previous studies on susceptibility prediction did not fully utilize the edge characteristics and spatial shape features of landslides. Moreover, the complex non-linear relationships between landslides and influencing factors make simple network structures unable to fully utilize feature information; thus, there remains space for improvements in relevant research [25,26].

Commonly used landslide susceptibility prediction methods in the past have utilized statistical methods (e.g., the certainty factors method and informativeness method) to generate landslide samples, which were then combined with machine learning or deep learning models to perform the assessment. These methods are mainly based on counting the number of landslides (area) within each classification by classifying the impact factors. The samples generated in this way are one-dimensional textual datasets, which lose the boundary information of landslides to a certain extent and cannot effectively incorporate the corresponding spatial information. In this study, a moving slider (size: 64×64) was used to traverse the landslide label and influence factor layers across the study area to produce a sample set, which was a 2D image dataset at this stage. Iterative pixel-by-pixel learning of 2D matrices using deep learning methods can maximize the use of edge characteristics and spatial shape features of landslides. This method can help further refine the final landslide susceptibility zoning results. The specific steps of this study are as follows: 1. We combine a topographic map, geological map, and remote sensing image of the area to interpret landslides in Wenchuan and build an interpretation database. 2. By traversing landslide vectors and evaluation factors at a fixed scale, a stable, diverse, and high-quality sample training set is constructed. 3. We achieve susceptibility predictions in Wenchuan by considering the spatial characteristics of landslides as area-based geographic objects (including spatial relation, spatial scale, and spatial shape features) and using deep learning models with a focus on spatial features (UPerNet, ViG, ViT). These predictions enable us to provide technical and theoretical support for the early identification, warning, and accurate prediction of landslides, thus strengthening environmental protection.

2. Materials

2.1. Study Area

The study area of Wenchuan is located in the northwestern Sichuan Basin, with coordinates of 30°45′N–31°43′N and 102°51′E–103°44′E, featuring a total area of 4084 km^2 characterized by highly developed faults and folds (Figure 1). The two major fracture zones of Maoxian–Wenchuan and Beichuan–Yingxiu extend across Wenchuan County from northeast to southwest, in which the Beichuan–Yingxiu fracture was associated with the "5.12" earthquake [27]. Before the 2008 earthquake, the Wenchuan region was already highly susceptible to landslides, with landslide hazards particularly prominent at riverbanks.

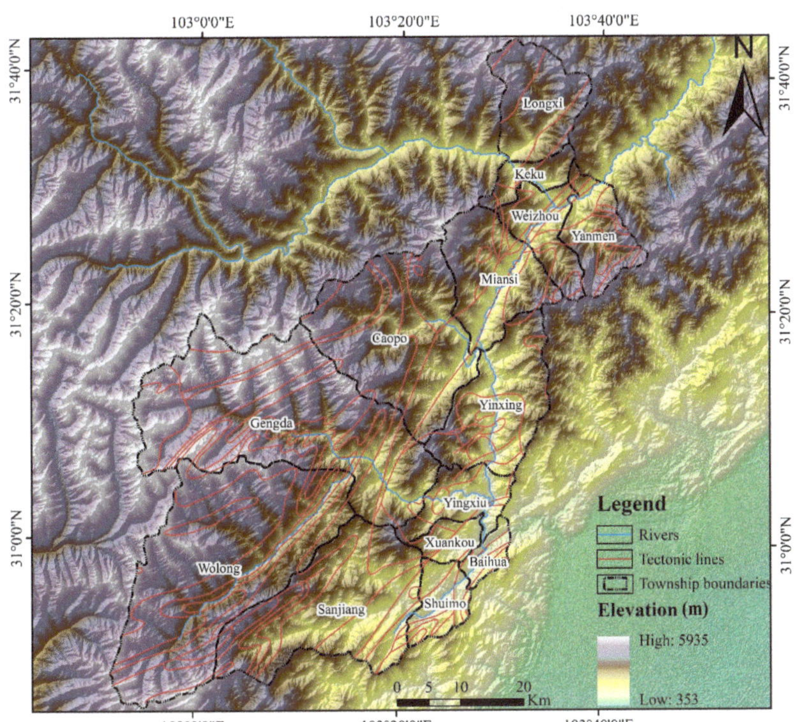

Figure 1. Overview map of Wenchuan County.

2.2. Experiment Data

Landslide susceptibility prediction based on the semantic segmentation of deep learning requires a landslide dataset for training. Since no landslide interpretation and influencing factor dataset for Wenchuan existed prior to this study, it was necessary to collect and build a dataset for the county.

2.2.1. Source of Data

Data were placed into two classes according to the needs of this study (Table 1): (1) remote sensing image and regional investigation data for the sample set on susceptibility prediction and (2) original data for the feature dataset of influencing factors on landslide susceptibility.

Table 1. Data source of influencing factors on landslide susceptibility prediction.

Name	Type	Spatial Resolution	Use	Source
Gaofen-1 satellite image	Raster	Panchromatic 2 m	Landslide visual interpretation	Gaofen Sichuan center for data and application
Sichuan geological hazard survey data	Point	-	Reference for landslide interpretation	Sichuan Provincial Department of Natural Resources
Fundamental geographic information	Line, polygon	1:50,000	Extracting factors of roadway, river, etc.	National Geomatics Center of China
Digital elevation model (DEM)	Raster	30 m	Extracting factors of slope, aspect, etc.	Geospatial Data Cloud
Geological data	Line, polygon	1:200,000	Extracting tectonic line factor	Sichuan Geological Survey Research Institute
Rainfall data from weather stations	Point	-	Extracting average annual rainfall factor	China Meteorological Administration

2.2.2. Production of Landslide Interpretation Database

The goal of the present study on landslide hazards was to develop a detailed, objective, and accurate landslide interpretation database, including the patterns and characteristics of spatial distributions. As reported in this section, we performed a remote sensing interpretation of landslides in Wenchuan for the last 10 years, combining detailed landslide investigation point sources and historical high-resolution remote sensing images with field investigations and on-site validation. The interpretation relied primarily on landslide exposure, vegetation destruction, and accumulation recognition, ultimately identifying 5944 landslides, as shown in Figure 2.

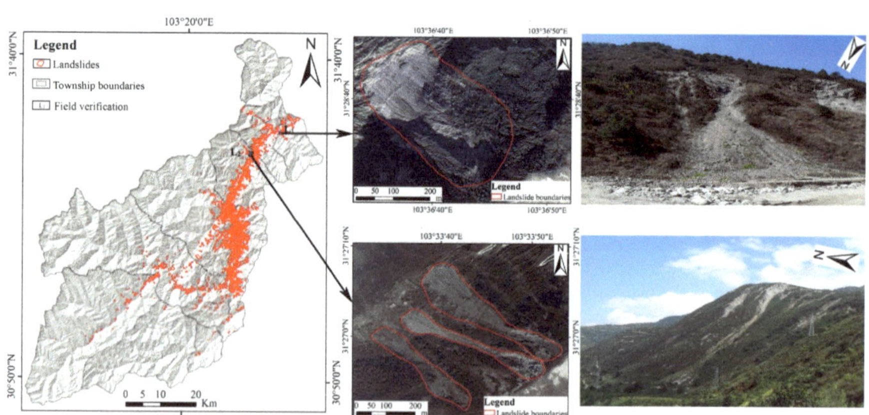

Figure 2. Interpreting Landslides in Wenchuan.

2.2.3. Factors That Influence Landslide Susceptibility

Topography, landforms, and other factors influence the occurrence of landslides. Moreover, a scientific and rational understanding of the factors influencing landslide occurrence in the geological environment is a precondition for conducting landslide susceptibility prediction [28]. Different combinations of influencing factors have different effects on the prediction results of landslide susceptibility, so it is necessary to analyze and compare the importance of factors. Thus, this study extensively investigated the geological environment and conditions that cause landslides in Wenchuan. Considering the actual situation in the study area, the 12 factors were chosen to establish landslide susceptibility prediction metrics: elevation, slope, aspect, topographic relief, plan curvature, profile curvature, tectonic lines, lithology, rivers, rainfall, land use, and roadways [29–31].

Deep learning models are often regarded as black-box models, meaning their internal decision-making process is opaque to us. Due to the complexity and non-linear nature of deep learning models, it is difficult to parse out the specific contribution of individual feature factors to the final prediction result directly from the models. In deep learning models, the importance of features is not directly visible because they are embedded in the complex computational processes of multi-layer neural networks. Traditional machine learning models tend to provide direct information about the importance of features. In this study, the importance of the evaluation factors was calculated based on the decision tree, and the calculated factor weights were normalized before proceeding with the model construction. The order of factor importance is shown in Figure 3. Therefore, the two factors of lithology and land use were not selected for modeling in this study.

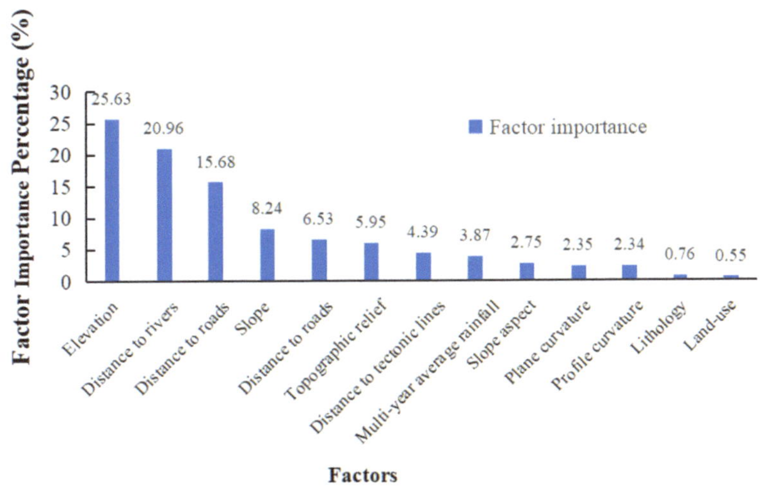

Figure 3. Ranking the importance of factors.

Elevation (Figure 4a) does not have a direct relationship with landslides; instead, the impact of elevation on landslides is usually reflected by other associated conditions. In addition, the landform and topography vary in different elevation ranges, greatly affecting the form of landslide faces [32].

Slope (Figure 4b) or slope angle refers to the angle between the slope and horizontal surfaces. Slope angle is one of the major factors influencing the stability of slopes and is directly related to the shear strength of sliding surfaces. When the slope exceeds the maximum stability angle of the gravitational stress on the soil, the rock and soil bodies will slide downwards [33]. The slope aspect reflects the direction corresponding to different slope bodies, and the difference between shaded and sunny slopes can play a different role in the development and distribution of landslides. As Wenchuan County is in the northern hemisphere, the convention is that the sunny slopes face south, and conversely, the shaded slopes face north.Sunny slopes have longer hours of sunlight than shady slopes, resulting in more solar radiation, more water evaporation from sunny slopes, and more weathering of slopes.

The slope aspect (Figure 4c) is the direction that the largest slope surface on the landform faces and is widely used in landslide susceptibility modeling due to its direct correlation with slope stability. This aspect is relevant to slope stability since it determines the slope's exposure time to the sun and the direction of rainfall. These elements lead to different moisture absorption and evaporation results on different slopes and change the weathering intensity on rock and soil bodies [34]. The slope aspect reflects the direction corresponding to different slope bodies, wherein shaded and sunny slopes can play different roles in the development and distribution of landslides. Wenchuan is located in the northern hemisphere; thus, sunny slopes face south, while shaded slopes face north. Sunny slopes experience longer hours of sunlight than shady slopes, resulting in greater solar radiation, more water evaporation from sunny slopes, and increased slope weathering.

Topographic relief (Figure 4d) in slope elements usually refers to the difference between maximum and minimum elevation, which impacts the possibility of landslide occurrence on a slope [35]. Less than 1% of the study area features a topographic relief greater than 349 m. Landslides are mainly concentrated between 29 and 394 m, with the proportion of landslides increasing as the topographic relief increases.

Curvature is the degree of bending or the rate of change to neighboring points at a certain point and includes components along both the horizontal and vertical directions [35]. Areas with curvature greater than 0 are convex slopes, while areas with curvature less than 0 are concave slopes. The component on the horizontal axis is extracted to determine plan curvature (Figure 4e,f). The density of landslide distribution rises overall with an increase in plan curvature. The curvature component on the vertical axis is known as profile curvature, which is the changing rate of the surface slope at that point. Similar to landslide distribution in ranges of plan curvature, the percentage of landslide distribution grows overall with an increase in profile curvature. Curvature quantitatively describes the degree of bending and deformation at a point on a terrain surface. Curvature breaks down the degree of distorted deformation into vertical and horizontal directions. The component of ground curvature in the vertical direction is called the profile, which measures the change in elevation in terms of the rate of change in the direction of the maximum drop in slope. The component of curvature in the horizontal direction is called plane curvature. Convex slopes are more prone to landslides because the bending deformation of convex slopes causes stress concentration within that slope, leading to slope instability. The proportion of landslide area in the study area is higher within the area of convex slopes than in the area of landslide hazards within concave slopes.

The rupture and fracture caused by tectonic movements result in overhanging surfaces and valleys, which provide conditions for landslide development and occurrence. In areas affected by tectonic and seismic movements, the spatial distribution of landslides is closely related to the tectonic pattern [36].

Rivers (Figure 4h) also impact landslides, mostly in terms of changes in the slope stress state and landslide mechanics. Dynamics related to river erosion and corrosion weaken the slope's stability and catalyze the occurrence of landslides [37]. The bank slopes of rivers, lakes (reservoirs), seas, and ditches, as well as canyon areas with large differences in topographic elevation, are prone to landslides due to the scouring and erosion of water currents, which make the bank slopes steeper and higher. In terms of spatial distribution, a large number of landslides in this study are located relatively close to the stream and river system.

Rainfall (Figure 4i) can increase excessive water saturation and the overall weight of the slope. The lateral pressure along the slope surface due to gravity and the weakened shear strength of rock and soil bodies results in a sliding surface that can lead to landslide hazards [38].

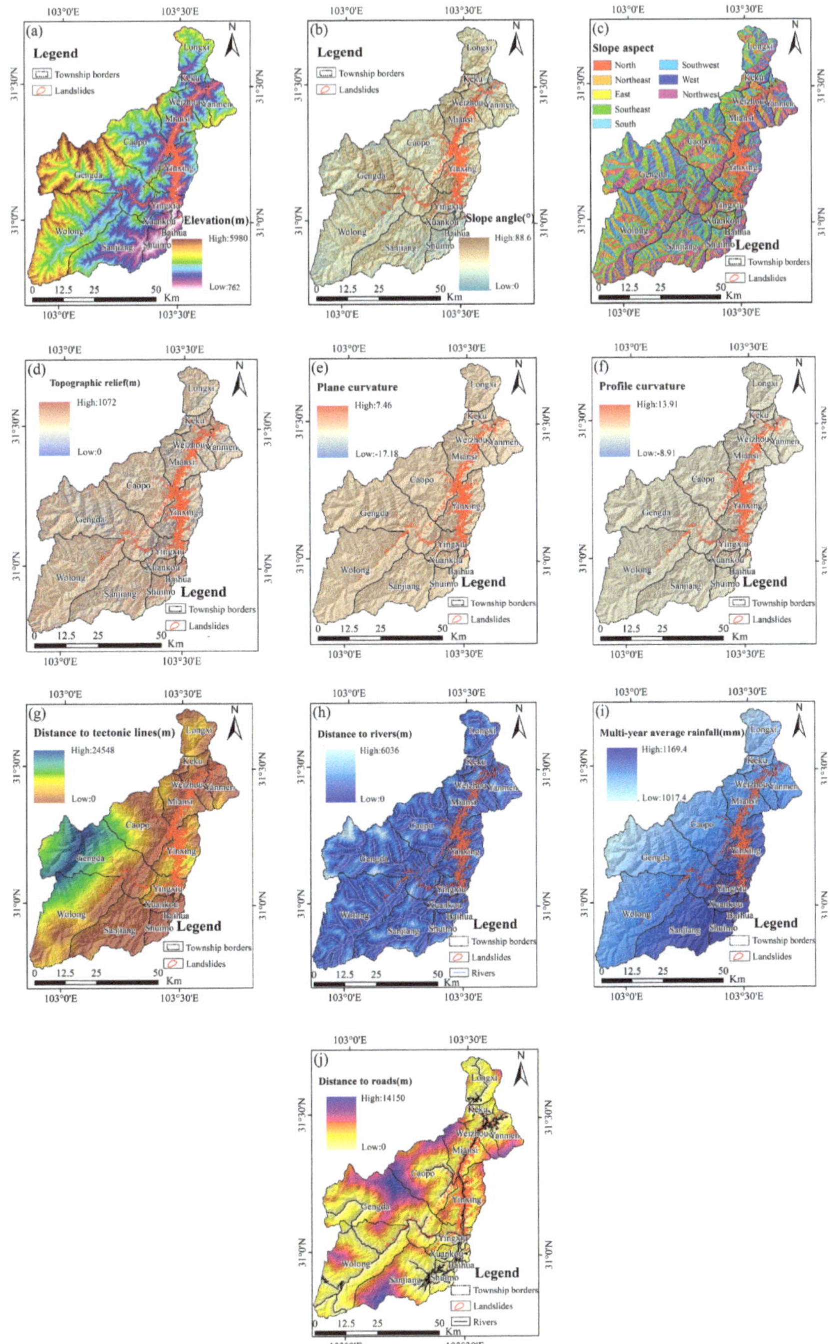

Figure 4. Influencing factors of landslide susceptibility in Wenchuan County: (**a**) Elevation, (**b**) Slope angle, (**c**) Slope aspect, (**d**) Topographic relief, (**e**) Plane curvature, (**f**) Profile curvature, (**g**) Distance to tectonic lines, (**h**) Distance to rivers, (**i**) Multi-year average rainfall, (**j**) Distance to road.

Lastly, the distance to roads (Figure 4j) can reflect the regional intensity of human engineering activities. Potential side-slope excavations and the earth's surface vibrations caused by vehicle transportation on roadways are additional potential factors influencing landslides. Existing studies suggest a centered landslide distribution along roadways and an inverse relationship between landslide density and distance to roads [39].

2.3. Production of Datasets on Susceptibility Prediction

Samples provide crucial data to support deep learning. Indeed, the quality of samples is critical to the practicality and precision of such models. Thus, the quality of samples should be strictly controlled. As reported in Section 2.2.2., a detailed landslide interpretation of Wenchuan was conducted and validated in the field. Landslide sample data were constructed from landslide attribute data (landslide or non-landslide data) and influencing factor attributes that affect landslide occurrence. The ultimate goal of the landslide susceptibility prediction model is to predict landslide attributes, with influencing factor attributes used as the input parameters. Since many factors in this study were derived from DEM (digital elevation model) data, all influencing factors used were uniformly sampled to a spatial resolution of 30 × 30 m [40,41], consistent with the DEM. This study is based on the interpretation results, which were transformed into raster data with the same spatial resolution as the influencing factor (a single pixel size of 30 × 30 m). Here, landslide data are labeled 1, and non-landslide data are labeled 0. To effectively utilize landslide margin and spatial shape features, a mobile block with 64 × 64 pixels [42,43] was introduced to build the prediction dataset; as can be seen from the cropping results of the block in Figure 5, the boundary information of the landslide can be well preserved.. In this way, we obtained a training set of 924 samples and a test set of 396 samples. In addition, 25% of the data in the training set was partitioned as a validation set to select training epochs and avoid overfitting.

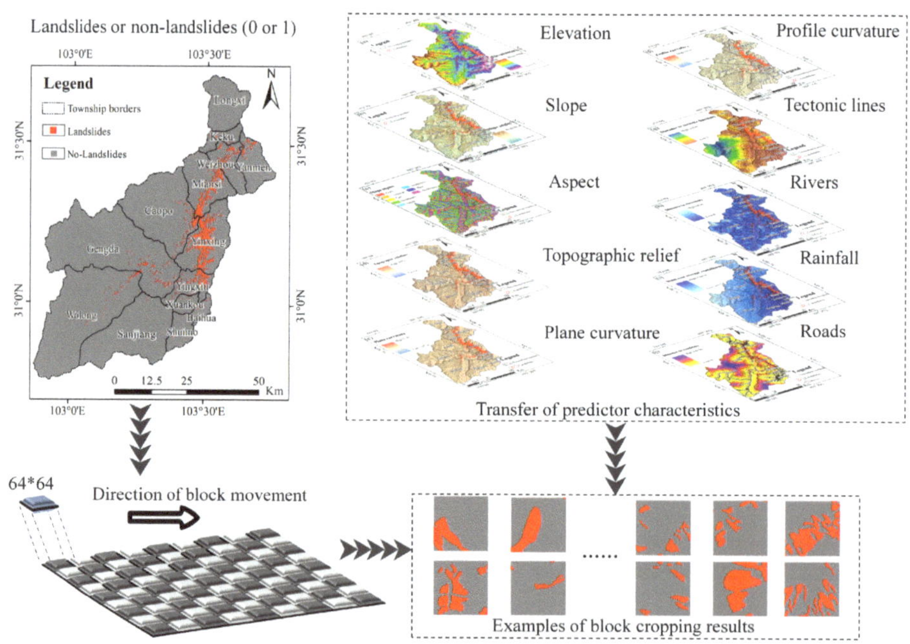

Figure 5. Production process of the prediction dataset in Wenchuan.

Online data augmentation, also known as in-place data augmentation, primarily includes the following stages [44]: (1) starting model training; (2) extracting a small batch of data from the training set according to assigned or random rules (batch size); (3) sending original small-batch data to the data magnifier and using the magnifier to perform geometrical transformation according to the designated rules; (4) using the data augmentation module input to augment small batch data input into the training model and performing training; (5) repeating stages 2 to 4 for the required training times. The process of online data augmentation is illustrated in Figure 6. During each augmentation step, the new data input into the model are very similar to the original data. To gradually improve the small-batch data, we augmented data based on geometric and image color transformation through the online data augmentation module.

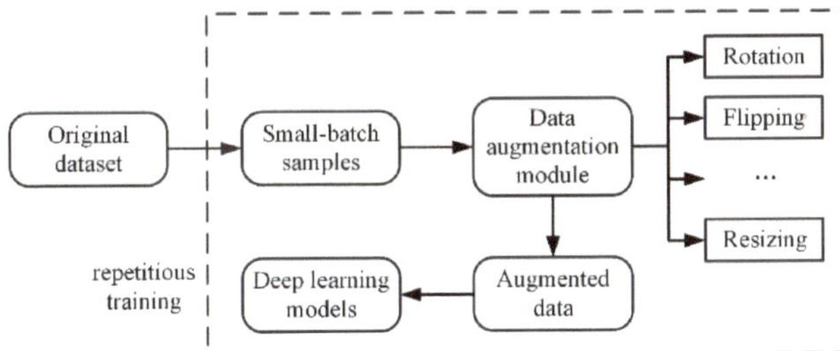

Figure 6. Online data augmentation process.

The geometric transformation of data usually involves flip, rotation, cropping, distortion, scaling, and other operations. Extensive experiments from previous studies suggest that flipping and distorting the image are valid data augmentation techniques for semantic segmentation [45]. Flipping is an operation similar to mirroring, while rotation involves a circular clockwise or counterclockwise movement. Moreover, it is important to rotate the image by 90 to 180 degrees to avoid inconsistent scaling among the image samples. Here, the flipped and rotated augmented images differed from the original images only in their directions and angles, as they retained the same size [46]. We selected random rotation and flipping for geometric data augmentation and expanded the sample dataset to three times its original size. A final training set of 2772 samples and a test set of 1188 samples were thus obtained. Labels were transformed according to the geometric transformation of the images to ensure consistency with the transformed images.

3. Methods

3.1. Description of the Applied Models

3.1.1. Unified Perceptual Parsing Network (UPerNet)

UPerNet [47] is based on the feature pyramid network (FPN) and uses a bottom-up process to sequentially extract image features, including small-scale detailed feature information and large-scale generalized feature information. The backbone network used in this study is the classical ResNet50. UPerNet connects high-level features that characterize the generalized information and low-level features that characterize the detailed information with top–down sidelobes through top–down paths, enabling features at all scales to contain rich information. The network architecture is shown in Figure 7. Since this susceptibility prediction study uses 10 impact factor layers, the input features correspond to 10 channels.

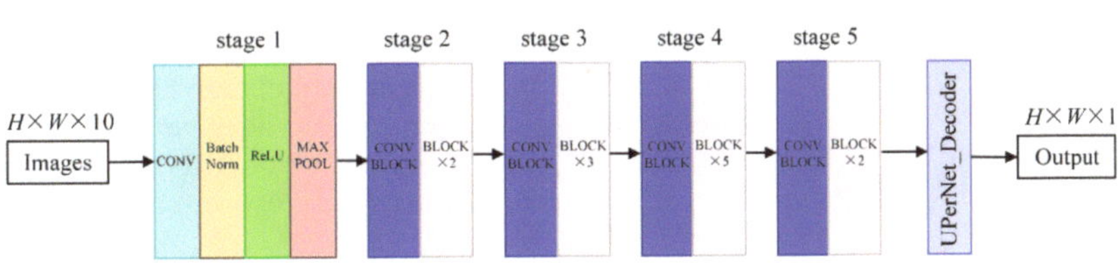

Figure 7. Structure of UPerNet.

3.1.2. Vision Transformer (ViT)

The classical Transformer model [48] uses a one-dimensional embedded token as input. Dosovitski et al. [49] proposed using a Vision Transformer (ViT) to process 2D images. Considering time complexity and performance, ViT is able to directly input feature images into the Transformer network to perform image classification tasks without relying on the CNN network, which demonstrates for the first time the possibility of using the Transformer in the field of computer vision. In the natural language domain, the transformer accepts word tokens as sequence input. Thus, ViT cuts the input image into fixed blocks of P × P size without overlapping and then converts the segmented feature subgraphs into one-dimensional sequences, i.e., patch embedding. The category and location information are then fed into the Transformer encoder as a learnable vector along with the sequence data using a method similar to BERT [50], which performs a self-attention operation. The final category vector output is used for downstream classification tasks. The ViT network structure is shown in Figure 8.

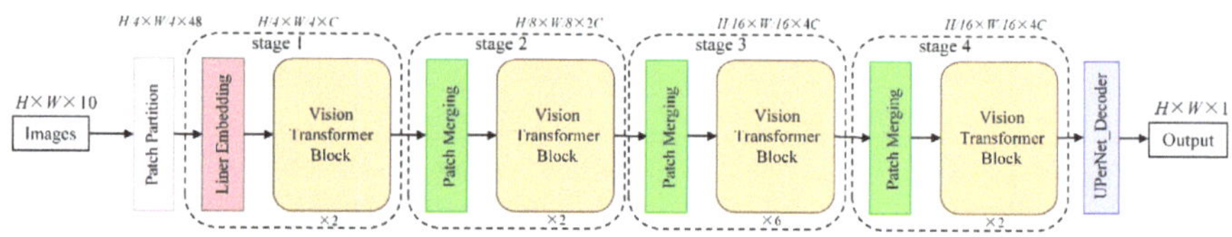

Figure 8. Structure of ViT.

3.1.3. Vision Graph Neural Networks (ViG)

The GNN is a neural network specialized in processing graph data. The GNN is most useful when handling complex graph data, including social networks, knowledge graphs, and molecular geometry. The core idea of the GNN is to update the output of nodes through iteration using the graph's structural information [51]. In each iteration step, every node agglomerates information on its neighboring pixels and updates its state accordingly. Through this process, the nodes can capture global structural information from the graph. In reality, many important datasets exist as graphs or networks. Thus, combining a neural network with graph analysis could resolve numerous problems [52]. A graph, also known as a topology, is an important concept in computer science. The two primary components of a graph are its vertices and edges, represented as follows:

$$G = (V, E), \tag{1}$$

where G is the graph, V is the set of vertices, and E is the set of edges connecting vertices.

Vision GNN (ViG) is an innovative neural network architecture [53] that combines the advantages of computer vision with GNN. The core idea of this approach is to convert traditional image data into graph structures and utilize GNNs to capture the complex patterns and relationships in images. In ViG, an image is first segmented into multiple local regions (or "nodes"). These nodes are connected as a graph by considering the spatial relationship (e.g., adjacency) between them. Through this transformation, pixels in an image are no longer isolated but instead organized into a graph structure that expresses the overall form and local details of the image. ViG next utilizes the graph convolution operation in GNNs to aggregate and update the feature information of the nodes. Each node can capture information about its neighboring nodes through multiple iterations and updates, resulting in a rich representation that contains global and local information about the image. The network structure is illustrated in Figure 9.

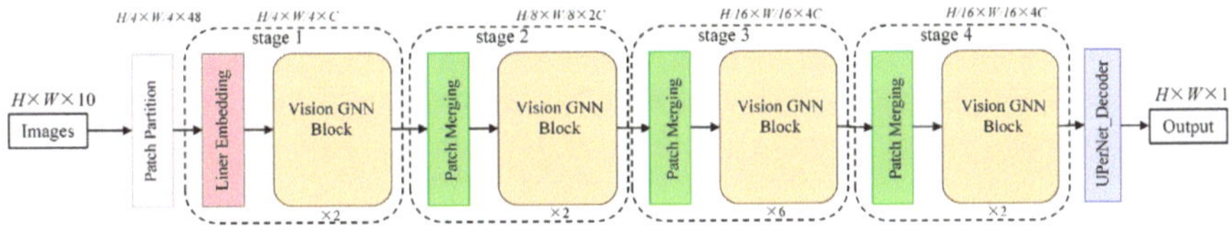

Figure 9. Structure of ViG.

3.2. Evaluation of Model Accuracy

To objectively validate the prediction results, we selected four validation factors, including pixel accuracy, overall accuracy, recall rate, and F1 score, based on the test images [54–56]. Then, the performance of different methods on landslide susceptibility prediction was assessed to choose the optimal prediction model. Here, pixel accuracy (*PA*) reflects the proportion of accurate predictions in all pixels predicted as landslides:

$$PA = \frac{TP}{TP + FP}. \tag{2}$$

Overall accuracy (*OA*) is the proportion of correct landslide and non-landslide predictions in all pixels predicted, as follows:

$$OA = \frac{TP + TN}{P + N}. \tag{3}$$

The recall rate (*Recall*) is the proportion of correctly predicted landslide pixels in all sample landslide pixels:

$$Recall = \frac{TP}{TP + FN}. \tag{4}$$

The *F1-score* considers both pixel accuracy and the recall:

$$F1\text{-}score = \frac{2 \times Recall \times PA}{Recall + PA}, \tag{5}$$

where P is the number of landslide pixels; N is the number of non-landslide pixels; TP refers to landslide pixels predicted as landslide pixels; TN refers to non-landslide pixels predicted as landslide pixels; FP refers to non-landslide pixels predicted as landslide pixels; and FN refers to landslide pixels predicted as non-landslide pixels.

4. Results

4.1. Application of the UPerNet, ViT, and ViG Models

Landslide susceptibility predictions using deep learning models involve processing large amounts of sample data alongside high algorithmic and time complexity, which demand significant computational resources. Consequently, the present experiment used a 13th Gen Intel® Core™ i7-13700KF CPU, an NVIDIA GeForce RTX 4090 GPU, and 64.0 GB of RAM. The software environment employed the deep learning framework pytorch-gpu-1.8.1 + CUDA 11.3 + CUDNN 8.2.1 for model construction, corresponding to Python version 3.7.

The augmented training set described in Section 3.2 was input into the UPerNet, ViT, and ViG models, with the best results obtained by optimizing the model parameters. The parameter settings of the different models are listed in Table 2, which also shows the optimal parameter weights.

All prediction samples were considered monochrome images and input into the established models. Landslide susceptibility index maps were generated by calculating the possibility of landslide occurrence in each pixel (Figures 10a, 11a and 12a). Using the equal distance method, calculated indices were categorized into five levels: very low landslide-susceptibility areas (0–0.2), low landslide-susceptibility areas (0.2–0.4), medium landslide-susceptibility areas (0.4–0.6), high landslide-susceptibility areas (0.6–0.8), and very high landslide-susceptibility areas (0.8–1). The landslide susceptibility zoning maps are illustrated in Figures 10b, 11b and 12b.

Table 2. Model Parameter Settings.

Models	Same parameters	Backbone
UPerNet	Epochs:100 Dropout: 0.5 Learning rate: 0.001 Batch_size: 16 Activation function: ReLU Optimizer: Adam Loss function: Binary CrossEntropy	7 × 7, 64, Conv 3 × 3, max pool ResNet50 _Block [2,2,2,2]
ViT		4 × 4 Patch 64 Embedding Transformer_Layer [2,2,6,2]
ViG		7 × 7, 64, Conv 3 × 3, max pool GNN_Layer [2,2,6,2]

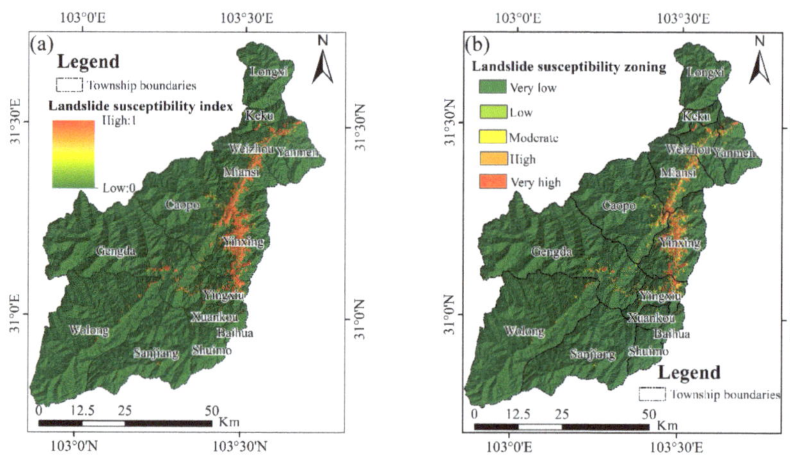

Figure 10. (**a**) UPerNet-based landslide susceptibility index map; (**b**) UPerNet-based landslide susceptibility zoning map.

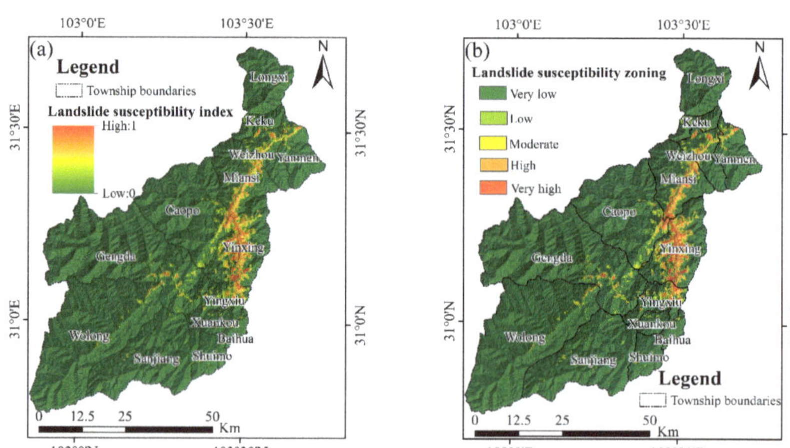

Figure 11. (**a**) ViT-based landslide susceptibility index map; (**b**) ViT-based landslide susceptibility zoning map.

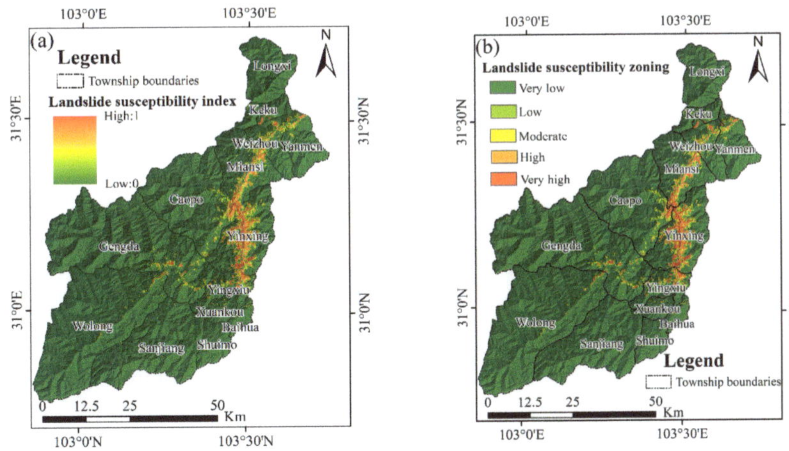

Figure 12. (**a**) ViG-based landslide susceptibility index map; (**b**) ViG-based landslide susceptibility zoning map.

4.2. Validation of the UPerNet, ViT, and ViG Models

The network training performance on the test set is presented in Table 3. Compared with UPerNet and ViG, the overall test results for ViT lacked competitiveness and offered lower evaluation results for all criteria. Although the transformer (TR) model has excellent performance in the natural language processing (NLP) field, it is less competitive in handling image or graph data. The ViT extracts feature representations with context relevance, which is effective for text data. However, for images or graphs, ViT may not fully capture the data's spatial–structural information or geometric features.

Table 3. Comparison of landslide susceptibility test sets from different models.

Models	PA	OA	Recall	F1-Score
UPerNet	0.78	0.98	0.77	0.77
ViT	0.66	0.97	0.72	0.69
ViG	0.72	0.97	0.74	0.73

UPerNet performs better than ViG on the test set for all criteria, with 6% higher pixel accuracy, 1% higher overall accuracy, 3% higher recall, and a 4% higher F1 score than ViG. UPerNet provides increased performance because it specializes in processing image data. By performing convolution on input images with the convolutional layer, UPerNet can effectively extract regional features from images and generate layered feature representations by stacking multiple convolutional layers. In comparison, ViG is primarily used to process graph data, such as in social networks and knowledge graphs. Although ViG can accurately process graph data, it may not effectively extract and utilize regional features in image data or UPerNet.

5. Discussion

5.1. Comparing Datasets with Different Data Volumes

Data augmentation is a key technology in the field of deep learning. Through a series of transformations and expansions of the original dataset, this method can effectively enlarge the training sample size and improve the model's generalization ability. In the training process of deep learning models, data augmentation can reduce the risk of overfitting and help the model learn more useful features to enhance the accuracy and stability of the prediction results. In this study, we expanded samples of the dataset to three times their original size. The data augmentation results show that UPerNet, ViG, and ViT all obviously improved in accuracy. This improvement benefited from data augmentation, which provided more diverse and abundant training samples and enabled the model to learn more useful features and patterns. Nevertheless, UPerNet presented the greatest improvements in accuracy.

Compared to models before data augmentation, UPerNet improved the PA, OA, Recall, and F1 score by 26%, 2%, 8%, and 18%, respectively; ViT improved them by 17%, 1%, 2%, and 11%; and ViG improved them by 18%, 1%, 5%, and 12%. Comparing the susceptibility prediction results of three models before and after data augmentation, ViG was found to provide more accurate susceptibility zoning than the other two models because ViG updates the representations of nodes by agglomerating information from the neighboring pixels, which enables the model to learn the dependency between nodes. For small samples, even with a limited sample size, ViG can still diversify representations of nodes through agglomeration.

After data augmentation, the susceptibility zoning predictions of UPerNet and ViG improved greatly. Table 4 and Figure 13 suggest that UPerNet offered slightly better improvements than ViG because of its strength in processing image data. UPerNet's convolution operations can effectively extract regional features from images and generate layered feature representations by stacking multiple convolutional layers. Data augmentation provides UPerNet with more image samples. The transformation and expansion of these samples additionally expose the model to different image patterns, further enhancing the feature extraction and prediction accuracy of UPerNet. ViG is excellent in processing graph data and can capture complex relationships and dependency patterns between nodes.

Table 4. Comparison of evaluation metrics before and after data enhancement for different models.

Models	Data Volumes	PA	OA	Recall	F1-Score
UPerNet	original	0.52	0.96	0.69	0.59
	3× data enhancement	0.78	0.98	0.77	0.77
ViT	original	0.49	0.96	0.70	0.58
	3× data enhancement	0.66	0.97	0.72	0.69
ViG	original	0.54	0.96	0.69	0.61
	3× data enhancement	0.72	0.97	0.74	0.73

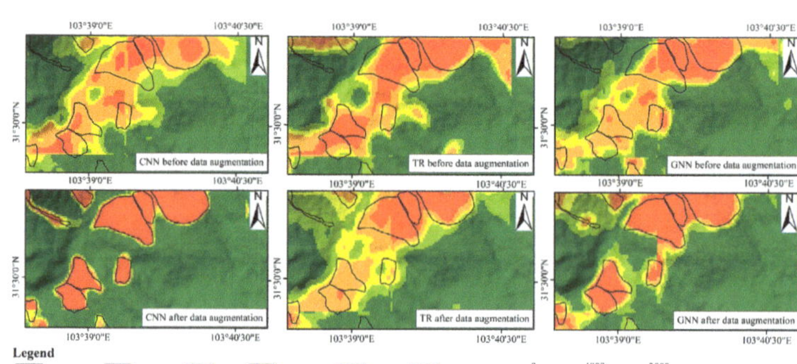

Figure 13. Comparison of susceptibility zones before and after data enhancement.

However, ViG might not be as effective as UPerNet when processing image data because the regional features and spatial structure of image data can be better extracted through convolution. ViG sometimes requires more complex transformations and representations when handling this type of data. Although data augmentation provides ViG with more graph data, ViG cannot be compared with UPerNet in terms of accuracy improvements because of its different processing mechanisms and data characteristics. Moreover, ViG might face challenges in computation efficiency and extensibility when processing large-sized graph data, which can also restrict its improvements in accuracy. While ViT has had success in the field of NLP, it may have limitations in processing image data.

Moreover, although data augmentation may help improve the accuracy of the ViT model, ViT may not be as efficient as UPerNet in extracting image features and processing spatial structures since it is not designed for image data. Furthermore, ViT has relatively high computation complexity and may require more computational resources and time when processing large image data. This factor may also affect the accuracy improvements of ViT.

5.2. Comparison of Models for Susceptibility Zones

In this study, areas within Wenchuan County were divided into five zones: relatively low landslide susceptibility, low landslide susceptibility, medium landslide susceptibility, high landslide susceptibility, and very high landslide susceptibility. The classified statistics for the susceptibility zoning results for landslide density are shown in Table 5.

Table 5. Statistical zoning results of the different models.

Models	Zoning Levels	Percentage of Landslide Area L_i (%)	Percentage of Susceptibility Zones S_i (%)	$R_i = L_i/S_i$
UPerNet	I	15.1	94.8	0.16
	II	2.6	0.9	2.89
	III	5.8	0.8	7.25
	IV	14.1	1.1	12.82
	V	62.4	2.4	26
ViT	I	7.8	90.3	0.09
	II	7.4	3.6	2.06
	III	10.9	2	5.45
	IV	29.7	2.3	12.91
	V	44.2	1.8	24.56
ViG	I	8.1	92.4	0.08
	II	6.2	2.3	2.69
	III	10.8	1.6	6.75
	IV	19.2	1.5	12.8
	V	55.7	2.2	25.32

The evaluation criteria for this method were as follows: The very high landslide-susceptibility zone represented the smallest proportion in the entire study area with the maximum proportion (number) of landslide test set samples, while the low landslide-susceptibility zone represented the greatest proportion in the area with the minimum proportion (number) of landslide test-set samples. The proportion of each landslide susceptibility zone in Table 5 shows that the deep learning prediction results under the new strategy all satisfy the needs of susceptibility zoning. In other words, the low landslide-susceptibility zone occupies a higher proportion of the area with a lower rate of landslides, while the very high landslide-susceptibility zone occupies a smaller proportion with a higher rate of landslides. The zoning results of the three models all satisfy reasonable susceptibility requirements: $R_I < R_{II} < R_{III} < R_{IV} < R_V$. The proportion of the very low landslide-susceptibility zone in all models was $S_{UPerNet} < S_{ViG} < S_{ViT}$; the zonal proportion of landslides was $L_{UPerNet} > L_{ViG} > L_{ViT}$; the proportion of the low landslide-susceptibility zone was $S_{UPerNet} > S_{ViG} > S_{ViT}$; and the zonal proportion of landslides was $L_{ViT} < L_{ViG} < L_{UPerNet}$.

With neighboring pixel collaboration, the UPerNet model offers the best results in zonal statistics. Under this model, the high landslide-susceptibility zone occupied only 2.4%

of the entire area with 62.4% of landslide hazards, whereas the low landslide-susceptibility zone represented 94.8% of the entire area region, covering most of Wenchuan County, with only 15.1% of landslide hazards. The zoning and hazard prediction results both demonstrate that the UPerNet model can identify very high landslide-susceptibility zones.

The contributions of neighboring pixel collaboration to landslide susceptibility prediction are twofold. First, this method assists deep learning models in capturing the spatial patterns of landslide occurrence. Landslides often happen under certain geological environmental conditions that are distributed according to certain spatial laws. By involving information from neighboring pixels, the model can better recognize such patterns and evaluate the possibilities for landslide occurrence. Figure 14 suggests that flat areas such as rivers usually have low relevance in landslide occurrence, while steep hills represent more susceptible zones. The UPerNet, ViT, and DNN models have strong feature extraction and learning capabilities. These models can fully utilize the advantages of neighboring pixel collaboration and achieve excellent results in landslide susceptibility zoning by better distinguishing between flat and susceptible regions.

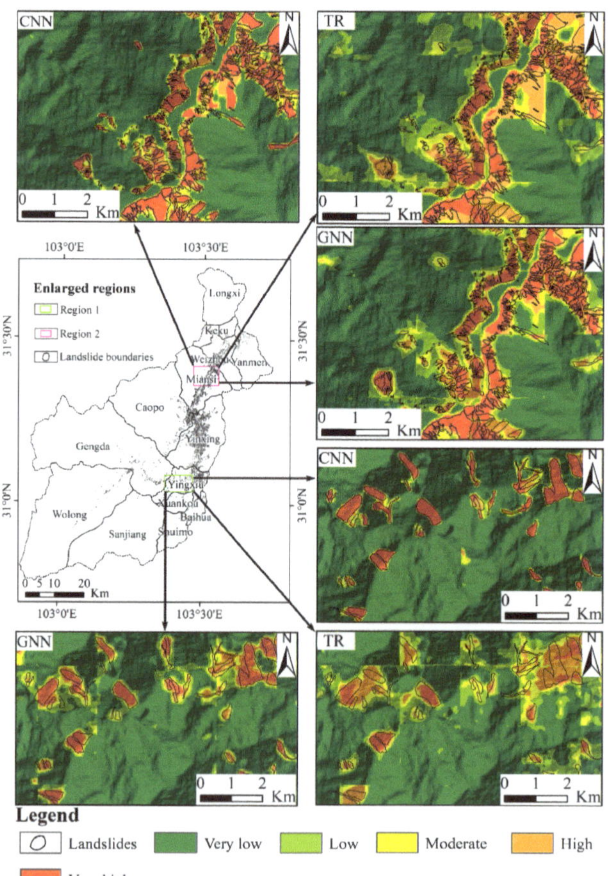

Figure 14. Comparison of Landslide Susceptibility Zoning Results for Old and New Strategies.

5.3. Comparison with the Traditional Method

Statistical techniques such as the informational method and the deterministic factor method [57–59] have been widely adopted in traditional studies of landslide susceptibility prediction to construct feature sample sets. These methods focus on calculating the specific quantitative relationships between landslides and various types of influencing factors to accurately assess the sensitivity of influencing factors to landslides. After combining sensitivity analysis with the spatial distribution of landslides, predictions of landslide susceptibility were derived via hierarchical superposition. This type of sample set is a two-dimensional text dataset that tends to ignore the precise boundary information of landslides. Even using the Transformer model [60], which is very advanced in the textual domain, could not achieve granular predictions of susceptibility.

To solve this problem, we proposed a novel strategy using neighboring pixel collaboration. Employing iterative pixel-by-pixel learning of 2D images with deep learning methods can utilize the maximum number of edge features and spatial shape features of landslides. A 2D image dataset was produced based on this strategy, which can retain the boundary information of landslides as completely as possible when predicting landslide susceptibility. We also used the Transformer model, which performs well in textual data modeling, to compute landslide susceptibility in Wenchuan County using the traditional strategy. Taking Miansi town as an example, we explored the landslide susceptibility prediction results using different strategies.

As shown in Figure 15, our new strategy combines different deep-learning models to label areas with flat terrain and a very low probability of landslide occurrence as very low susceptibility zones. At the same time, very high susceptibility zones accounted for a very low percentage of the study area. This outcome reflects the ultimate goal of landslide susceptibility prediction: to focus on areas that are extremely susceptible to landslides at the lowest possible monitoring cost. In contrast, the traditional strategy employed a continuous zoning approach to analyze the influencing factors, thus closely linking the river neighborhoods to the landslide susceptibility areas of the lower mountains and achieving the relevant data mining objectives. However, our new method offered significant advantages in preserving landslide boundary information, providing new perspectives and strategies for landslide susceptibility prediction.

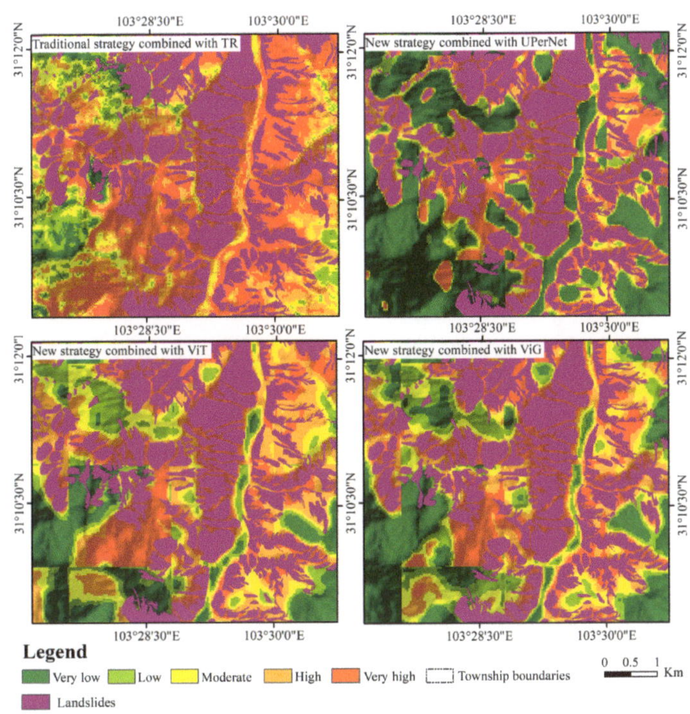

Figure 15. Comparison with the traditional method (traditional strategy combined with TR).

6. Conclusions

This study proposed a new strategy using UPerNet, ViT, and ViG for neighboring pixel collaboration to effectively utilize landslide features such as margins and spatial shapes in landslide susceptibility prediction. This strategy efficiently combines the strengths of deep learning in image processing, sequence modeling, and data processing. We selected Wenchuan County as the study area, conducted various experiments, and obtained the following conclusions.

We designed an online data augmentation module that successfully expanded the sample size of the dataset to three times its original size. This expansion greatly improved the accuracy of the prediction models. Compared to the models before data augmentation, UperNet improved the PA, OA, Recall, and F1-score, respectively, by 26%, 2%, 8%, and 18%; ViT improved them by 17%, 1%, 2%, and 11%; and ViG improved them by 18%, 1%, 5%, and 12%.

The UPerNet model offered the best performance in susceptibility zoning prediction. The very high landslide-susceptibility zone predicted by UPerNet contained 62.4% of landslides, which was considerably better than the results from ViG (55.7%) and ViT (29.7%) because the regional features and spatial structures of the image data could be better extracted via convolution.

By using the latest machine learning and statistical models and incorporating neighboring pixel collaborative analysis, the proposed model could more accurately predict landslide susceptibility. These predictions identified flat areas such as rivers and distinguished between areas with high and very high landslide susceptibility.

Author Contributions: Conceptualization, X.W., S.C. and D.W.; methodology, X.W., W.L. and J.D.; software, D.W. and L.X.; validation, T.S. and J.D.; formal analysis, X.S.; investigation, X.W.; resources, D.W.; data curation, X.W.; writing—original draft preparation, X.W. and M.Z.; writing—review and editing, M.Z.; visualization X.W.; supervision, X.W.; project administration, X.W.; funding acquisition, L.X. All authors have read and agreed to the published version of the manuscript.

Funding: This research was funded by the National Natural Science Foundation of China, grant number 52208006.

Data Availability Statement: The datasets presented in this article are not readily available because the data are part of an ongoing study. Requests to access the datasets should be directed to Xiao Wang.

Conflicts of Interest: Author Xiaochuan Song was employed by the company Sichuan 402 Surveying and Mapping Technology Corp. Tiegang Sun was employed by the company China Building Materials Southwest Survey and Design Co., Ltd. The remaining authors declare that the research was conducted in the absence of any commercial or financial relationships that could be construed as a potential conflict of interest.

References

1. Yong, C.; Jinlong, D.; Fei, G.; Bin, T.; Tao, Z.; Hao, F.; Li, W.; Qinghua, Z. Review of landslide susceptibility assessment based on knowledge mapping. *Stoch. Environ. Res. Risk Assess.* **2022**, *36*, 2399–2417. [CrossRef]
2. Chen, X.; Chen, W. GIS-based landslide susceptibility assessment using optimized hybrid machine learning methods. *Catena* **2021**, *196*, 104833. [CrossRef]
3. Segoni, S.; Pappafico, G.; Luti, T.; Catani, F. Landslide susceptibility assessment in complex geological settings: Sensitivity to geological information and insights on its parameterization. *Landslides* **2020**, *17*, 2443–2453. [CrossRef]
4. Bui, D.T.; Tsangaratos, P.; Nguyen, V.-T.; Liem, N.V.; Trinh, P.T. Comparing the prediction performance of a Deep Learning Neural Network model with conventional machine learning models in landslide susceptibility assessment. *Catena* **2020**, *188*, 104426. [CrossRef]
5. Hodasová, K.; Bednarik, M. Effect of using various weighting methods in a process of landslide susceptibility assessment. *Nat. Hazards* **2020**, *105*, 481–499. [CrossRef]
6. Yang, C.; Liu, L.-L.; Huang, F.; Huang, L.; Wang, X.-M. Machine learning-based landslide susceptibility assessment with optimized ratio of landslide to non-landslide samples. *Gondwana Res.* **2023**, *123*, 198–216. [CrossRef]
7. Pourghasemi, H.R.; Sadhasivam, N.; Amiri, M.; Eskandari, S.; Santosh, M. Landslide susceptibility assessment and mapping using state-of-the art machine learning techniques. *Nat. Hazards* **2021**, *108*, 1291–1316. [CrossRef]

8. Wu, R.; Zhang, Y.; Guo, C.; Yang, Z.; Tang, J.; Su, F. Landslide susceptibility assessment in mountainous area: A case study of Sichuan–Tibet railway, China. *Environ. Earth Sci.* **2020**, *79*, 1–16. [CrossRef]
9. Zhang, Y.-X.; Lan, H.-X.; Li, L.-P.; Wu, Y.-M.; Chen, J.-H.; Tian, N.-M. Optimizing the frequency ratio method for landslide susceptibility assessment: A case study of the Caiyuan Basin in the southeast mountainous area of China. *J. Mt. Sci.* **2020**, *17*, 340–357. [CrossRef]
10. Chen, L.; Guo, H.; Gong, P.; Yang, Y.; Zuo, Z.; Gu, M. Landslide susceptibility assessment using weights-of-evidence model and cluster analysis along the highways in the Hubei section of the Three Gorges Reservoir Area. *Comput. Geosci.* **2021**, *156*, 104899. [CrossRef]
11. Yi, Y.; Zhang, W.; Xu, X.; Zhang, Z.; Wu, X. Evaluation of neural network models for landslide susceptibility assessment. *Int. J. Digit. Earth* **2022**, *15*, 934–953. [CrossRef]
12. Alsabhan, A.H.; Singh, K.; Sharma, A.; Alam, S.; Pandey, D.D.; Rahman, S.A.S.; Khursheed, A.; Munshi, F.M. Landslide susceptibility assessment in the Himalayan range based along Kasauli—Parwanoo road corridor using weight of evidence, information value, and frequency ratio. *J. King Saud Univ.-Sci.* **2022**, *34*, 101759. [CrossRef]
13. Sur, U.; Singh, P.; Meena, S.R. Landslide susceptibility assessment in a lesser Himalayan road corridor (India) applying fuzzy AHP technique and earth-observation data. *Geomat. Nat. Hazards Risk* **2020**, *11*, 2176–2209. [CrossRef]
14. Chanu, M.L.; Bakimchandra, O. Landslide susceptibility assessment using AHP model and multi resolution DEMs along a highway in Manipur, India. *Environ. Earth Sci.* **2022**, *81*, 1–11. [CrossRef]
15. Okalp, K.; Akgün, H. Landslide susceptibility assessment in medium-scale: Case studies from the major drainage basins of Turkey. *Environ. Earth Sci.* **2022**, *81*. [CrossRef]
16. Di Napoli, M.; Marsiglia, P.; Di Martire, D.; Ramondini, M.; Ullo, S.L.; Calcaterra, D. Landslide Susceptibility Assessment of Wildfire Burnt Areas through Earth-Observation Techniques and a Machine Learning-Based Approach. *Remote Sens.* **2020**, *12*, 2505. [CrossRef]
17. Rong, G.; Li, K.; Su, Y.; Tong, Z.; Liu, X.; Zhang, J.; Zhang, Y.; Li, T. Comparison of Tree-Structured Parzen Estimator Optimization in Three Typical Neural Network Models for Landslide Susceptibility Assessment. *Remote Sens.* **2021**, *13*, 4694. [CrossRef]
18. Dias, H.C.; Gramani, M.F.; Grohmann, C.H.; Bateira, C.; Vieira, B.C. Statistical-based shallow landslide susceptibility assessment for a tropical environment: A case study in the southeastern Brazilian coast. *Nat. Hazards* **2021**, *108*, 205–223. [CrossRef]
19. Deng, N.; Li, Y.; Ma, J.; Shahabi, H.; Hashim, M.; de Oliveira, G.; Chaeikar, S.S. A comparative study for landslide susceptibility assessment using machine learning algorithms based on grid unit and slope unit. *Front. Environ. Sci.* **2022**, *10*, 1009433. [CrossRef]
20. Ge, Y.; Liu, G.; Tang, H.; Zhao, B.; Xiong, C. Comparative analysis of five convolutional neural networks for landslide susceptibility assessment. *Bull. Eng. Geol. Environ.* **2023**, *82*, 377. [CrossRef]
21. Pandey, A.; Sarkar, M.S.; Palni, S.; Parashar, D.; Singh, G.; Kaushik, S.; Chandra, N.; Costache, R.; Singh, A.P.; Mishra, A.P.; et al. Multivariate statistical algorithms for landslide susceptibility assessment in Kailash Sacred landscape, Western Himalaya. *Geomat. Nat. Hazards Risk* **2023**, *14*, 2227324. [CrossRef]
22. Zou, Q.; Jiang, H.; Cui, P.; Zhou, B.; Jiang, Y.; Qin, M.; Liu, Y.; Li, C. A new approach to assess landslide susceptibility based on slope failure mechanisms. *Catena* **2021**, *204*, 105388. [CrossRef]
23. Tekin, S.; Çan, T. Slide type landslide susceptibility assessment of the Büyük Menderes watershed using artificial neural network method. *Environ. Sci. Pollut. Res.* **2022**, *29*, 47174–47188. [CrossRef] [PubMed]
24. Chen, T.; Wang, Q.; Zhao, Z.; Liu, G.; Dou, J.; Plaza, A. LCFSTE: Landslide Conditioning Factors and Swin Transformer Ensemble for Landslide Susceptibility Assessment. *IEEE J. Sel. Top. Appl. Earth Obs. Remote Sens.* **2024**, *17*, 6444–6454. [CrossRef]
25. Depicker, A.; Jacobs, L.; Delvaux, D.; Havenith, H.-B.; Mateso, J.-C.M.; Govers, G.; Dewitte, O. The added value of a regional landslide susceptibility assessment: The western branch of the East African Rift. *Geomorphology* **2020**, *353*, 106886. [CrossRef]
26. Azarafza, M.; Azarafza, M.; Akgün, H.; Atkinson, P.M.; Derakhshani, R. Deep learning-based landslide susceptibility mapping. *Sci. Rep.* **2021**, *11*, 24112. [CrossRef]
27. Yin, Y.; Wang, F.; Sun, P. Landslide hazards triggered by the 2008 Wenchuan earthquake, Sichuan, China. *Landslides* **2009**, *6*, 139–152. [CrossRef]
28. Li, Y.; Lin, F.; Lou, L.; Li, J.; Xu, Z.; Zhu, S.; Luo, X.; Huo, G.; Zhao, Q. Performance analysis of landslide susceptibility assessment under different factor-filtering models. *Arab. J. Geosci.* **2021**, *14*, 1160. [CrossRef]
29. Wang, X.; Zhang, S.; Zhang, H.; Wang, D.; Bai, M.; Li, W.; Li, S.; Sun, T.; Wang, Y. Prediction of landslide susceptibility in Wenchuan County based on pixel-level samples. *Bull. Eng. Geol. Environ.* **2023**, *82*, 203. [CrossRef]
30. Tang, C.; Zhu, J.; Liang, J. Emergency assessment of seismic landslide susceptibility: A case study of the 2008 Wenchuan earthquake affected area. *Earthq. Eng. Eng. Vib.* **2009**, *8*, 207–217. [CrossRef]
31. Bai, S.B.; Wang, J.; Thiebes, B.; Cheng, C.; Chang, Z.Y. Susceptibility assessments of the Wenchuan earthquake-triggered landslides in Longnan using logistic regression. *Environ. Earth Sci.* **2013**, *71*, 731–743. [CrossRef]
32. Su, Y.; Chen, Y.; Lai, X.; Huang, S.; Lin, C.; Xie, X. Feature adaptation for landslide susceptibility assessment in "no sample" areas. *Gondwana Res.* **2024**, *131*, 1–17. [CrossRef]
33. Hong, H. Landslide susceptibility assessment using locally weighted learning integrated with machine learning algorithms. *Expert Syst. Appl.* **2024**, *237*, 121678. [CrossRef]
34. He, Y.; Zhao, Z.A.; Zhu, Q.; Liu, T.; Zhang, Q.; Yang, W.; Zhang, L.; Wang, Q. An integrated neural network method for landslide susceptibility assessment based on time-series InSAR deformation dynamic features. *Int. J. Digit.* **2023**, *17*, 2295408. [CrossRef]

35. Meng, S.; Shi, Z.; Li, G.; Peng, M.; Liu, L.; Zheng, H.; Zhou, C. A novel deep learning framework for landslide susceptibility assessment using improved deep belief networks with the intelligent optimization algorithm. *Comput. Geotech.* **2024**, *167*, 106106. [CrossRef]
36. Jiang, Y.; Wang, W.; Zou, L.; Cao, Y. Regional landslide susceptibility assessment based on improved semi-supervised clustering and deep learning. *Acta Geotech.* **2023**, *19*, 509–529. [CrossRef]
37. Yingze, S.; Yingxu, S.; Xin, Z.; Jie, Z.; Degang, Y. Comparative analysis of the TabNet algorithm and traditional machine learning algorithms for landslide susceptibility assessment in the Wanzhou Region of China. *Nat. Hazards* **2024**, *120*, 7627–7652. [CrossRef]
38. Sun, J.; Yan, T.; Hu, J.; Ma, C.; Gao, J.; Xu, H. Slope-scale landslide susceptibility assessment based on coupled models of frequency ratio and multiple regression analysis with limited historical hazards data. *Nat. Hazards* **2023**, *120*, 1–23. [CrossRef]
39. Li, Y.; Ming, D.; Zhang, L.; Niu, Y.; Chen, Y. Seismic Landslide Susceptibility Assessment Using Newmark Displacement Based on a Dual-Channel Convolutional Neural Network. *Remote Sens.* **2024**, *16*, 566. [CrossRef]
40. Wei, X.; Gardoni, P.; Zhang, L.; Tan, L.; Liu, D.; Du, C.; Li, H. Improving pixel-based regional landslide susceptibility mapping. *Geosci. Front.* **2024**, *15*, 101782. [CrossRef]
41. Wu, J.; Zhang, Y.; Yang, L.; Zhang, Y.; Lei, J.; Zhi, M.; Ma, G. Identifying the essential influencing factors of landslide susceptibility models based on hybrid-optimized machine learning with different grid resolutions: A case of Sino-Pakistani Karakorum Highway. *Environ. Sci. Pollut. Res.* **2023**, *30*, 100675–100700. [CrossRef] [PubMed]
42. He, T.; Zhou, H.; Xu, C.; Hu, J.; Xue, X.; Xu, L.; Lou, X.; Zeng, K.; Wang, Q. Deep Learning in Forest Tree Species Classification Using Sentinel-2 on Google Earth Engine: A Case Study of Qingyuan County. *Sustainability* **2023**, *15*, 2741. [CrossRef]
43. Ye, M.; Ruiwen, N.; Chang, Z.; He, G.; Tianli, H.; Shijun, L.; Yu, S.; Tong, Z.; Ying, G. A Lightweight Model of VGG-16 for Remote Sensing Image Classification. *IEEE J. Sel. Top. Appl. Earth Obs. Remote Sens.* **2021**, *14*, 6916–6922. [CrossRef]
44. Upadhyay, A.K.; Bhandari, A.K. Semi-Supervised Modified-UNet for Lung Infection Image Segmentation. *IEEE Trans. Radiat. Plasma Med Sci.* **2023**, *7*, 638–649. [CrossRef]
45. Maslej-Krešňáková, V.; Sarnovský, M.; Jacková, J. Use of Data Augmentation Techniques in Detection of Antisocial Behavior Using Deep Learning Methods. *Future Internet* **2022**, *14*, 260. [CrossRef]
46. Shorten, C.; Khoshgoftaar, T.M. A survey on Image Data Augmentation for Deep Learning. *J. Big Data* **2019**, *6*, 60. [CrossRef]
47. Hou, Q.; Zhang, L.; Cheng, M.-M.; Feng, J. Strip pooling: Rethinking spatial pooling for scene parsing. *arXiv* **2020**, arXiv:2003.13328.
48. Al-Najjar, H.A.H.; Pradhan, B.; Kalantar, B.; Sameen, M.I.; Santosh, M.; Alamri, A. Landslide Susceptibility Modeling: An Integrated Novel Method Based on Machine Learning Feature Transformation. *Remote Sens.* **2021**, *13*, 3281. [CrossRef]
49. Dosovitskiy, A.; Beyer, L.; Kolesnikov, A.; Weissenborn, D.; Zhai, X.; Unterthiner, T.; Dehghani, M.; Minderer, M.; Heigold, G.; Gelly, S.; et al. An image is worth 16X16 words: Transformers for image recognition at scale. *arXiv* **2020**, arXiv:2010.11929.
50. Devlin, J.; Chang, M.-W.; Lee, K.; Toutanova, K. BERT: Pre-training of deep bidirectional transformers for language understanding. *arXiv* **2018**, arXiv:1810.04805.
51. Kuang, P.; Li, R.; Huang, Y.; Wu, J.; Luo, X.; Zhou, F. Landslide Displacement Prediction via Attentive Graph Neural Network. *Remote Sens.* **2022**, *14*, 1919. [CrossRef]
52. Babu, M.V.S.; Ashokkumar, N.; Joshi, A.; Deshpande, P.S.; Keshta, I.; Maaliw, R.R. RETRACTED ARTICLE: Spatio–temporal attention based real-time environmental monitoring systems for landslide monitoring and prediction. *Spat. Inf. Res.* **2023**, *32*, 207. [CrossRef]
53. Han, K.; Wang, Y.; Guo, J.; Tang, Y.; Wu, E. Vision GNN: An Image is Worth Graph of Nodes. *arXiv* **2022**, arXiv:2206.00272.
54. Kohno, M.; Higuchi, Y. Landslide Susceptibility Assessment in the Japanese Archipelago Based on a Landslide Distribution Map. *ISPRS Int. J. Geo-Inf.* **2023**, *12*, 37. [CrossRef]
55. Miao, F.; Ruan, Q.; Wu, Y.; Qian, Z.; Kong, Z.; Qin, Z. Landslide Dynamic Susceptibility Mapping Base on Machine Learning and the PS-InSAR Coupling Model. *Remote Sens.* **2023**, *15*, 5427. [CrossRef]
56. Zhang, Y.; Deng, L.; Han, Y.; Sun, Y.; Zang, Y.; Zhou, M. Landslide Hazard Assessment in Highway Areas of Guangxi Using Remote Sensing Data and a Pre-Trained XGBoost Model. *Remote Sens.* **2023**, *15*, 3350. [CrossRef]
57. Zhao, Z.; Liu, Z.Y.; Xu, C. Slope Unit-Based Landslide Susceptibility Mapping Using Certainty Factor, Support Vector Machine, Random Forest, CF-SVM and CF-RF Models. *Front. Earth Sci.* **2021**, *9*, 589630. [CrossRef]
58. Wang, X.; Li, S.; Liu, H.; Liu, L.; Liu, Y.; Zeng, S.; Tang, Q. Landslide susceptibility assessment in Wenchuan County after the 5.12 magnitude earthquake. *Bull. Eng. Geol. Environ.* **2021**, *80*, 5369–5390. [CrossRef]
59. Bai, S.; Lu, P.; Thiebes, B. Comparing characteristics of rainfall- and earthquake-triggered landslides in the Upper Minjiang catchment, China. *Eng. Geol.* **2020**, *268*, 105518. [CrossRef]
60. Zhao, Z.; Chen, T.; Dou, J.; Liu, G.; Plaza, A. Landslide Susceptibility Mapping Considering Landslide Local-Global Features Based on CNN and Transformer. *IEEE J. Sel. Top. Appl. Earth Obs. Remote Sens.* **2024**, *17*, 7475–7489. [CrossRef]

remote sensing

Article

A Novel Attention-Based Generalized Efficient Layer Aggregation Network for Landslide Detection from Satellite Data in the Higher Himalayas, Nepal

Naveen Chandra [1], Himadri Vaidya [2,3], Suraj Sawant [4] and Sansar Raj Meena [5,6,*]

1 Geomorphology, Environmental, and Engineering Geology, Wadia Institute of Himalayan Geology, Dehradun 248001, Uttarakhand, India; naveenchandra@wihg.res.in
2 Department of Computer Science and Engineering, Graphic Era Hill University, Dehradun 248002, Uttarakhand, India; himadrivaidya@gehu.ac.in
3 Graphic Era Deemed to be University, Dehradun 248002, Uttarakhand, India
4 Department of Computer Science, COEP Technological University, Pune 411005, Maharashtra, India; sts.comp@coep.ac.in
5 Machine Intelligence and Slope Stability Laboratory, Department of Geosciences, University of Padova, 35139 Padova, Italy
6 Center for Remote Sensing, Department of Earth and Environment, Boston University, Boston, MA 02215, USA
* Correspondence: sansarraj.meena@unipd.it

Abstract: Landslide event detection poses a significant challenge in the remote sensing community, especially with the advancements in computer vision technology. As computational capabilities continue to grow, the traditional manual and partially automated methods of landslide recognition from remote sensing data are transitioning towards automatic approaches using deep learning algorithms. Moreover, attention models, encouraged by the human visual system, have emerged as crucial modules in diverse applications including natural hazard assessment. Therefore, we suggest a novel and intelligent generalized efficient layer aggregation network (GELAN) based on two prevalent attention modules, efficient channel attention (ECA) and convolutional block attention module (CBAM), to enrich landslide detection techniques from satellite images. CBAM and ECA are separately integrated into GELAN at different locations. The experiments are conducted using satellite images of the Nepal Himalayan region. Standard metrics such as precision, recall, F-score, and mAP (mean average precision) are considered for quantitative evaluation. GELANc+CBAM (F-score = 81.5%) demonstrates the best performance. This study underscores the suitability of the proposed approach in up-to-date inventory creation and accurate landslide mapping for disaster recovery and response efforts. Moreover, it contributes to developing early prediction models for landslide hazards.

Keywords: attention; CBAM; ECA; GELAN; landslides; YOLOv9

Citation: Chandra, N.; Vaidya, H.; Sawant, S.; Meena, S.R. A Novel Attention-Based Generalized Efficient Layer Aggregation Network for Landslide Detection from Satellite Data in the Higher Himalayas, Nepal. *Remote Sens.* **2024**, *16*, 2598. https://doi.org/10.3390/rs16142598

Academic Editors: Federico Raspini, Chao Zhou and Kunlong Yin

Received: 22 May 2024
Revised: 3 July 2024
Accepted: 8 July 2024
Published: 16 July 2024

1. Introduction

Landslides represent critical geohazards influenced by climatic, seismic, and human-induced factors, resulting in significant loss of life, infrastructure, and economic resources [1–3]. With current weather patterns, urban extension, and population growth, landslides are expected to rise, particularly in mountainous regions worldwide [4,5]. Hence, effective mitigation strategies and risk reduction measures are essential for humankind. Extensive monitoring, forecasting, and detection form the foundation of landslide risk management [1,6]. Monitoring involves tracking the displacement of landslides over time, while forecasting, or prediction, involves timely estimation of potential events. Detection provides detailed and precise information about landslide occurrences. Continuous monitoring is crucial for predicting landslides, while detection facilitates the identification of key parameters, such as location and scale, essential for reducing their cascading effects. Furthermore, accurate detection of landslide events is vital for post-hazard recovery efforts and preparing up-to-date inventories. Additionally, detecting landslide events aids in mapping vulnerable areas, which is essential for sustainable planning. So, there is a growing trend towards rapidly and precisely detecting landslide events.

The traditional approach to landslide extraction relied upon ground-based measurements [7,8], which were time-consuming, labor-intensive, and less effective in emergency response situations [7]. However, with advancements in remote sensing technology, high-resolution images have become vital in landslide hazard assessment studies [1].

Four primary techniques [8] leverage remote sensing data for landslide detection: visual interpretation, pixel-based, object-oriented, and artificial intelligence (AI) methods. Although relying on expert knowledge, visual interpretation methods are time-consuming and may not meet the requirements for rapid response actions [7]. Pixel-based approaches overcome the restrictions of visual analysis by employing binary algorithms [8] to classify pixels in an image into specific categories (e.g., landslide and background). However, accurately distinguishing objects with spectral appearances similar to landslides can be challenging. Object-oriented procedures rely on multiscale segmentations that consider image primitives such as texture, shape, and spectrum [9]. These techniques require empirical settings (threshold criteria) and may struggle with the rapid segmentation of large geographical areas using high-resolution satellite imagery [10]. Keyport et al. (2018) [11] conducted a comparative study to infer the capabilities and limitations of both methods for landslide detection. Furthermore, AI-based techniques have witnessed remarkable advancements across multiple fields, including hazard exploration. Sequentially, remote sensing-centered big data [12,13] have provided auxiliary support to AI initiatives.

Additionally, machine learning (ML) algorithms have shown significant progress in landslide prevention and assessment studies [4]. Specifically, logistic regression [14], support vector machines [15], random forest [16], and decision trees [17] have been prominent. With advancements in computational tools, particularly graphical processing units (GPUs), deep learning (a subdiscipline of ML) algorithms have demonstrated exceptional performance across various fields, including object detection [18], image segmentation [19], and scene classification [20]. Moreover, deep learning has been successfully applied in geohazard analysis, including earthquakes, avalanches, and landslides [6].

Previously, various deep learning algorithms, particularly convolutional neural networks (CNNs), with diverse architectures have been employed for landslide information recognition [21], specifically Mask-RCNN [9,21] and U-Net [22–24]. Currently, YOLO (you only look once) models have gained popularity in the areas of computer vision [25]. For landslide detection, a series of YOLO models, namely, YOLOv3 [8], YOLOv4 [26], YOLOv5 [27], YOLOv7 [28], and YOLOv8 [29], have been augmented and applied. In addition, attention modules [30], inspired by human visual mechanisms, have been integrated into CNN architectures to enhance detection accuracy, specifically 3D-SCAM [31], SW-MSA [32], YOLO-SA [2], and LA-YOLO-LLL [33]. Despite the notable contributions by researchers, identifying a suitable deep learning-based model for landslide detection remains a challenge. In addition, evaluating the performance of attention models fused within a one-stage algorithm is in the preliminary stage and is, hence, crucial. We suggest an attention-driven generalized efficient layer aggregation network (GELAN) to leverage its strength and efficiency concerning landslide detection using satellite images with complex environments to address these concerns. The major contributions of this research work are threefold:

1. Developing an intelligent and optimized GELAN for automated landslide mapping in the Himalayan terrain.
2. Incorporating attention-based cognitive models, mainly CBAM and ECA, within GELAN to augment landslide hazard analysis.
3. Providing a comprehensive model evaluation and comparison considering earth observational images of different geological and geomorphological settings.

This paper is structured as follows: Section 2 provides the details of the study area and dataset, while Section 3 outlines the overall methodology. Section 4 presents the results and Section 5 contains the discussion. Lastly, Section 6 includes the conclusion.

2. Study Area and Dataset

The study area is in Nepal, the Greater Himalayas (Figure 1), comprising 275 GeoTIFF-8 bit Landsat images of 416 × 416 pixels. This region is prone to landslides due to various factors such as seismic activity, topographical conditions, seasonal rainfall, and human activities such as deforestation [34,35]. The Nepal region has experienced significant economic losses and numerous fatalities from natural disasters such as earthquakes and landslides. For instance, the Gorkha earthquake, with a magnitude of 7.8, occurred on 25 April 2015, followed by aftershocks on 12 May 2015, resulting in extensive damage and approximately 9000 fatalities [34,36]. The land cover pattern of the study site includes bare soil, vegetation, water bodies, rocks, and urban areas, making the extraction of landslide information from this complex environment particularly challenging. The dataset is available at https://zenodo.org/record/3675410#.ZBv3UZBxD8 (accessed on 15 March 2024). Figure 2 represents sample images of the Nepal landslide detection database.

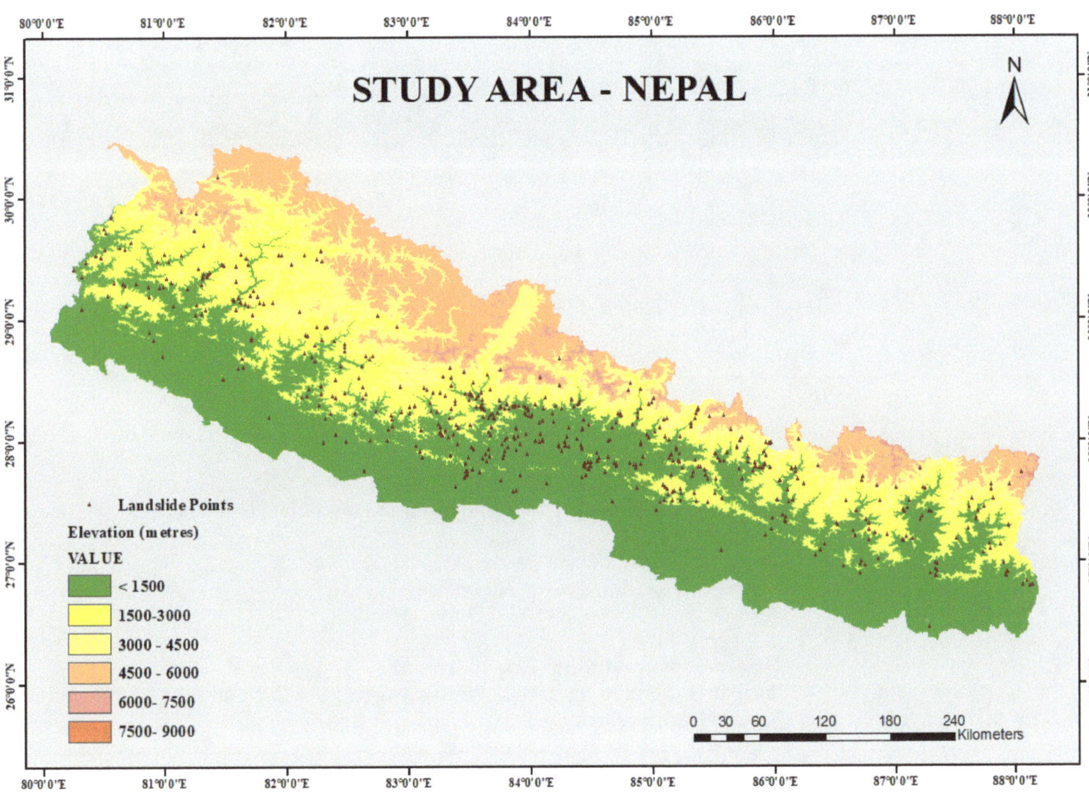

Figure 1. Location map: Nepal (study area) and the landslide event.

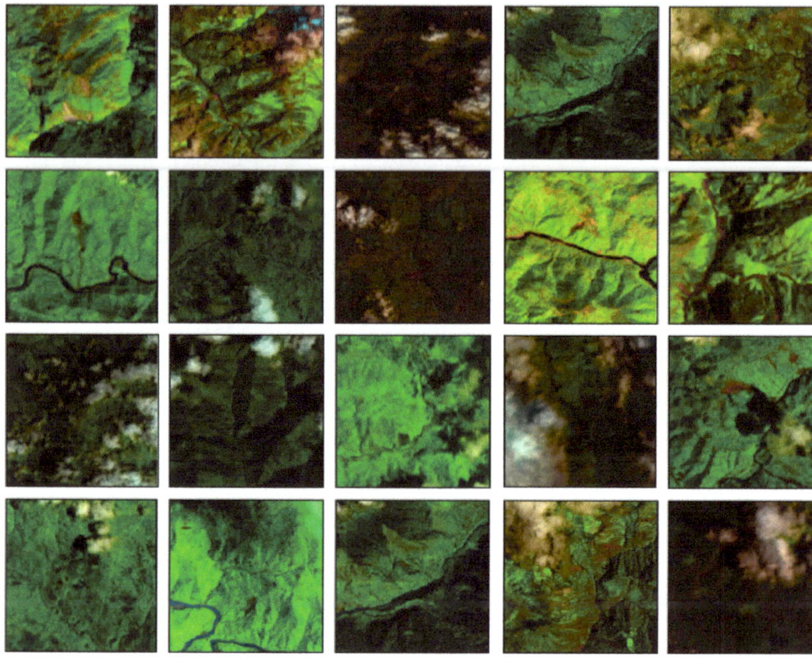

Figure 2. Examples from Nepal landslide detection database.

3. Methodology

This section outlines the technique for automated landslide recognition, which encompasses four key stages: (1) data preparation, (2) model development, (3) experimental settings, and (4) evaluation measures. Algorithm 1 provides the pseudocode for the proposed method.

Algorithm 1 Pseudocode of the novel GELAN based on an attention mechanism.

Input: Labeled landslide-based satellite imagery
Output: Predicted bounding box coordinates
Annotation transformation: Convert annotations into YOLO format.
Data formation: Organize data into training, validation, and test sets.
Model training:
    GELANc, and GELANc+attention
    GELANe, and GELANe+attention
    YOLOv9c, YOLOv9e, and YOLOv9c+attention
for each epoch **do**
    Hyperparameter tuning: Fine-tune image size, batch size, etc.
    Calculate the loss function (classification, bounding box, and confidence), and update the network's parameters.
    Model validation: Evaluate detection performance on the validation dataset.
end for
Save the model weights with the best performance (`best_weight.pt`).
Model testing: Apply the trained network to detect landslides in test images.
Predicted bounding box coordinates for detected landslides.

3.1. Data Preparation

The dataset initially contained annotations represented as masks in .png format. To align with our requirements for the YOLO format, we transformed the annotations into five elements: object class ID, X_center, Y_center, height, and width. Algorithm 2 represents the procedure of annotation conversion.

Algorithm 2 Procedure for annotation conversion.

1: **INPUT** Satellite images with landslide events, and ground truth images in binary format.
2: **OUTPUT** Annotations in YOLO format.
3: Read ground truth images (mask).
4: Convert images to grayscale.
5: Calculate the height (h) and width (w) of the image.
6: Contour estimation method: Identify the edges of the ground truth image.
7: **for** each identified contour **do**
8: Get the minimum and maximum coordinates of the bounding box:
9: Minimum x-coordinate ($xmin$), Minimum y-coordinate ($ymin$), Maximum x-coordinate ($xmax$), Maximum y-coordinate ($ymax$).
10: **end for**
11: Non-maximum suppression to remove overlapping bounding boxes.
12: **for** each remaining bounding box **do**
13: Calculate normalized bounding box coordinates:

$$x_{\text{center}} = \frac{x_{\text{min}} + x_{\text{max}}}{2w}$$

$$y_{\text{center}} = \frac{y_{\text{min}} + y_{\text{max}}}{2h}$$

$$\text{normalized-height} = \frac{y_{\text{max}} - y_{\text{min}}}{h}$$

$$\text{normalized-width} = \frac{x_{\text{max}} - x_{\text{min}}}{w}$$

14: Create a line in the YOLO annotation file in the format:

$$\langle\text{class}\rangle\,\langle x_{\text{center}}\rangle\,\langle y_{\text{center}}\rangle\,\langle\text{normalized-width}\rangle\,\langle\text{normalized-height}\rangle$$

15: **end for**
16: Repeat step 7 for all remaining bounding boxes.
17: Save YOLO annotations.

Moreover, we created a YAML file comprising essential information, specifically the file paths for the training, validation, and testing sets and the number and names of classes. For experimentation, we subdivided the annotated data into training (70%), validation (20%), and testing (10%) sets.

3.2. Model Development

This section illustrates the background of the original YOLOv9 algorithm and describes the applied attention modules, particularly CBAM and ECA. Subsequently, a detailed explanation of the proposed GELAN+attention network including training details and evaluation measures will be given.

3.2.1. YOLOv9 Algorithm

Neural networks usually come across the challenge of information loss as the input data undergo numerous layers of feature extraction and spatial transformation, leading to the loss or degradation of the original information. Hence, to report this concern, the YOLOv9 model [37] uses PGI (programmable gradient information) and GELAN (generalized efficient layer aggregation network) for effective key feature extraction.

PGI is a supporting framework aimed at managing the propagation of gradient information across several semantic levels to enrich the detection performance of the network. PGI consists of three primary modules: (1) the main branch, (2) the auxiliary reversible branch, and (3) multilevel auxiliary information. During the inference stage, only the main branch is accountable for both forward- and backpropagation. As the network becomes deeper, an information bottleneck may occur, leading to loss functions that fail to produce valuable gradients. In such cases, reversible functions, employed by the auxiliary reversible branch, become active, preserving information integrity and minimizing information degradation in the main branch. Further, multilevel auxiliary information addresses the problem of errors commencing from the deep supervision process, improving the learning capacity of the model through the introduction of supplementary information at multiple levels.

During model training, to boost information integration and propagation efficiency, the YOLOv9 model presents an innovative lightweight network architecture called GELAN. It assimilates ELAN (efficient layer aggregation network) and CSPNet (cross-stage partial network) to effectively combine network information, thereby minimizing information loss during propagation and improving interlayer information communication. This architecture is well suited for hazard assessment with restricted computing resources because of its lesser parameter count and computational complexity. Additionally, based on the parameter count, four variants of YOLOv9 are released, namely, v9-S, v9-M, v9-C, and v9-E, pretrained on the MS-COCO dataset, providing a balance between speed and accuracy. However, considering the size of the database and the available computational resources, this study concentrates on GELANc, GELANe, YOLOv9c, and YOLOv9e only.

3.2.2. CBAM

CBAM [38], an effective attention-based module, can be integrated into neural networks with a remarkably low parameter count. As illustrated in Figure 3, CBAM consists of a channel attention module (CAM) and a spatial attention module (SAM). To generate feature maps by multiplying weights for adaptive refinement, CBAM computes attention weights across two dimensions. CAM executes two vital operations: global average pooling (GAP) and maximum global pooling, while SAM executes maximum and average pooling. This dual-module framework aids the estimation of attention weights for the precise refinement of feature maps. In GAP, each feature map of the convolutional layer is spatially averaged to produce a single value. Unlike traditional pooling operations like max pooling, which select the maximum value within a region, GAP computes the average value across the entire feature map. This process encompasses three stages: (1) Spatial average calculation, where the values of all the neurons are averaged together for each feature map in the convolutional layer. (2) The spatial average operation, which reduces the spatial dimensions of the feature maps to a single value per channel. (3) Channel-wise aggregation, where the averaged values from each channel are concatenated to form the output of the pooling operation. GAP is generally applied at the end of the convolutional layers in a CNN, before the fully connected layers. GAP also introduces translation invariance, making the network less sensitive to small spatial variations in the input and enhancing the network's performance in challenging tasks like landslide detection.

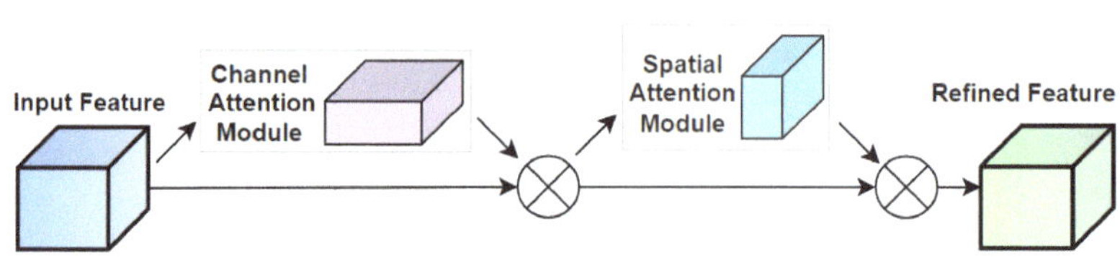

Figure 3. Architecture of CBAM.

Through convolution and pooling operations, CBAM derives the 1D channel attention $X_{\text{Ch\_A}} \in \mathbb{R}^{C\times 1\times 1}$ and 2D spatial attention $X_{\text{Sp\_A}} \in \mathbb{R}^{1\times H\times W}$ maps, given an input feature map $Z \in \mathbb{R}^{C\times H\times W}$. The inclusive attention is depicted by Equations (1) and (2).

$$Z' = X_{\text{Ch\_A}}(Z) \otimes Z \tag{1}$$

$$Z'' = M_{\text{Sp\_A}}(Z') \otimes Z' \tag{2}$$

where $\otimes$ signifies element-wise multiplication and Z″ denotes the refined output.

Moreover, Equations (3) and (4) define the weights of CAM and SAM, respectively.

$$\text{CAM}_{\text{Weights}} = \sigma\Big(\text{Multilayer}_{\text{perceptron}}\Big(\text{Max}_{\text{pooling}}(Z)\Big) + \text{Multilayer}_{\text{perceptron}}\Big(\text{Average}_{\text{pooling}}(Z)\Big)\Big) \tag{3}$$

$$\text{SAM}_{\text{Weights}} = \sigma\Big(\text{Convolution}\Big(\text{concatenation}\Big(\text{Max}_{\text{pooling}}(Z), \Big(\text{Average}_{\text{pooling}}(Z)\Big)\Big)\Big)\Big) \tag{4}$$

3.2.3. ECA Network

Efficient channel attention (ECA) [39] is a mechanism designed to enhance the representation power of CNNs by focusing on important channels within feature maps. It aims to capture long-range dependencies between channels while maintaining computational efficiency, making it suitable for real-time applications such as landslide event extraction employing large-scale datasets. ECA, a lightweight framework related to SENet [39], readily integrates into CNNs to streamline the model's complexity while preserving the original dimensionality.

Initially, the input feature map is transformed using a 1D convolutional layer. This transformation enables the network to learn channel-wise relationships. Further, the transformed feature map is then passed through a GAP layer, capturing the global context of each channel to recalibrate the feature map. This is achieved by applying a sigmoid activation function followed by element-wise multiplication with the original feature map. The sigmoid function scales the channel-wise importance scores between 0 and 1, effectively emphasizing informative channels while suppressing less relevant ones.

To ensure computational efficiency, ECA employs a kernel-wise operation, where each channel is processed independently. This reduces the computational overhead compared to traditional attention mechanisms that consider pairwise interactions between channels. The convolutional layers' output forms a 4D tensor, which serves as input to the ECA network, encompassing four features, namely, the number of channels (Ch), height (He), width (Wi), and batch size (Bs). Similarly, the output of ECA-Net remains a 4D tensor. Its architecture consists of three modules: (1) the global feature descriptor, (2) adaptive neighborhood interaction, and (3) broadcast scaling. The GAP (global average pooling) method involves processing the input tensor by calculating the average of all pixels within an explicit feature map, followed by a reduction to a single pixel. Afterward, a tensor represented as $C \times 1 \times 1$ endures a 1D striding convolution. Further, the adaptive estimation of kernel size K_{Size} is estimated by Equations (5) and (6), where γ and 'X' indicate the predefined hyperparameters. The architecture of the ECA model is given in Figure 4.

$$Z'' = M_{\text{Sp\_A}}(Z') \otimes Z' \tag{5}$$

$$Ch = \varnothing(K_{\text{Size}}) = 2^{(\gamma \cdot K_{\text{Size}} - X)} \tag{6}$$

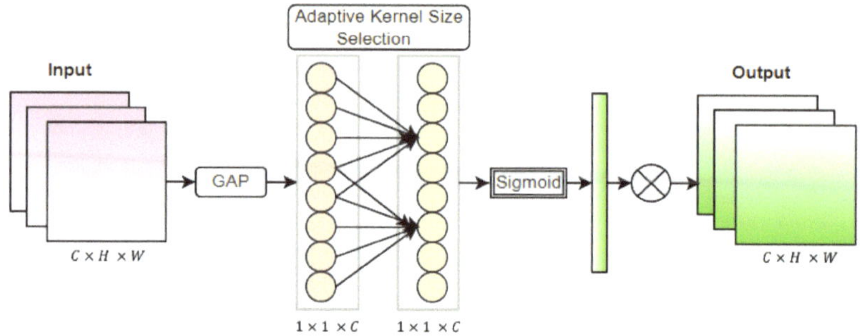

Figure 4. Architecture of ECA model.

3.2.4. Proposed Architecture

The proposed architecture encompasses the backbone and head. The backbone includes a series of convolutional layers that progressively extract hierarchical features from the input data. These convolutional layers allow the model to capture complex patterns and relationships within the data. These layers apply convolutional filters to the input, learning spatial patterns and extracting features. The first layer has a -1 input channel and outputs 64 channels (indicating the layer will generate 64 feature maps), with a kernel size of 3×3 and a stride of 2 for downsampling. The output feature maps from the first convolutional layer (with 64 channels) serve as an input to the second convolutional layer (with 128 channels). This process continues through the entire backbone. The architecture of the backbone is aimed to efficiently aggregate network information while minimizing information loss during propagation. This is achieved through the integration of specialized modules such as CSPNet and ELAN, resulting in RepNCSPELAN, which enhances inter-layer information interaction and preserves information integrity. The output feature maps from the previous convolutional layer (with 128 channels) are fed into the RepNCSPELAN block. Average pooling (a down layer) downsamples feature maps spatially, reducing their resolution while preserving some spatial information to create additional feature maps at different scales. Later, the pooled features are processed by convolution, resulting in further downsampled feature maps (with 256 channels). Hence, the backbone progressively reduces the spatial resolution of the feature maps while extracting increasingly complex features. The backbone follows a similar pattern for the remaining layers, and the number of output channels typically increases as the network progresses (256, 512, 512). This allows the model to capture both fine-grained and high-level information from the input image, which is crucial for accurate landslide detection.

Further, the neck receives the feature maps extracted by the backbone and refines them for landslide detection. The SPPELAN block performs spatial pyramid pooling (SPP), which partitions the input feature maps into grids of different sizes and extracts features from each region. This allows the network to capture features at different scales within the input. The upsampled features are concatenated with the features from the backbone, creating a richer representation at a higher resolution. Concatenation combines information from different scales, allowing the network to leverage both high-level semantic information and low-level spatial details for landslide detection.

The concatenated features are processed by the RepNCSPELAN blocks (similar to those in the backbone), potentially involving group normalization, channel splitting, and residual connections. Further, these features are enhanced by the attention module. Attention, a fundamental cognitive ability of human beings, enables them to prioritize relevant information while filtering out irrelevant details. Inspired by this aspect of human vision, attention modules show significant potential when integrated within a CNN to enhance its performance. Therefore, we deliberately integrated two substantial attention modules, CBAM and ECA, within GELAN. Particularly, an attention module is added in the neck after each RepNCSPELAN layer. The architecture of the head follows repeated upsampling, concatenation, RepNCSPELAN block, attention to the remaining layers, and processing features at different scales. Figure 5 represents the algorithmic framework of the proposed GELAN+attention model. The addition of the attention module enhances feature representation as it allows the model to dynamically focus on the most relevant features within the input data. By assigning higher weights to important features and suppressing less relevant ones, the attention mechanism boosts the representation of key information, leading to more discriminative feature maps. Further, the attention layer facilitates efficient integration of information across different layers of the network. By selectively attending to relevant features from multiple layers, the model can effectively combine information from various levels of abstraction, leading to richer and more informative representations. The attention mechanism enables the model to adaptively learn which features to prioritize based on the input data. This characteristics allows the model to selectively consider different parts of the input, depending on their importance to the task, leading to more flexible and adaptive feature learning. The attention mechanism highlights the regions of the input that are most influential for the predictions, enabling increased interpretability and supporting model debugging and refinement. Integrating attention models within GELAN empowers the network to become more robust towards input data, consequently improving its ability to generalize across different conditions and environments.

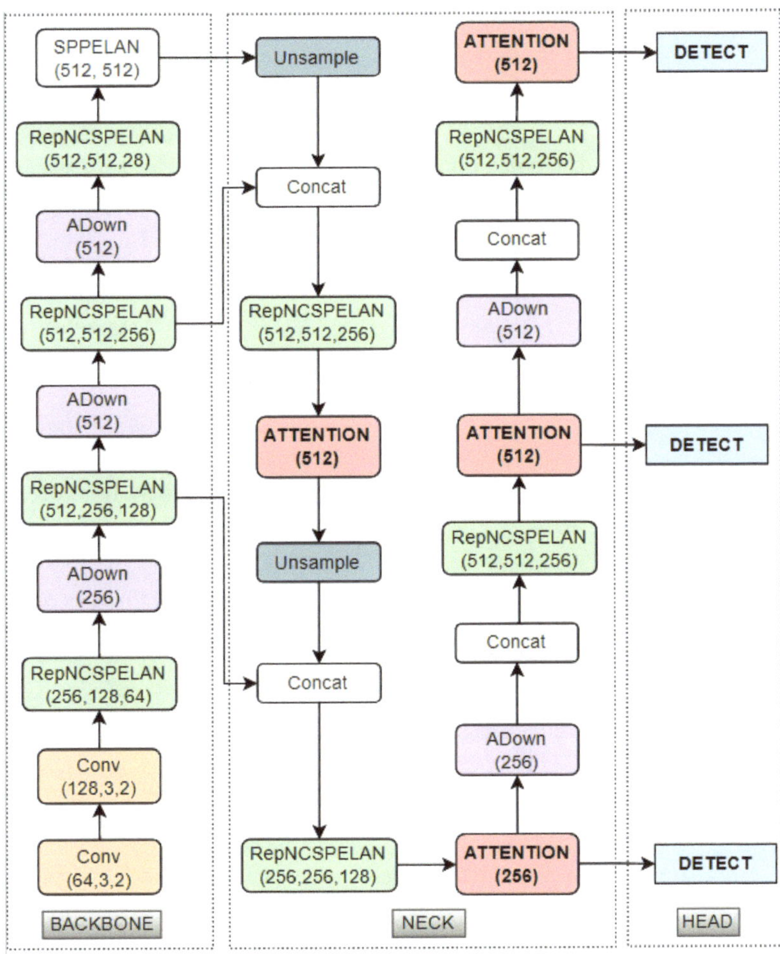

Figure 5. Schematic representation of the proposed architecture.

Lastly, the detection layer performs the complex task of landslide detection. It takes the processed features and predicts the bounding boxes for potential objects in each grid cell of the feature maps and the class probabilities for each bounding box. The proposed architecture improves accuracy and facilitates real-time landslide information extraction from remotely sensed data.

3.3. Experimental Settings and Evaluation Measures

For experimentation, an image size of 416 × 416 pixels is chosen as the input to the network. The architecture is optimized to effectively process images of the selected size, capturing intricate details in the satellite image and ensuring efficient processing by the network. The batch size is set to 4 due to memory constraints, allowing for efficient training. Although the aim is to train the model for a maximum of 500 epochs, early stopping criteria are adopted to avoid inefficient utilization of computing resources and time. For optimization, the training units utilize a stochastic gradient descent (SGD) optimizer that uniformly scales gradients in all directions and facilitates faster convergence by collecting gradients from previous data points. Furthermore, GELAN uses an SGD optimizer by default, which is therefore employed in this research work. The initial learning rate is set at 0.01, with an initial momentum factor of 0.937 and an initial weight decay of 0.0005. These parameters are chosen to ensure effective training and optimization of the model. The training environment comprises an Intel® Xeon-CPU E3-1231 v3 @ 3.40 GHz CPU and NVIDIA GeForce RTX 3080Ti GPU running on the Ubuntu 18.04 operating system. The deep learning frameworks used are PyTorch 1.7 and CUDA 11.4. The model's performance evaluation encompasses both quantitative and qualitative assessments. For quantitative evaluation, four metrics, precision, recall, f-score, and mean average precision (mAP), are computed using Equations (7)–(10). Simultaneously, qualitative evaluation involves visual analysis of the outcomes to assess the model's capability.

$$\text{Precision} = \frac{\text{True positive}}{\text{True positive} + \text{False positive}} \tag{7}$$

$$\text{Recall} = \frac{\text{True positive}}{\text{True positive} + \text{False negative}} \tag{8}$$

$$\text{F-Score} = \frac{2 \times \text{Precision} \times \text{Recall}}{\text{Precision} + \text{Recall}} \tag{9}$$

$$\text{mAP} = \sum_{i=0}^{n} \left(\frac{1}{n} \int_0^1 \text{Precision}(\text{Recall})\, d(\text{Recall}) \right) \tag{10}$$

4. Results

This section presents the experimental findings of this study. Initially, quantitative outcomes based on the metrics outlined in Section 3.3 are provided. Subsequently, the qualitative or visual representations of the detected landslide events by the proposed model are described.

4.1. Quantitative Evaluation

The outcomes of quantitative measures, specifically f-score, precision, recall, and mAP, of each attention-induced GELAN are given in Table 1. The computed f-score of GELANe and GELANc is 78.0% and 79.4%, respectively. Integration of the ECA module within GELANe showed an improvement of +2.6%. Further, the f-score significantly increased by +2.8% considering CBAM. In addition, an enhancement of +1.8% and +2.1% is noted when the ECA model and CBAM are added within the architecture of GELANc, respectively. The GELANc+CBAM outperformed other models and demonstrated the highest accuracy (81.5%).

Table 1. Experimental results of the proposed model.

Model	GFLOP	Params	P	R	F-Score	mAP@50	Time (h)	GPU (GBs)
GELANe	190.8	58.00	88.3	69.9	78.0	75.2	1.814	3.74
GELANe + ECA	190.8	58.00	86.4	75.5	80.6	76.2	1.973	3.76
GELANe + CBAM	191.5	58.85	84.2	77.6	80.8	75.0	1.869	4.06
GELANc	102.5	25.41	86.3	73.6	79.4	78.3	1.037	1.95
GELANc + ECA	102.5	25.41	90.2	73.8	81.2	77.8	1.027	1.95
GELANc + CBAM	120.6	31.06	85.7	77.7	81.5	75.2	1.301	2.15

A difference of 3.3% is noted between the maximum (GELANc) and minimum (GELANe + CBAM) mAP@0.50, while it is inferred to be equivalent (75.2%) for GELANe and GELANc + CBAM. Moreover, the precision reached 90.2% (GELANc + ECA), and recall is estimated between 66.9% (GELANe) and 77.7% (GELANc + CBAM, almost similar to GELANe + CBAM). A minor deviation (+0.2%) is observed in the recall of GELANc and GELANc + ECA. Furthermore, the computational time (in hours (h)) is an important assessment factor and is hence considered in this research work. Table 1 shows the execution time of each model, and it is inferred that GELANe + ECA exhibits the maximum execution time (1.973 h), while GELANc + ECA requires the minimum time (1.027 h). In addition, the amount of GPU consumed (1.95 GB–4.06 GB) by each model is indicated. Moreover, Table 1 illustrates the computed GFLOPS (giga floating-point operations per second), with the highest noted at 190.8 (GELANe) and the lowest at 102.5 (GELANc). GFLOPS measures the computational complexity of a model, indicating the number of floating-point operations (FLOPs) required to process a single forward pass of the model, expressed in billions (giga) of operations. The FLOPs signify the total number of multiplications and additions required by all the operations in the model. GFLOPS represents an estimate of the computational resources needed to run the model. Params refers to the number of trainable parameters in the model. These include weights (coefficients that are learned from the data during training) and biases (additional parameters in each layer that shift the activation function) across all layers in the network. In our experiments, the maximum (58.85 million (m)) and minimum (25.41 m) Params are computed by GELANe+CBAM and GELANc, respectively. Hence, GFLOPS and Params are critical factors for evaluating the model's performance as they provide insights into the computational and memory requirements, influencing the choice of model for specific applications.

Additionally, Figures 6 and 7 illustrate the estimated taring and validation loss functions, respectively. Particularly the bounding box regression loss (box loss), distributional focal loss (dfl loss), and classification loss are computed for the suggested networks (GELANe + ECA, GELANe + CBAM, GELANc + ECA, and GELANc + CBAM). The x axis represents the number of epochs, whereas the y axis indicates the computed loss values during the training and validation stages.

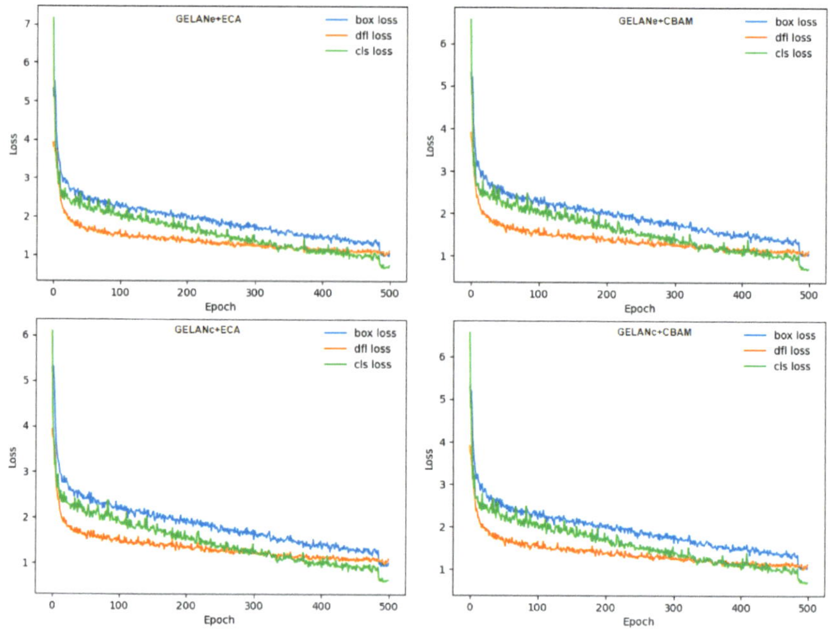

Figure 6. Results of the training loss function by the GELAN + attention model.

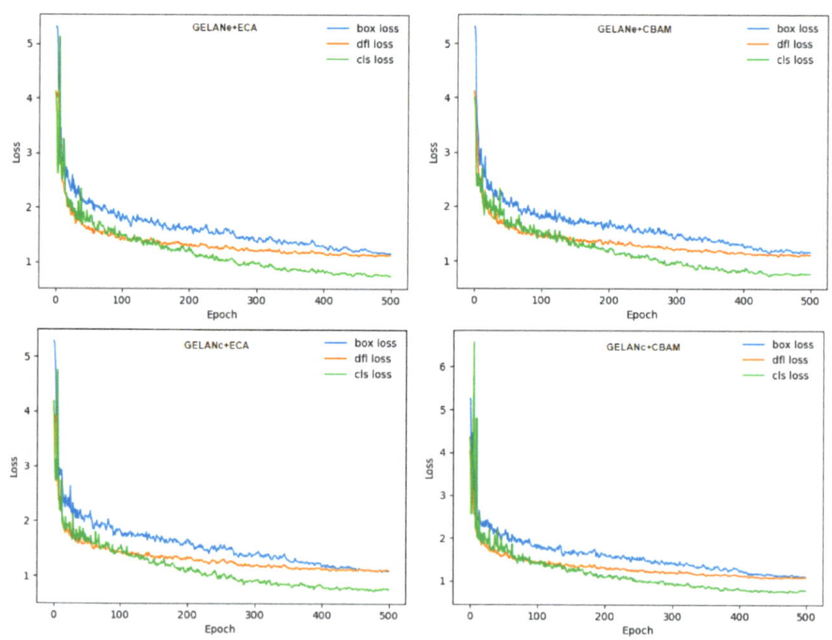

Figure 7. Results of the validation loss function by the GELAN + attention model.

4.2. *Qualitative Evaluation*

Visual analysis of the detected landslide by the proposed model is essential for validating model accuracy, interpreting results, and applying the findings in practical scenarios. It enhances the model's reliability, supports effective communication and decision making, and is critical in both immediate disaster response and long-term risk mitigation strategies.

The overall impact of this study is significantly enhanced by integrating visual analysis into the workflow. Presenting the results as image patches offers significant advantages. Image patches enable a detailed and focused examination of specific areas, enhancing the clarity and resolution of the detected landslides. The results facilitate precise comparisons, highlight specific sites, and support communication efficiency with stakeholders by providing a concise and targeted representation of the results. Additionally, handling smaller patches is computationally efficient and aligns with the technical requirements of machine learning model training and validation. Therefore, using image patches alongside full-area visualizations provides a robust and comprehensive approach to landslide detection analysis. The YOLO models create the labels before target prediction. Figure 8 represents the generated labels for detecting landslides. The visual representation of the identified landslides by GELANc + CBAM and GELANe + CBAM are depicted through bounding boxes in Figure 8. The dataset comprises images with complex backgrounds/landscapes, along with cloud cover, where the manual interpretation is difficult and poses a challenge for landslide recognition; yet, the model effectively predicts them. The proposed algorithm demonstrates the ability to distinguish between landslides of varying extents. Due to the high detection correctness, the model proved its robustness, subsequently yielding enhanced prediction outcomes.

Figure 8. Detected landslides by the proposed method: Labels (R1), GELANc + CBAM (R2), GELANc + ECA (R3).

5. Discussion

The proposed deep learning-based algorithm identifies the presence of landslides successfully. However, to infer its suitability, we considered the additional benchmark dataset. Moreover, we examined the effectiveness of integrating an additional attention-based model into GELAN to assess its capabilities. Subsequently, we compared our results with previously proposed studies and examined the performance of state-of-the-art YOLO variants for recognizing landslides. Additionally, we underscore the importance of the patience factor. Finally, the limitations, prospects, and applications of this study are outlined.

5.1. Model's Performance on Benchmark Dataset

To assess the effectiveness of the suggested model, we included an additional dataset with geological and geomorphological characteristics distinct from those of the Himalayan terrain, specifically the popular benchmark Bijie landslide detection database. The quantitative outcomes (f-score, precision, and recall) are shown in Table 2. A competitive f-score is achieved for GELANc (98.0%) and GELANe (98.5%). By integrating ECA within GELANc (improvement = +1.5%) and GELANe (improvement = +1.0%), the computed f-score reached 99.5%. Similarly, GELANe+CBAM (99.0%) and GELANc + CBAM (98.0%) demonstrated a challenging f-score. Recently, Du et al., 2024 [29] and Yang et al., 2024 [33] employed a similar dataset for their experiments based on YOLO models. Comparing our results with them (considering the average f-score), we noted a substantial improvement of +5.15% compared to Du et al., 2024 [29] and a remarkable progress of +10.25% in the case of Yang et al., 2024 [33]. Moreover, a noteworthy improvement of +19.75% is estimated when compared with the results of Tanatipuknon et al., 2021 [40]. This comparison unveils the novelty of the proposed algorithms.

Table 2. Results of the additional dataset (Bijie landslide detection database).

Reference	Model	F-Score (%)	P (%)	R (%)
Our work	GELANe	98.5	97.9	99.1
	GELANe + ECA	99.5	99.9	99.1
	GELANe + CBAM	99.0	98.1	99.9
	GELANc	98.0	98.1	98.0
	GELANc + ECA	99.5	99.9	99.1
	GELANc + CBAM	98.0	98.0	98.1
Tanatipuknon et al., 2021 [40]	Faster R-CNN	79.0	86.0	77.0
Du et al., 2024 [29]	LBE_YOLO	88.5	90.6	86.5
Yang et al., 2024 [33]	YOLOv4 + VGG16	94.0	96.1	91.9
	YOLOv4 + ECA	92.4	93.5	91.4
	YOLOv4 + CBAM	93.1	94.0	92.2
	YOLOv4 + LPFRF	94.9	95.5	94.2

5.2. Additional Attention Model

We further explored an additional attention model to identify the most suitable one for landslide identification. Particularly, the global attention mechanism (GAM) [41], a widely recognized model, enhances the utility of global dimension-interactive features to reduce information dispersion. GAM innovatively transforms the channel and spatial components of CBAM. GELANc + GAM (f-score = 76.7%, mAP@0.50 = 73.2%) demonstrated better performance when compared with GELANe + GAM (f-score = 70.3%, mAP@0.50 = 67.4%). The recall computed by GELANc + GAM and GELANe + GAM is 71.8% and 60.2%, respectively. However, the estimated precision of GELANe + GAM is estimated to be higher (+2.2 units) than GELANc + GAM (P = 82.4%). Further, the computational efficiency of GELANc + GAM (time = 1.301 h, memory = 2.15 GB, GFLOPS = 120.6, and Params = 31.06) is also noted to be better than GELANe + GAM (time = 2.095 h, memory = 4.21 GB, GFLOPS = 208.9, and Params = 63.65). In addition, YOLOv9-C+GAM revealed a competitive f-score (77.5%) and mAP@0.50 (74.3%). The accurateness computed by the GAM is assessed to be less than the CBAM and ECA modules.

5.3. Comparative Analysis

To exhibit the effectiveness of the attention mechanism integrated into GELAN, the results are compared with previously proposed methods for landslide detection from the Nepal region, particularly Bragagnolo et al., 2021 [34], Chen et al., 2018 [36], and Yu et al., 2020 [35]. Comparison and evaluation are based on computed precision, recall, and f-score of our study's best model. Table 3 indicates the notable improvement in f-score (+14.5% to +55.5%), precision (+24.7 to +55.7 %), and recall (+3.7% to +54.7%), thus affirming its novelty and efficacy.

Table 3. Comparison of our model with previous studies.

Reference	F-Score (%)	P (%)	R (%)
GELANc + CBAM (our work)	81.5	85.7	77.7
Bragagnolo et al., 2021 [34]	67.0	61.0	74.0
Improvement	+14.5	+24.7	+3.7
Yu et al., 2020 [35]	60.0	55.0	65.0
Improvement	+21.5	+30.7	+12.7
Chen et al., 2018 [36]	26.0	30.0	23.0
Improvement	+55.5	+55.7	+54.7

We also evaluated the performance of YOLOv9 with seven algorithms, specifically YOLOv5 small (s), YOLOv5 medium (m), YOLOv6 nano (n), YOLOv6s, YOLOv6m, YOLOv7 tiny (t), and YOLOv8n. The obtained mAP at the intersection over the union threshold of 0.5 (mAP@0.50) for all models is depicted in Figure 9. Notably, GELANc achieved the highest mAP@0.50 at 78.3% followed by YOLOv9e (76.3%), which is almost equivalent to YOLOv9e (76.0%). In addition, the mAP@0.50 of GELANe reached 75.2%. The computed mAP@0.50 of the considered YOLO variants for comparison is significantly lower (+2.6 to +4.5) than YOLOv9. However, the marginal difference (0.5 units) is noted in the case of YOLOv8n and GELANe. This comparison highlights the effectiveness of the YOLOv9 network.

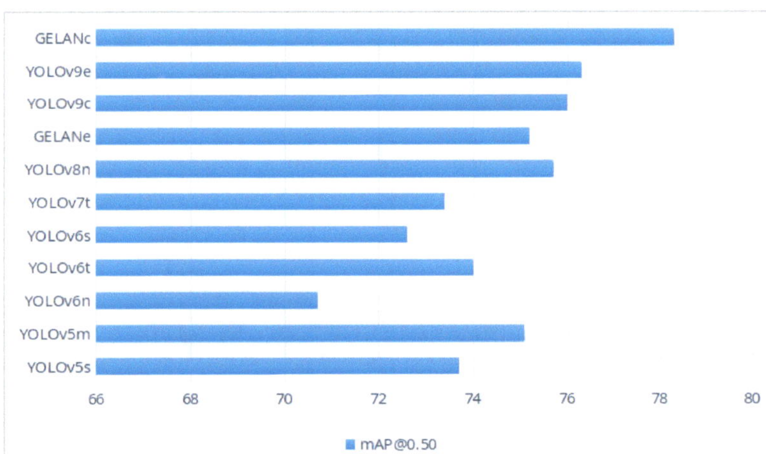

Figure 9. Comparison of YOLOv9 with the state-of-the-art algorithms.

The results of the proposed model (GELANe + ECA, GELANe + CBAM GELANc + ECA, GELANc + CBAM) are compared with state-of-the-art models, particularly YOLOv6n, YOLOv6t, YOLOv6s, and YOLOv7t, considering mAP@0.50–0.95 matrices (refer to Figure 10). The maximum improvement of +7.3% is noted in the case of GELANc+ECA when compared with YOLOv6n. In addition, the progress of +6.6% and +6.1% is estimated when compared with YOLOv6s and YOLOv6t, respectively. Further, the mAP@0.50–0.95 improved by +3.4% in comparison with YOLOv7t. GELANc+CBAM shows an improvement of +4.9% over YOLOv6n, and the enhancements are +4.2%, +3.7%, and +1.0% when compared to YOLOv6s, YOLOv6t, and YOLOv7t, respectively. Additionally, GELANe+ECA improves mAP@0.50–0.95 by +4.4% over YOLOv6n. The improvements are +3.7%, +3.2%, and +0.5% compared to YOLOv6s, YOLOv6t, and YOLOv7t, respectively. Furthermore, enrichment of +3.1%, +2.4%, and +1.9% is noted when the mAP@0.50–0.95 of GELANe+CBAM is compared with YOLOv6n, YOLOv6s, and YOLOv6t, respectively, while a marginal difference (0.8%) is observed between the mAP@0.50–0.95 of

GELANe+CBAM and YOLOv7t. These comparisons highlight the significant enhancements in mAP@0.50–0.95 achieved by the proposed models, demonstrating their superior performance over the popular object detection models.

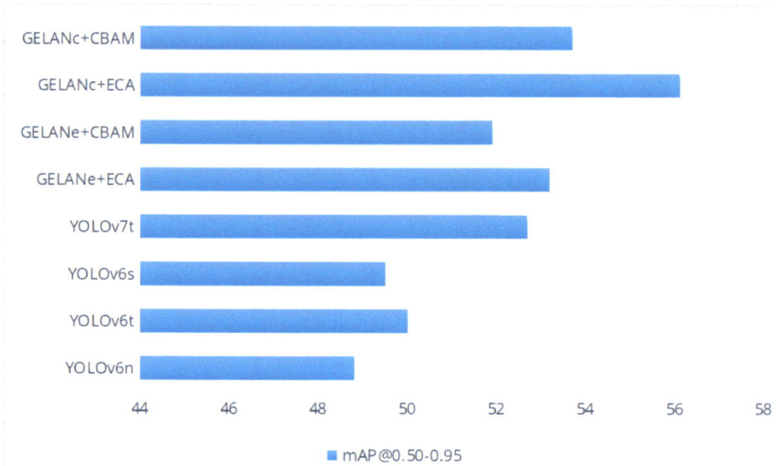

Figure 10. Comparison of the proposed method with the state-of-the-art models.

5.4. Significance of Early Stopping Factor

During the execution and implementation of YOLO algorithms, early stopping is a technique utilized to improve the efficiency of the network. It involves continuously monitoring evaluation metrics during training and halting the training process if no improvement in accuracy is observed after a certain number of epochs, known as the patience parameter. Patience is a user-defined integer that specifies the number of iterations or epochs the training process can endure without any advancement in evaluation metrics before terminating early. This supports maintaining a balance between training time and network capability. Choosing a suitable patience value is crucial for optimizing outcomes. Setting the patience value too low may prematurely end model training, potentially missing out on performance improvements. On the other hand, setting it too high can unnecessarily prolong training, resulting in ineffective use of computational resources/hardware and time. There is no standard formula for determining the optimal patience value; it relies on factors such as dataset, model complexity, speed, and machine configuration. Finding the optimal patience value requires trial experimentation and fine-tuning. The default patience value for the YOLOv9 network is 100, as utilized in this research work (patience = 100). Patience parameters are flexible and can be disabled (set patience = 0) and enabled easily. In our experiments, only GELANe terminated early (at 468 epochs), saving the best results at 367 epochs, and the other five models (GELANc, GELANe + ECA, GELANc + ECA, GELANe + CBAM, and GELANc + CBAM) were trained for the defined number of epochs.

5.5. Limitations, Prospects, and Applications

The development of intelligent and fully automated landslide event recognition models by deep learning algorithms is still ongoing, presenting opportunities for advancement. Hence, we suggest an attention-based GELAN for detecting landslide hazards of varying sizes from complex backgrounds. We explore the implications of different attention modules and compare their performance with state-of-the-art YOLO models to showcase their detection capabilities. However, our study is constrained by the dataset and computational resources. The experimentation database comprises a limited number of satellite images, which impacts the accuracy of the suggested network. Additionally, the applied deep learning models are executed on a machine with limited GPU access. For future research work, we aim to enhance the accuracy of the model network by increasing the database size. Furthermore, the model can be evaluated using unmanned aerial vehicle images. The proposed approach holds the potential for quick and accurate mapping of landslide scenes, supporting effective rescue and recovery actions. Additionally, the findings of this study can be valuable in time-constrained landslide mapping for up-to-date inventory generation. Moreover, accurately identifying the spatial characteristics of landslides is essential for developing a landslide forecasting system.

6. Conclusions

This research work introduces an attention-embedded GELAN algorithm for detecting landslide events. Specifically, CBAM and ECA are discretely integrated into the neck of the original GELAN architecture. The proposed algorithm addresses the challenges associated with the correct landslide information extraction from fragile and complex landscapes, mainly the Himalayas. The results demonstrate that GELANc + CBAM yields the best performance. Overall, the findings are novel and encouraging, indicating the model's suitability for prompt hazard mapping to facilitate fast recovery and response efforts. Future research directions may explore the perspective of GELAN and attention models considering unmanned aerial vehicle-based images for landslide identification. This study offers an advanced solution to overcome the challenges allied with near-real-time landslide hazard mapping.

Author Contributions: N.C.: conceptualization, model execution, and manuscript writing; H.V.: implementation and manuscript editing; S.S.: manuscript editing; and S.R.M.: manuscript writing and editing. All authors have read and agreed to the published version of the manuscript.

Funding: This research work is supported by the Department of Science and Technology, Science and Engineering Research Board, New Delhi, India, under Grant No: EEQ/2022/000812.

Data Availability Statement: The data presented in this study can be downloaded from https://zenodo.org/record/3675410#.ZBv3UZBxD8 and http://gpcv.whu.edu.cn/data/Bijie_pages.html.

Acknowledgments: The authors would like to thank the directors of the Wadia Institute of Himalayan Geology (WIHG), Dehradun, for their continuous encouragement. The contribution number for this research work is WIHG/0372.

Conflicts of Interest: The authors declare no conflict of interest.

References

1. Casagli, N.; Intrieri, E.; Tofani, V.; Gigli, G.; Raspini, F. Landslide detection, monitoring and prediction with remote-sensing techniques. *Nat. Rev. Earth Environ.* **2023**, *4*, 51–64. [CrossRef]
2. Cheng, L.; Li, J.; Duan, P.; Wang, M. A small attentional YOLO model for landslide detection from satellite remote sensing images. *Landslides* **2021**, *18*, 8. [CrossRef]
3. Mohan, A.; Singh, A.K.; Kumar, B.; Dwivedi, R. Review on remote sensing methods for landslide detection using machine and deep learning. *Trans. Emerg. Telecommun. Technol.* **2021**, *32*, e3998. [CrossRef]
4. Ma, Z.; Mei, G. Machine Learning for Landslides Prevention: A Survey. *Neural Comput. Appl.* **2020**, *30*, 10881–10907. [CrossRef]
5. Xu, Y.; Ouyang, C.; Xu, Q.; Wang, D.; Zhao, B.; Luo, Y. CAS Landslide Dataset: A Large-Scale and Multisensor Dataset for Deep Learning-Based Landslide Detection. *Sci. Data* **2024**, *11*, 12. [CrossRef] [PubMed]
6. Ma, Z.; Mei, G. Deep learning for geological hazards analysis: Data, models, applications, and opportunities. *Earth-Sci. Rev.* **2021**, *223*, 103858. [CrossRef]
7. Liu, P.; Wei, Y.; Wang, Q.; Xie, J.; Chen, Y.; Li, Z.; Zhou, H. A Research on Landslides Automatic Extraction Model Based on the Improved Mask R-CNN. *Isprs Int. J. Geo-Inf.* **2021**, *10*, 168. [CrossRef]
8. Han, Z.; Fang, Z.; Li, Y.; Fu, B. A novel Dynahead-Yolo neural network for the detection of landslides with variable proportions using remote sensing images. *Front. Earth Sci.* **2023**, *10*, 1077153. [CrossRef]
9. Lu, P.; Stumpf, A.; Kerle, N.; Casagli, N. Object-Oriented Change Detection for Landslide Rapid Mapping. *IEEE Geosci. Remote Sens. Lett.* **2011**, *8*, 4. [CrossRef]
10. Blaschke, T.; Feizizadeh, B.; Holbling, D. Object-Based Image Analysis and Digital Terrain Analysis for Locating Landslides in the Urmia Lake Basin, Iran. *IEEE J. Sel. Top. Appl. Earth Obs. Remote Sens.* **2014**, *7*, 4806–4817. [CrossRef]
11. Keyport, R.N.; Oommen, T.; Martha, T.R.; Sajinkumar, K.S.; Gierke, J.S. A comparative analysis of pixel- and object-based detection of landslides from very high-resolution images. *Int. J. Appl. Earth Obs. Geoinf.* **2018**, *64*, 1–11. [CrossRef]

12. Ma, Y.; Wu, H.; Wang, L.; Huang, B.; Ranjan, R.; Zomaya, A.; Jie, W. Remote sensing big data computing: Challenges and opportunities. *Future Gener. Comput. Syst.* **2015**, *51*, 47–60. [CrossRef]
13. Zhang, X.; Zhou, Y.; Luo, J. Deep learning for processing and analysis of remote sensing big data: A technical review. *Big Earth Data* **2021**, *6*, 527–560. [CrossRef]
14. Lombardo, L.; Mai, P.M. Presenting logistic regression-based landslide susceptibility results. *Eng. Geol.* **2018**, *244*, 14–24. [CrossRef]
15. Huang, Y.; Zhao, L. Review on landslide susceptibility mapping using support vector machines. *Catena* **2018**, *165*, 520–529. [CrossRef]
16. Hu, Q.; Zhou, Y.; Wang, S.; Wang, F.; Wang, H. Improving the Accuracy of Landslide Detection in "Off-site" Area by Machine Learning Model Portability Comparison: A Case Study of Jiuzhaigou Earthquake, China. *Remote Sens.* **2019**, *11*, 2530. [CrossRef]
17. Fang, Z.; Wang, Y.; Duan, G.; Peng, L. Landslide Susceptibility Mapping Using Rotation Forest Ensemble Technique with Different Decision Trees in the Three Gorges Reservoir Area, China. *Remote Sens.* **2021**, *13*, 238. [CrossRef]
18. Zhao, Z.-Q.; Zheng, P.; Xu, S.; Wu, X. Object Detection with Deep Learning: A Review. *arXiv* **2019**, arXiv:1807.05511. Available online: http://arxiv.org/abs/1807.05511 (accessed on 7 July 2024).
19. Minaee, S.; Boykov, Y.Y.; Porikli, F.; Plaza, A.J.; Kehtarnavaz, N.; Terzopoulos, D. Image Segmentation Using Deep Learning: A Survey. *IEEE Trans. Pattern Anal. Mach. Intell.* **2021**, *44*, 3523–3542. [CrossRef]
20. Zeng, D.; Liao, M.; Tavakolian, M.; Guo, Y.; Zhou, B.; Hu, D.; Pietikäinen, M.; Liu, L. Deep Learning for Scene Classification: A Survey. *arXiv* **2021**, arXiv:2101.10531. Available online: http://arxiv.org/abs/2101.10531 (accessed on 7 July 2024).
21. Ullo, S.; Mohan, A.; Sebastianelli, A.; Ahamed, S.; Kumar, B.; Dwivedi, R.; Sinha, G.R. A New Mask R-CNN-Based Method for Improved Landslide Detection. *IEEE J. Sel. Top. Appl. Earth Obs. Remote Sens.* **2021**, *14*, 3799–3810. [CrossRef]
22. Devara, M.; Maurya, V.K.; Dwivedi, R. Landslide extraction using a novel empirical method and binary semantic segmentation U-NET framework using sentinel-2 imagery. *Remote Sens. Lett.* **2024**, *15*, 326–338. [CrossRef]
23. Ghorbanzadeh, O.; Crivellari, A.; Ghamisi, P.; Shahabi, H.; Blaschke, T. A comprehensive transferability evaluation of U-Net and ResU-Net for landslide detection from Sentinel-2 data (case study areas from Taiwan, China, and Japan). *Sci. Rep.* **2021**, *11*, 14629. [CrossRef] [PubMed]
24. Meena, S.R.; Nava, L.; Bhuyan, K.; Puliero, S.; Soares, L.P.; Dias, H.C.; Floris, M.; Catani, F. HR-GLDD: A globally distributed dataset using generalized DL for rapid landslide mapping on HR satellite imagery. *Earth Syst. Sci. Data Discuss.* **2022**, *15*, 3283–3298. [CrossRef]
25. Diwan, T.; Anirudh, G.; Tembhurne, J.V. Object detection using YOLO: Challenges, architectural successors, datasets and applications. *Multimed. Tools Appl.* **2023**, *82*, 9243–9275. [CrossRef] [PubMed]
26. Li, B.; Li, J. Methods for landslide detection based on lightweight YOLOv4 convolutional neural network. *Earth Sci. Inform.* **2022**, *15*, 765–775. [CrossRef]
27. Mo, P.; Li, D.; Liu, M.; Jia, J.; Chen, X. A Lightweight and Partitioned CNN Algorithm for Multi-Landslide Detection in Remote Sensing Images. *Appl. Sci.* **2023**, *13*, 8583. [CrossRef]
28. Liu, Q.; Wu, T.; Deng, Y.; Liu, Z. SE-YOLOv7 Landslide Detection Algorithm Based on Attention Mechanism and Improved Loss Function. *Land* **2023**, *12*, 1522. [CrossRef]
29. Du, Y.; Xu, X.; He, X. Optimizing Geo-Hazard Response: LBE-YOLO's Innovative Lightweight Framework for Enhanced Real-Time Landslide Detection and Risk Mitigation. *Remote Sens.* **2024**, *16*, 534. [CrossRef]
30. Niu, Z.; Zhong, G.; Yu, H. A review on the attention mechanism of deep learning. *Neurocomputing* **2021**, *452*, 48–62. [CrossRef]
31. Ji, S.; Yu, D.; Shen, C.; Li, W.; Xu, Q. Landslide detection from an open satellite imagery and digital elevation model dataset using attention boosted convolutional neural networks. *Landslides* **2020**, *17*, 1337–1352. [CrossRef]
32. Jia, L.; Leng, X.; Wang, X.; Nie, M. Recognizing landslides in remote sensing images based on enhancement of information in digital elevation models. *Remote Sens. Lett.* **2024**, *15*, 224–232. [CrossRef]
33. Yang, Y.; Miao, Z.; Zhang, H.; Wang, B.; Wu, L. Lightweight Attention-Guided YOLO with Level Set Layer for Landslide Detection from Optical Satellite Images. *IEEE J. Sel. Top. Appl. Earth Obs. Remote Sens.* **2024**, *17*, 3543–3559. [CrossRef]
34. Bragagnolo, L.; Rezende, L.R.; da Silva, R.V.; Grzybowski, J.M.V. Convolutional neural networks applied to semantic segmentation of landslide scars. *Catena* **2021**, *201*, 105189. [CrossRef]
35. Yu, B.; Chen, F.; Xu, C. Landslide detection based on contour-based deep learning framework in case of national scale of Nepal in 2015. *Comput. Geosci.* **2020**, *135*, 104388. [CrossRef]
36. Chen, F.; Yu, B.; Li, B. A practical trial of landslide detection from single-temporal Landsat8 images using contour-based proposals and random forest: A case study of national Nepal. *Landslides* **2018**, *15*, 453–464. [CrossRef]
37. Wang, C.-Y.; Yeh, I.-H.; Liao, H.-Y.M. YOLOv9: Learning What You Want to Learn Using Programmable Gradient Information. *arXiv* **2024**, arXiv:2402.13616. Available online: http://arxiv.org/abs/2402.13616 (accessed on 7 July 2024).
38. Woo, S.; Park, J.; Lee, J.-Y.; Kweon, I.S. CBAM: Convolutional Block Attention Module. *arXiv* **2018**, arXiv:1807.06521. Available online: http://arxiv.org/abs/1807.06521 (accessed on 7 July 2024).
39. Wang, Q.; Wu, B.; Zhu, P.; Li, P.; Zuo, W.; Hu, Q. ECA-Net: Efficient Channel Attention for Deep Convolutional Neural Networks. *arXiv* **2020**, arXiv:1910.03151. Available online: http://arxiv.org/abs/1910.03151 (accessed on 7 July 2024).

40. Tanatipuknon, A.; Aimmanee, P.; Watanabe, Y.; Murata, K.T.; Wakai, A.; Sato, G.; Hung, H.V.; Tungpimolrut, K.; Keerativittayanun, S.; Karnjana, J. Study on Combining Two Faster R-CNN Models for Landslide Detection with a Classification Decision Tree to Improve the Detection Performance. *J. Disaster Res.* **2021**, *16*, 4. [CrossRef]
41. Liu, Y.; Shao, Z.; Hoffmann, N. Global Attention Mechanism: Retain Information to Enhance Channel-Spatial Interactions. *arXiv* **2021**, arXiv:2112.05561. Available online: http://arxiv.org/abs/2112.05561 (accessed on 7 July 2024).

remote sensing

MDPI

Article

Advancing Volcanic Activity Monitoring: A Near-Real-Time Approach with Remote Sensing Data Fusion for Radiative Power Estimation

Giovanni Salvatore Di Bella [1,*], Claudia Corradino [1], Simona Cariello [1,2], Federica Torrisi [1] and Ciro Del Negro [1]

1 Istituto Nazionale di Geofisica e Vulcanologia, Sezione di Catania, Osservatorio Etneo, Piazza Roma 2, 95125 Catania, Italy; claudia.corradino@ingv.it (C.C.); simona.cariello@ingv.it (S.C.); federica.torrisi@ingv.it (F.T.); ciro.delnegro@ingv.it (C.D.N.)

2 Department of Electrical, Electronic and Computer Engineering, University of Catania, Viale Andrea Doria 6, 95125 Catania, Italy

* Correspondence: giovanni.dibella@ingv.it

Abstract: The global, near-real-time monitoring of volcano thermal activity has become feasible through thermal infrared sensors on various satellite platforms, which enable accurate estimations of volcanic emissions. Specifically, these sensors facilitate reliable estimation of Volcanic Radiative Power (VRP), representing the heat radiated during volcanic activity. A critical factor influencing VRP estimates is the identification of hotspots in satellite imagery, typically based on intensity. Different satellite sensors employ unique algorithms due to their distinct characteristics. Integrating data from multiple satellite sources, each with different spatial and spectral resolutions, offers a more comprehensive analysis than using individual data sources alone. We introduce an innovative Remote Sensing Data Fusion (RSDF) algorithm, developed within a Cloud Computing environment that provides scalable, on-demand computing resources and services via the internet, to monitor VRP locally using data from various multispectral satellite sensors: the polar-orbiting Moderate Resolution Imaging Spectroradiometer (MODIS), the Sea and Land Surface Temperature Radiometer (SLSTR), and the Visible Infrared Imaging Radiometer Suite (VIIRS), along with the geostationary Spinning Enhanced Visible and InfraRed Imager (SEVIRI). We describe and demonstrate the operation of this algorithm through the analysis of recent eruptive activities at the Etna and Stromboli volcanoes. The RSDF algorithm, leveraging both spatial and intensity features, demonstrates heightened sensitivity in detecting high-temperature volcanic features, thereby improving VRP monitoring compared to conventional pre-processed products available online. The overall accuracy increased significantly, with the omission rate dropping from 75.5% to 3.7% and the false detection rate decreasing from 11.0% to 4.3%. The proposed multi-sensor approach markedly enhances the ability to monitor and analyze volcanic activity.

Keywords: satellite remote sensing; volcanic radiative power; volcano monitoring; data fusion

Citation: Di Bella, G.S.; Corradino, C.; Cariello, S.; Torrisi, F.; Del Negro, C. Advancing Volcanic Activity Monitoring: A Near-Real-Time Approach with Remote Sensing Data Fusion for Radiative Power Estimation. *Remote Sens.* **2024**, *16*, 2879. https://doi.org/10.3390/rs16162879

Academic Editors: Federico Raspini, Chao Zhou and Kunlong Yin

Received: 19 July 2024
Revised: 4 August 2024
Accepted: 5 August 2024
Published: 7 August 2024

1. Introduction

Technological advancements in satellite remote sensing have transformed our perception and understanding of volcanic processes. The continuous monitoring of volcanoes and the study of volcanic hazardous phenomena are of great interest and are constantly evolving [1,2]. This progress has led to the development of various sensors, both dedicated and non-dedicated, to measure volcanic thermal anomalies from space [3,4]. Among these, the Moderate Resolution Imaging Spectroradiometer (MODIS), the Sea and Land Surface Temperature Radiometer (SLSTR), and the Visible Infrared Imaging Radiometer Suite (VIIRS) onboard polar satellites, as well as the Spinning Enhanced Visible and InfraRed Imager (SEVIRI) on geostationary satellites, are widely used for volcano monitoring. High spatial resolution satellite sensors also play a key role in monitoring the behavior of a volcano, such as the MultiSpectral Instrument (MSI) on Sentinel-2 (from ESA) and the Operational Land Imager (OLI) and the Thermal Infrared Sensor (TIRS) on Landsat-8 (from NASA & USGS) [5–7].

Implementing new algorithms that enhance precision and reliability in real-time observation of high-temperature volcanic features is fundamental to improving our understanding of volcanic processes, identifying renewed volcanic activity, forecasting eruptions, and assessing hazards [8,9]. Thermal infrared sensors on board satellite platforms allow for continuous monitoring of active volcanoes worldwide, providing accurate estimates of volcanic thermal emissions [10]. Several volcanic hotspot detection algorithms have been proposed, utilizing different spatial [11], temporal [12], and spectral contexts [13] of the satellite sensors. These detection algorithms can be classified into four main categories based on the information used to identify an anomalous pixel: contextual (e.g., the VAST algorithm [14]), temporal (e.g., the RAT algorithm [15]), spectral (e.g., MODVOLC [16]), and hybrid (e.g., the Okmok algorithm [17]). Further classification is possible based on the criteria used to identify hotspots. Most algorithms apply intensity thresholds to one or more bands, which can be predetermined (e.g., the NHI algorithm [18,19]), statistical (e.g., the HOTSAT algorithm [20,21]), or based on machine learning approaches [6,22].

Several volcanic monitoring satellite systems operate in near real time (NRT), such as MIROVA [23], MODVOLC, HOTSAT, HOTVOLC [24] and FASTVRP [25], utilizing traditional algorithms to automatically provide radiance information from volcanic anomalous pixels worldwide. These systems process infrared images acquired by satellite sensors to calculate relevant Volcanic Radiative Power (VRP), which measures the heat radiated by volcanic activity, analogous to the Fire Radiative Power (FRP) [26,27] measure used for fire detection. While these algorithms generally perform well in detecting intense volcanic activity [28,29], they may have limitations, particularly in detecting more subtle thermal anomalies. These limitations arise from the low spatial resolution of satellite sensors, which average incoming radiance information over a large area into a single value. Consequently, spatially-confined volcanic emissions might not significantly affect the averaged pixel value, making it appear similar to pixels covering heated slopes.

In more advanced techniques, the replication of visual system processes has been introduced to account for spatial features, such as the shapes of volcanic thermal anomalies [30]. This approach enhances the detection of volcanic anomalies by using geometrical features to overcome the intrinsic limitations of the instruments. Recent studies have demonstrated the importance of using spatial features that replicate the focus mechanism, particularly in the context of the ASTER (Advanced Spaceborne Thermal Emission and Reflection Radiometer) multispectral thermal infrared (TIR) dataset using statistical [31] and deep learning [32] algorithms. Regardless of the spatial resolution, this approach appears to be feasible for every satellite sensor. Typically, each satellite sensor has its own specific volcanic hotspot detection algorithm, primarily relying on intensity. To operationally utilize all available satellite data, data fusion is needed to provide VRP estimates as soon as new data becomes available.

Here, we introduce a new algorithm for multi-sensor VRP estimates, named the Remote Sensing Data Fusion (RSDF) algorithm. This algorithm employs a novel volcanic hotspot detection method that exploits both intensity and spatial features. We describe and demonstrate the operation of the RSDF algorithm by analyzing the eruptive activity at the Etna and Stromboli volcanoes from 2021 to 2023 [33,34]. Specifically, we first focus on two similar polar satellite sensors, SLSTR and MODIS, demonstrating enhanced performance compared to the SLSTR Level 2 products, distributed by Copernicus Dataspace [35], and the MODIS fire detection product, available online on the Fire Information for Resource Management System (FIRMS) [36]. We then show the potential of using this approach with the polar-orbiting VIIRS and SEVIRI sensors. Improved performance over single sensor existing algorithms is shown for the SLSTR, MODIS, VIIRS, and SEVIRI sensors.

2. Materials

2.1. Case Studies: Etna and Stromboli Volcanoes

We investigated the Etna and Stromboli volcanoes, both characterized by frequent eruptive events that produced exemplary volcanic heat sources during the analyzed period from 2021 to 2023.

Mount Etna, located on the east coast of Sicily, is one of the most active volcanoes in the world, known for its persistent activity in summit craters—consisting of degassing and explosive phenomena associated with fast-moving lava flows—and recurrent effusive eruptions from vents located on the flanks of the volcano, producing lava flows that can extend for several kilometers [37,38]. During the investigated period, Etna experienced numerous paroxysmal events, particularly in early 2021, with powerful lava fountains, ash plumes reaching several kilometers into the atmosphere, and extensive lava flows that significantly altered the landscape [39]. Notably, the series of paroxysmal events in February and March 2021 were among the most intense, resulting in widespread ash fall and disruptions to local communities and air traffic [6,40,41].

Similarly, Stromboli, situated in the Aeolian Islands, is renowned for its continuous mild explosive activity, often referred to as "Strombolian" eruptions [42,43]. From 2021 to 2023, Stromboli exhibited consistent eruptive behavior with frequent explosive bursts that ejected incandescent lava fragments, ash, and volcanic gases. The volcano's activity included several significant episodes, such as the July 2021 eruption which featured a major explosive event, producing high-energy pyroclastic flows that reached the sea and prompted evacuation warnings for nearby residents and tourists. Additionally, its volcanic activity in late September 2023 was marked by heightened explosive activity, creating substantial volcanic heat sources and providing valuable data for our analysis [44–46].

The frequent and varied eruptive behaviors of these volcanoes during the study period provided an excellent opportunity to test and validate our RSDF algorithm's ability to detect and analyze volcanic thermal activity across different sensors.

2.2. Satellite Sensors

The Sea and Land Surface Temperature Radiometer (SLSTR) is on board the Sentinel-3A and 3B satellites. It is designed to monitor temperatures of Earth's terrestrial and aquatic surfaces with a precision of under 0.2 K. The SLSTR features nine spectral channels ranging from 0.55 μm to 12.0 μm, including three in the visible and near-infrared spectrum, three in the short-wave infrared, one in the mid-infrared, and two in the thermal infrared. SLSTR data are captured in both nadir and oblique views, offering spatial resolutions of 1 km and 0.5 km for VIS/SWIR channels, across swaths of 1400 km and 750 km, respectively. The algorithm extracts information using the S7 (Medium InfraRed, MIR) and S8 (Thermal InfraRed, TIR) bands for brightness temperature, switching to Fire Channels F1 and F2 when saturation temperatures of 312 K and 500 K are exceeded. Data are provided in netCDF 4 format, covering both nadir and oblique views, with comprehensive metadata for each pixel [47].

The Moderate Resolution Imaging Spectroradiometer (MODIS) operates on NASA's Terra and Aqua satellites, providing global coverage every 1–2 days across 36 spectral bands from 0.4 μm to 14.4 μm. It boasts a high 12-bit radiometric sensitivity with spatial resolutions of 250 m, 500 m, and 1 km, covering a 2330-km swath in a 55-degree scanning pattern from a 705 km orbit. Its optical system includes a rotating double-sided scan mirror and off-axis telescope, cooled to 83 K for infrared bands. MODIS Level 1 data, available since February 2000, are provided in a geographic projection at 1 km resolution. The radiance values are extracted from the MIR (channel 22) and TIR (channel 32) bands, switching to channel 21 when the MIR band's saturation threshold of 328 K is exceeded. Data are encoded as 16-bit unsigned integers, requiring conversion for analysis, and are accessible via the LAADS web platform [48].

The Visible Infrared Imaging Radiometer Suite (VIIRS), deployed on the Suomi NPP and JPSS satellites, captures detailed data on Earth's land cover, surface temperatures, and atmospheric parameters across various spectral bands. It provides global coverage, collecting data from diverse regions during each orbit. A notable feature is its ability to observe at night using low-light bands, useful for monitoring urban lighting and nocturnal fires. VIIRS L1B granules, available from the LAADS-DAAC system, include MIR (channel I4) and TIR (channel I5) brightness temperature data at 375 m resolution. These bands have saturation temperatures of 367 K and 380 K, respectively, and switch to M13 (MIR) and M15 (TIR) bands at 750 m resolution when saturated. After resampling into a UTM grid, a matrix of 67 by 67 pixels is obtained, due to the sensor's 750-m resolution [48].

The Spinning Enhanced Visible and Infrared Imager (SEVIRI), carried on EUMETSAT's Meteosat Second Generation satellites, provides data at 15-min intervals with spatial resolutions of 3–5 km and 1 km for the High-Resolution Visible channel. It supports weather forecasting, climate monitoring, and environmental research by assessing parameters like cloud properties, sea surface temperatures, and land surface temperature. SEVIRI data are transmitted in 12 spectral channels through high-rate transmissions, with particular focus on the IR3.9 (MIR) and IR10.8 (TIR) bands. These data are resampled into UTM coordinates, forming a 12 × 12 km grid, aiding in the monitoring of severe weather events such as storms, wildfires, and hurricanes [47].

A summary of the main characteristics of the four satellite sensors utilized is shown in Table 1.

Table 1. Characteristics of the SLSTR, MODIS, VIIRS, and SEVIRI sensors.

	SLSTR (Sentinel-3)	**MODIS (AQUA/TERRA)**	**VIIRS (S-NPP/N20)**	**SEVIRI (MSG)**
Orbit altitude (km)	814	705	824	38.500
Type of Satellites	Polar	Polar	Polar	Geostationary
Equator crossing time	10:00 LT	10:30 LT/13:30 LT	13:30 LT/12:40 LT	
Spatial Resolution (km)	1	1	0.75–0.375	3
Temporal Resolution	1–2 days	1–2 days	12 h	10–15 min
Spectral coverage of thermal bands (μm)	3.700–12.000	3660–14.385	3.550–12.488	3.480–13.400
ID MIR Band	S7/F1	22/21	I04/M-13	IR-039
Spectral Range (μm)	3.700/3.700	3.940–4.001/3.929–3.989	3.973–4.128	3.480–4.360
Tmax	312 K/500 K	331 K/500 K	367 K/634 K	335 K
ID TIR Band	S8/F1	31	I05/M-15	I-108
Spectral Range (μm)	10.850/12.000	10.780–11.280	10.263–11.263	11.000–13.400
Tmax	323 K/400 K	400 K	380 K/343 K	

2.3. SLSTR and MODIS Level 2 Products and Fire Radiative Power

The Sentinel-3 SLSTR Level 2 databases are available for open access via Copernicus dataspace. This platform facilitates the download of volcanic anomaly data from both the Etna and Stromboli volcanoes within the same time window. We extract Fire Radiative Power (FRP) values from the same volcanic anomalies obtained through an Active Fire Detection algorithm [27,35,49]. Originally designed to identify and describe vegetation fires occurring on land surfaces, the algorithm was also adapted to detect areas with elevated temperatures related to active volcanism and intense industrial heat sources. Fire detection employs varying fixed thresholds to identify confirmed fire pixels in near-nadir view data, adjusting for daytime and nighttime conditions. For each image, FRP values of individual hotspots are collected and summed to derive the total FRP value for each acquired image. NASA's MODIS active fire products marked the introduction of remotely sensed fire datasets [36,50,51]. Utilizing moderate spatial resolution (~1 km), these data sets originate from advanced "fire-capable" sensors on satellite platforms [27]. By accessing MODIS Level 2 products at https://firms.modaps.eosdis.nasa.gov/, accessed on 18 January 2024, it is possible to download .csv files containing FRP values of detection alerts for a given period and geographical area [36].

3. Methods

The RSDF algorithm detects volcanic hotspots and estimates the emitted VRP for each image acquired. Information is extracted from multispectral satellite sensor images through spectral and spatial analysis, exploiting the potential of a bank of Gabor filters [31] known for their effectiveness in improving detection sensitivity and reducing the effects of clouds (Figure 1).

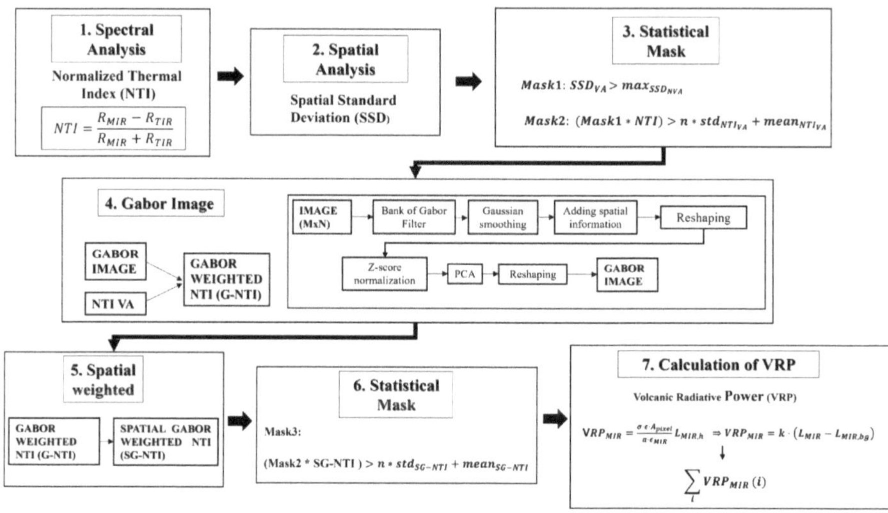

Figure 1. (**1**) Derivation of the Normalized Thermal Index (NTI) obtained by combining the radiance of the MIR and the radiance of the TIR. (**2**) Application of the Spatial Standard Deviation (SSD) filter to each pixel in the image. (**3**) Definition of two statistical masks, Mask1 and Mask2, to identify "potential" and "true" hotspots, applied on the SSD and NTI of the volcanic area (VA). (**4**) Application of Gabor filter to extract the significant features of the image, resulting in a matrix called Gabor Weighted NTI (G-NTI). (**5**) Highlighting hotspots in the crater area and defining the Spatial Gabor Weighted NTI (SG-NTI). (**6**) Application of a statistical mask to the previously extracted matrix. (**7**) Calculation of the final VRP.

3.1. Spectral Analysis

The initial step involves extracting Level 1 data from each satellite sensor, including geolocation, date, time, brightness temperature, and radiance specific to the relevant spectral band. Geolocation data is linked to the target volcano's position. Different satellites yield distinct Level 1 data based on sensor instruments. The algorithm focuses on the MIR and TIR bands, specifically targeting the MIR range of 3.973–4.128 μm and the TIR range of 10.263–11.263 μm. Band selection is influenced by the saturation temperature, which particularly affects the MIR band. The choice of these two bands is related to their frequency response; for low temperatures, the spectral radiance is similar for both bands, but for high temperatures, the MIR band shows a significant increase in spectral radiance compared to the TIR band. The next step involves performing a spectral analysis, deriving the Normalized Thermal Index (NTI) [16]. The NTI is calculated as follows, exploiting the differences between the MIR and TIR bands:

$$NTI = \frac{L_{MIR} - L_{TIR}}{L_{MIR} + L_{TIR}} \tag{1}$$

where L_{MIR} and L_{TIR} are the radiance recorded by the MIR and TIR bands, respectively. The NTI demonstrates heightened sensitivity to hot surfaces, leading to a substantial increase in L_{MIR} readings. This enhanced sensitivity allows for the precise detection of thermal anomalies in the image, facilitating the automatic identification of active hotspot pixels.

3.2. Spatial Analysis

Following the discovery of the NTI matrix, a spatial analysis is conducted using a Spatial Standard Deviation (SSD) filter applied to each pixel in the image. The SSD involves subtracting the mean (or median) of eight neighboring pixels from the NTI values matrix to mitigate artifacts such as seas or clouds [20]. The resulting SSD is then cropped and divided into volcanic areas (VAs) and non-volcanic areas (NVAs). The former is centered on the volcano's crater, halved from the original matrix, while the non-volcanic area is obtained by subtracting the volcanic area.

3.3. Statistical Mask

The first mask (Mask1) is defined by the maximum SSD value in the non-volcanic area (SSD_{NVA}) serving as a threshold. It is compared to the SSD values of each pixel in the volcanic area (SSD_{VA}), marking pixels exceeding the maximum SSD value of the non-volcanic area ($maxSSD_{NVA}$) as "potential hotspots", expressed as

$$SSD_{VA} > maxSSD_{NVA} \tag{2}$$

The second mask (Mask2) is applied to differentiate and identify the "true hotspots" among the potential ones. The condition for Mask2 is

$$NTI_{VA} > meanNTI_{VA} + n * stdNTI_{VA} \tag{3}$$

where $meanNTI_{VA}$ and $stdNTI_{VA}$ are, respectively, the mean of NTI and the standard deviation of NTI of the volcanic area (VA). n serves as a parameter determining the permissible deviation of the integrated MIR pixel temperature from the mean value. For polar satellites (MODIS, VIIRS and SLSTR), a distinction between daytime and nighttime data is necessary due to the significant influence of sunlight on daytime data values. Therefore, n = 10 is chosen for daytime data, while n = 5 is selected for nighttime data for MODIS and SLSTR sensors. Regarding VIIRS, given the different spatial resolution compared to previous sensors and the size of the volcanic area, n = 8 was adopted for daytime and n = 12 for nighttime. As for geostationary satellites (SEVIRI), as they are at a low spatial resolution they are not heavily affected by solar radiation, and thus a unique n value is choosen, set to 1.

3.4. Gabor Image

We opted to employ a bank of Gabor filters to enhance the performance of the detection algorithm [31]. These filters are derived from the product of a Gaussian function and an oriented sinusoidal plane wave. Gabor filters highlight local spatial features in images by focusing on the spatial frequency content around each pixel. When used in conjunction with thermal intensities, they enable the detection of anomalies based on both temperature and spatial 'texture'. This is especially crucial for identifying subtle anomalies characterized by low intensities, such as groups of warm pixels near volcanic crater vents.

The use of the Gabor filter on the selected image aims to extract the most significant features from MIR images. This task is achieved through the multi-step procedure described in Figure 1. Firstly, candidate features are retrieved using a bank of Gabor filters, Gaussian low-pass filter smoothing of the Gabor image, and spatial information. This results in a feature matrix with 24 Gabor features plus 2 spatial features per pixel. After normalization, i.e., the matrix is Z-score normalized to achieve zero mean and unit variance, Principal Component Analysis (PCA) is applied [31]. This step allows for the retrieving and selecting of only the most informative features, reducing the risk of overlap or redundancy.

The output is ultimately an image highlighting the most significant regions relative to the surrounding context. The choice to generate the Gabor image from the MIR band is due to the need to highlight thermal anomalies and avoid the cloud effects that influence the TIR band, which may affect the NTI. By multiplying the Gabor image with the NTI matrix obtained in the previous steps, we obtain the Gabor Weighted NTI (G-NTI).

3.5. Spatial Weighted

We generate a Spatial Gabor Weighted NTI (SG-NTI), aiming to highlight hotspots in the crater area and reduce the effect of any non-volcanic hotspots due to the uncertainty outside the volcanic zone.

3.6. Statistical Mask

In this step, similarly to Step 3, a mask is computed based on the SG-NTI matrix

$$Mask3 : SG - NTI > n * std_{SG-NTI} + mean_{SG-NTI} \quad (4)$$

where $mean_{SG-NTI}$ and std_{SG-NTI} are respectively the mean of SG-NTI and the standard deviation of SG-NTI.

3.7. Calculation of VRP

The final binary mask identifying real volcanic hotspots is the product of the three previously computed masks (Figure 2). Once we identify and localize hotspot pixels (i.e., Mask = Mask1 × Mask2 × Mask3), we calculate the VRP mask using

$$VRP_{MIR} = \frac{\sigma \cdot \epsilon \cdot A_{pixel}}{a \cdot \epsilon_{MIR}} L_{MIR,h} \Rightarrow VRP_{MIR} = k \cdot \left(L_{MIR} - L_{MIR,bg}\right) \quad (5)$$

where A_{pixel} is the pixel ground sampling area (m^2) and $L_{MIR} - L_{MIR,bg}$ is the excess of MIR radiance [6,22]. The background is the average value of the MIR band of the hotspots surrounding the "true hotspots" detected by the output of the total mask. The constant k is specific for each sensor and depends directly on the parameter a ($k_{MODIS} = 1.89 \times 10^7$ $Wm^{-2}sr^{-1}\mu m^{-1}K^{-4}$, $k_{SLSTR} = 1.70 \times 10^7$ $Wm^{-2}sr^{-1}\mu m^{-1}K^{-4}$, $k_{SEVIRI} = 3.70 \times 10^7$ $Wm^{-2}sr^{-1}\mu m^{-1}K^{-4}$, $k_{VIIRS375} = 2.48 \times 10^7$ $Wm^{-2}sr^{-1}\mu m^{-1}K^{-4}$, and $k_{VIIRS750} = 1.11 \times 10^7$ $Wm^{-2}sr^{-1}\mu m^{-1}K^{-4}$).

The summation of the VRP values is then computed to estimate the total VRP value of the analyzed image:

$$\sum_i VRP_{MIR}(i) \quad (6)$$

3.8. Eruptive Parameters

The time series of VRP (Volcanic Radiative Power) is necessary for characterizing three eruptive parameters: the Time-Averaged Discharge Rate (TADR [m^3/s]), the volume of lava flows, and the Volcanic Radiative Energy (VRE [J]). The TADR refers to the radiant density approach [52], connecting it directly to the VRP in cubic meters per second. This relationship is established through a single best-fit coefficient known as 'radiant density' (c_{rad}):

$$TADR = \frac{VRP}{c_{rad}}$$

where:

$$c_{rad} = \frac{6.45 * 10^{25}}{X_{SiO_2}^{10.4}}$$

where $X_{SiO_2}^{10.4}$ represents the concentration of silicon dioxide adjusted by a factor of 10.4. Values of the parameter c_{rad} used for Etna and Stromboli are shown in Table 2.

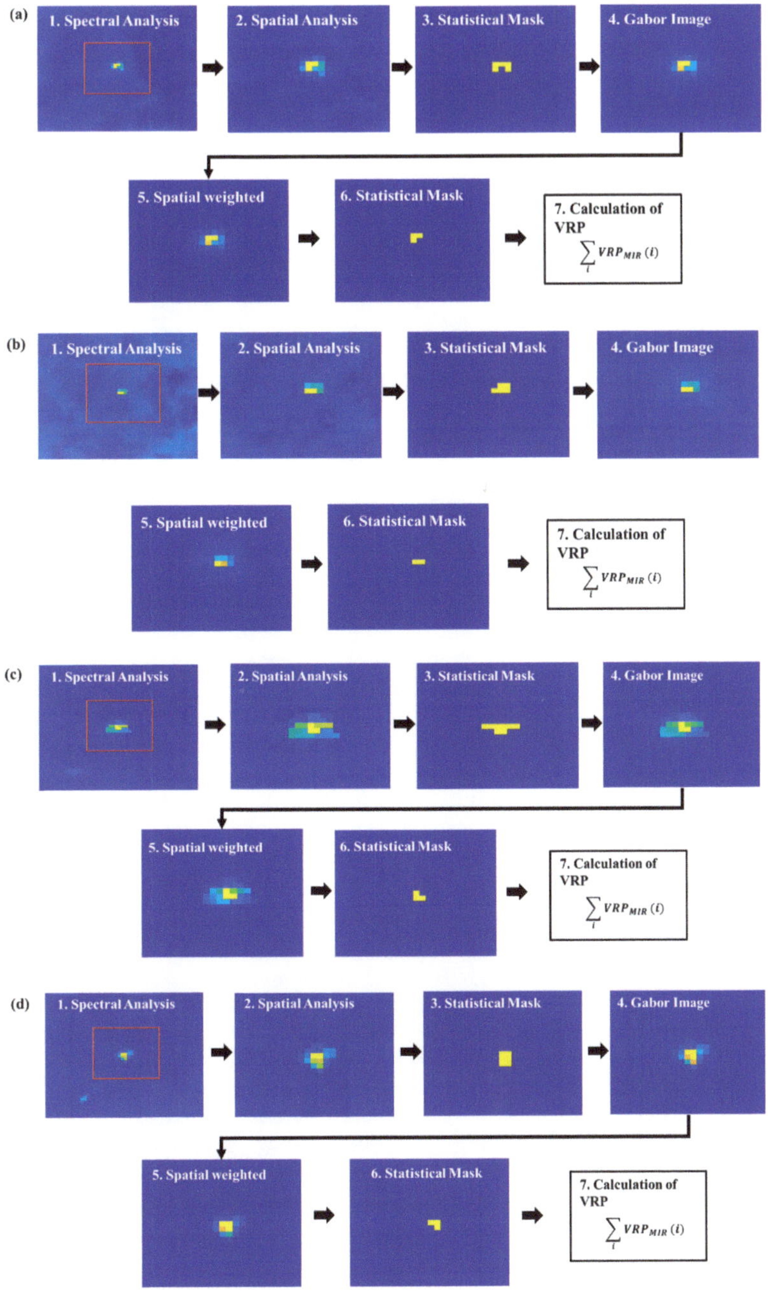

Figure 2. Workflow image of the RSDF algorithm. Study cases: (**a**) Etna on 2 December 2023 at 01:10 UTC, MODIS sensor; (**b**) Etna on 15 January 2023 at 20:46 UTC, SLSTR sensor; (**c**) Stromboli on 3 October 2023 at 13:10 UTC, MODIS sensor; (**d**) Stromboli on 23 October 2023 at 09:08 UTC, SLSTR sensor.

Table 2. Values of the parameter c_{rad} for Etna and Stromboli.

	Etna	Stromboli
c_{rad}	2.16×10^8	1.65×10^8
c_{radmax}	1.08×10^8	8.25×10^8
c_{radmin}	3.24×10^8	2.47×10^8

Due to the uncertainty in deriving the TADR parameter, we calculated its maximum, minimum, and mean values using three different values of c_{rad} [J/m^3]. Estimating c_{rad} is vital for understanding lava dynamics. With a ±50% uncertainty, minimum and maximum TADR values are computed for accuracy [53,54]. The volume of the lava flow is given by [55]:

$$TADR_{mean} = \frac{TADR_i + TADR_{i-1}}{2} \Rightarrow Volume_i = TADR_{mean} \cdot \Delta t$$

The last parameter is the Volcanic Radiated Energy (VRE), which represents the total energy emitted by a volcano, including thermal infrared radiation. The VRE is measured in Joules [J]. It serves as a crucial parameter for monitoring and understanding volcanic activity, enabling the analysis of eruption intensity and lava volume through remote sensing techniques such as satellite-based sensors. The calculation of VRE involves trapezoidal integration of VRP/FRP time series [56].

3.9. Uncertainties and Limits

The algorithm for Volcanic Radiative Power (VRP) estimation faces limitations, notably the MIR radiance method's constraint to emitter temperatures above 600 K. This restricts its accuracy for colder emitters, yet it maintains reliability for warmer ones. Challenges like atmospheric clouds, volcanic clouds, fires, and satellite viewing angles introduce noise during acquisitions, generating false hotspots and masking potential thermal anomalies. Clouds complicate satellite observations, making it challenging to discern real hotspots from false positives linked to cloud presence. Introducing the cloud coverage index (i_{cloud}) quantifies cloud presence within each pixel, aiding in understanding the impact of clouds on the image under investigation, expressed as

$$i_{cloud} = \frac{N_{cloudVA}}{N_{VA}}$$

where $N_{cloudVA}$ is the number of cloudy pixels inside the volcanic area and N_{VA} is the total number of pixels inside the volcanic area. The detection of cloudy pixels, using satellite data, is achieved by considering atmospheric and surface conditions to calculate the probability of clear sky for each pixel. A cloud coverage index of 1 denotes that the image is categorized as cloudy, while an index of 0 indicates the absence of identified cloudy pixels within the volcanic region. We consciously avoided modifying datasets through image inspections or filters to eliminate cloudy or geometrically challenging scenes. This decision aims to ensure a precise comparison between time series datasets, maintaining the integrity of the original data and allowing for a more accurate assessment of cloud impact on volcanic observations over time.

Another significant consideration is the satellite viewing geometry, meaning that the acquired images could have high zenith and azimuth angle values. In volcanic satellite monitoring, this poses several challenges: geometric distortion of images, reduced spatial resolution affecting small volcanic features, thermal measurement inaccuracies due to atmospheric interference, and compromised data quality from increased atmospheric dispersion and absorption. These factors hinder accurate identification of hotspots in deep craters and areas at the base of craters. Using the information on zenith and azimuth angles contained in the data provided by the satellite sensors, it is possible to calculate the value of the scan angle. When this value often exceeds 70/80, it indicates a certain degree of projected thermal anomaly distortion, visible as an increase in the number of pixels with the same exact value. However, the validation of this uncertainty in the algorithm's results, as with cloudiness, has not undergone filtering to maintain its integrity.

4. Results

4.1. Etna

The RSDF algorithm identified 764 alerts (N_{alert}) between 2021 and 2023, recognized as satellite acquisitions with thermal anomalies in the volcanic area by the SLSTR sensor. This is nearly two times the 456 alerts identified by the Level 2 data processing. Considering the 2545 SLSTR overpasses (N_{pass}), there is a noticeable contrast in alert detection frequency (f% = N_{alert}/N_{pass}). The algorithm registers a frequency of 30% (N_{alert} = 764), a marked difference compared to the 18% seen in the SLSTR product level 2 (N_{alert} = 445).

Similar to SLSTR, the results of MODIS active fire products, provided by NASA, indicate insufficient sensitivity for continuous volcanic monitoring. The RSDF algorithm provides a N_{alert} = 858 with f = 17%, while the Level 2 product provides N_{alert} = 13 with a frequency of 0.002%, both data related to N_{pass} = 5171, as achieved by the TERRA and AQUA satellites.

The temporal series of VRP from both sensors reveal a pattern that allows the identification of four volcanic eruptive phases, each representing a period of heightened volcanic activity over the analyzed timeframe.

Phase 1 spans from 15 February 2021 to 3 April 2021. Subsequent phases include phase 2, from 17 May 2021 to 28 September 2021, phase 3, from 12 May 2022 to 15 June 2022, and phase 4, from 27 November 2022 to 6 February 2023 (Figure 3).

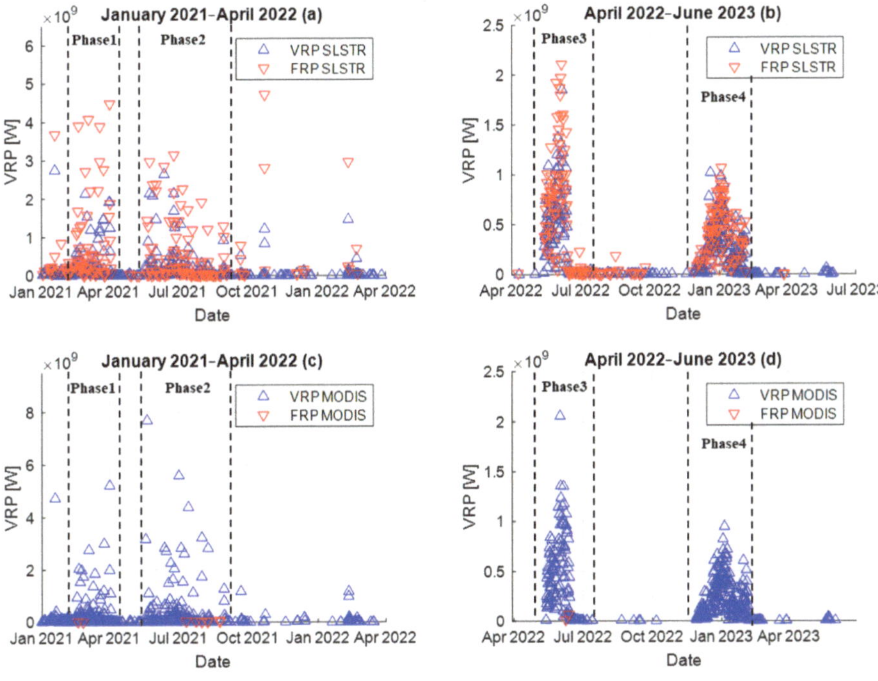

Figure 3. Time series of the Etna volcano. The panels show VRP calculated respectively from the RSDF Algorithm SLSTR (blue triangles) SLSTR Level 2 (red triangles), the RSDF Algorithm MODIS (blue triangles), MODIS Level 2 (red triangles). (**a**,**c**) shows data from January 2021 to April 2022, (**b**,**d**) from April 2022 to June 2023.

In the period of February to May 2021, and generally throughout 2021, Mount Etna exhibited exceptionally high volcanic activity, validated by the occurrence of 57 paroxysmal events [39]. These events significantly influence VRP calculations, with SLSTR showing VRP values ranging from 400 MW to 1000 MW. Notably, in phase 1, there is a peak VRP of 2139 MW detected on 28 February 2021, at 09:40 UTC.

Regarding MODIS, it exhibits very high peaks in the early part of 2021, with the series' maximum reaching 7684 MW on 21 May 2021, at 01:40 UTC. It is worth noting the absence of MODIS active fire products measurements. The period from April 2022 to June 2023 remains consistent in terms of magnitude. The highest peak during this timeframe for SLSTR was observed on 6 June 2022, at 09:36 UTC, with a VRP value of 1848 MW. On the other hand, MODIS recorded its peak on 30 May 2022, at 20:25 UTC, featuring a VRP value of 2054 MW.

An interesting observation is that in the months following intense volcanic activity, low-level thermal anomalies are rarely detected by Level 2 and MODIS active fire products data processing. In contrast, the implemented RSDF algorithm demonstrates enhanced accuracy and sensitivity in identifying these anomalies, enabling continuous monitoring of thermal anomalies throughout all phases of volcanic activity, including the presence of low-intensity thermal anomalies inside the volcanic crater.

Generally, higher Fire Radiative Power (FRP) values from the Sea and Land Surface Temperature Radiometer (SLSTR) are found compared to Volcanic Radiative Power (VRP). This correlation is due to the Radiative Surface Detection Function (RSDF) having lower false detection rates, resulting in lower total VRP values. This is especially true during periods of greater volcanic intensity, such as the 2021 lava fountain period on Etna. During such times, RSDF proves to be more accurate in detecting the number of hotspots, which tend to lower the VRP value.

Statistical analysis serves a dual purpose: it validates existing hypotheses regarding alert numbers and establishes thresholds capable of distinguishing between different volcanic thermal regimes. In long-term monitoring, the temporal complementarity of datasets is crucial, facilitating a point-to-point comparison of the diverse values obtained.

The distribution of alerts (Figure 4) confirms the RSDF algorithm's heightened sensitivity in detecting hotspots with low VRP values, evident for all values below a logVRP threshold of approximately 8.5 MW for SLSTR and 7.8 MW for MODIS. The absence of alerts in the Level 2 product dataset and MODIS active fire products further emphasizes this sensitivity. However, a similar trend is observed above the threshold, as demonstrated earlier, indicating the occurrence of high-intensity thermal volcanic events.

This analysis of the alert distribution for MODIS highlights the lack of thermal observations in the MODIS active fire products. The use of a probabilistic plot, where any alteration in the curve's slope is associated with a change in thermal regime, confirms that surpassing the previously defined thresholds causes the logVRP points to deviate from the theoretical line, manifesting extraordinary behavior that is attributed to phenomena such as lava fountains and paroxysms. This trend is also evident in the datasets obtained via processing SLSTR datasets. Consistently, this analysis holds true for the MODIS dataset processed by the RSDF algorithm, emphasizing the low reliability and sensitivity of the MODIS active fire products. Comparing RSDF performance, computed using the Etna and Stromboli study cases, with MIROVA outcomes for the same volcanoes, we find an improvement in terms of omitted detections (3.7% vs. 10%) and false detections (4.3% vs. 5%) [28]. The two metrics calculated for the two volcanoes refer to different datasets compared to this work [28]. By utilizing the calculation of average VRP values on a weekly scale (VRPw) for each sensor in each dataset, it is noted that the absence of intermediate values between two significant eruptive events confirms the previously shown results (Figure 5). This underscores the reliability and sensitivity demonstrated by the RSDF algorithm. From Figure 5, it is observed that the trends of MODIS and SLSTR processed through the RSDF algorithm are nearly coincident and overlapping, indicating considerable numerical compatibility. In the graph, the deviation of individual values of SLSTR and

MODIS, processed via RSDF, is attributable to the different revisit times of the two sensors and the uncertainty conditions present for each satellite pass.

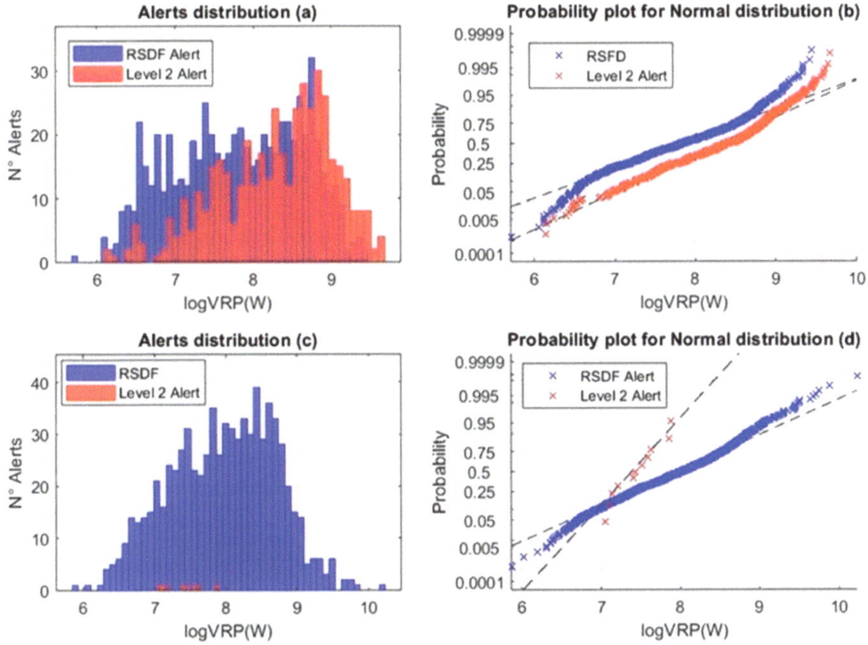

Figure 4. Histograms (**a**,**c**) and probability plots (**b**,**d**) for Etna datasets. (**a**,**c**) Histograms display data distribution related to VRP (and FRP) in logarithmic scale; (**a**) blue bars represent the distribution of SLSTR—RSDF algorithm processed data, and red bars represent SLSTR Level 2 product data; (**c**) blue bars represent the distribution of MODIS–RSDF algorithm processed data, and red bars represent MODIS active fire products. (**b**,**d**) Probability plots for normal distribution of RSDF algorithm processed data (blue), and Level 2 product data (red). The dashed grey lines represent the reference lines of the theoretical distributions, and the black dashed line in (**b**) corresponds to the slope change associated with the transition between regimes of background and high thermal activity.

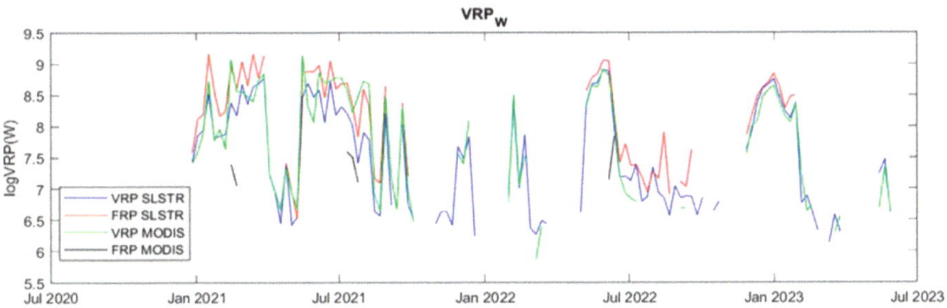

Figure 5. Stacked time series of VRPw (weekly mean) retrieved for the SLSTR–RSDF algorithm processed data (blue), SLSTR Level 2 product data (red), MODIS–RSDF algorithm processed data (green), and SLSTR Level 2 product data (black) at the Etna volcano, displayed on a logarithmic scale.

4.2. Stromboli

The application of the RSDF algorithm to a volcano under non-ideal conditions, such as the small island of Stromboli with its proximity to a sea that introduces noise during processing, demonstrates robust sensitivity for continuous monitoring over an extensive time scale. The algorithm detected 605 alerts out of 2545 overpasses, rendering the detection of thermal activity consistent, with an alarm frequency of 24%. For the same number of overpasses, the Level 2 product presents only 25 alerts (N_{alert}) and a frequency (f%) of 0.01% highlighting the insufficient sensitivity of the Level 2 product for sustained monitoring.

Turning to MODIS, the RSDF algorithm registers 570 alerts (N_{alert}) with a frequency (f%) of 11%. The relatively low frequency and alert count, compared to the total detections ($N_{pass} = 5171$), is attributed to the presence of the sea, which introduces noise into the system and negatively influences the calculation of VRP. Even less precise and reliable are the MODIS active fire products, which report only 11 alerts, considering the same number of overpasses (N_{pass} = 5171) and a frequency of 0.002%.

Examining Stromboli's volcanic activity from 2021 to 2023 reveals a dynamic, three-year period marked by a continuum of volcanic events, ranging from ordinary to extraordinary and encapsulated in three distinct phases. The first phase (13 May 2021–27 May 2021) witnesses pyroclastic flows and lava overflows. The second phase (11 May 2022–8 June 2022) encompasses a significant explosion on May 13, resulting in substantial material emission. The final phase spans from 27 November 2023 to 11 December 2023. While the VRP data from Level 2 and MODIS active fire products are slightly over a dozen units, the RSDF algorithm manages to detect peaks during the first eruptive cycle. Notably, on 19 May 2021 at 20:22 UTC, the maximum recorded VRP value is 959 MW. In the case of MODIS, the values are slightly lower, with a peak VRP of 997 MW on 19 May 2021 at 12:55 UTC, corresponding to the occurrence of a pyroclastic flow and overflow that reached the sea (Figure 6).

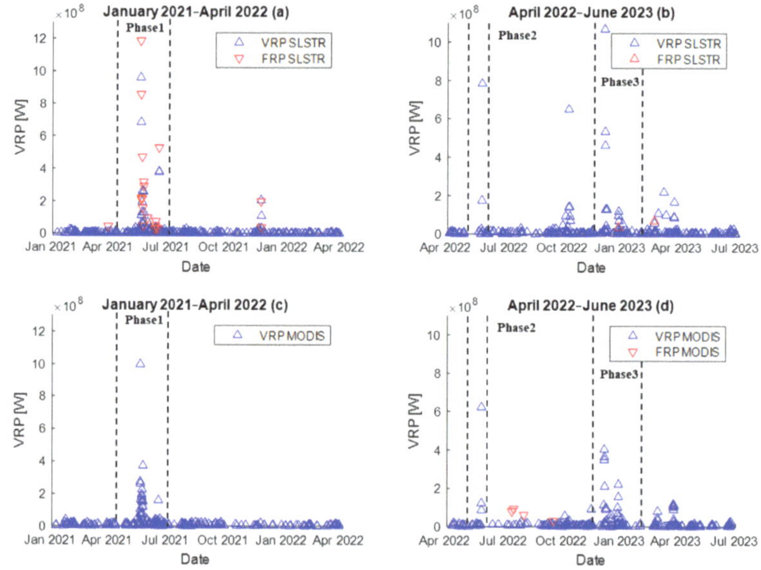

Figure 6. VRP time series of the Stromboli volcano. The panels show VRP calculated respectively from the RSDF Algorithm SLSTR (blue triangles) SLSTR Level 2 (red triangles), the RSDF Algorithm MODIS (blue triangles), MODIS Level 2 (red triangles). (**a**,**c**) shows data from January 2021 to April 2022, (**b**,**d**) from April 2022 to June 2023.

The utilization of Level 2 product and MODIS active fire product databases for both sensors proves inadequate for continuous monitoring of thermal anomalies in Stromboli. Consequently, leveraging the SWIR bands becomes crucial, capturing large thermal anomalies with very high VRP/FRP values. In this study, for a consistent and complementary comparison, the results of the processing are obtained exclusively using the MIR and TIR bands.

In general, the VRP values are lower than in the case of Etna, strongly influenced by the numerous paroxysmal events of 2021. Statistically analyzing the distribution of SLSTR sensor alerts reveals that below a logVRP threshold of 7.3 MW, the Level 2 product exhibits a complete absence of detections. The RSDF algorithm demonstrates an increasing trend up to an 8 MW threshold before a decline in the number of alerts corresponding to extraordinary volcanic events (Figure 7a). At this specific threshold, observed in the probabilistic plot, there is a marked change in slope, with points deviating from the theoretical distribution line.

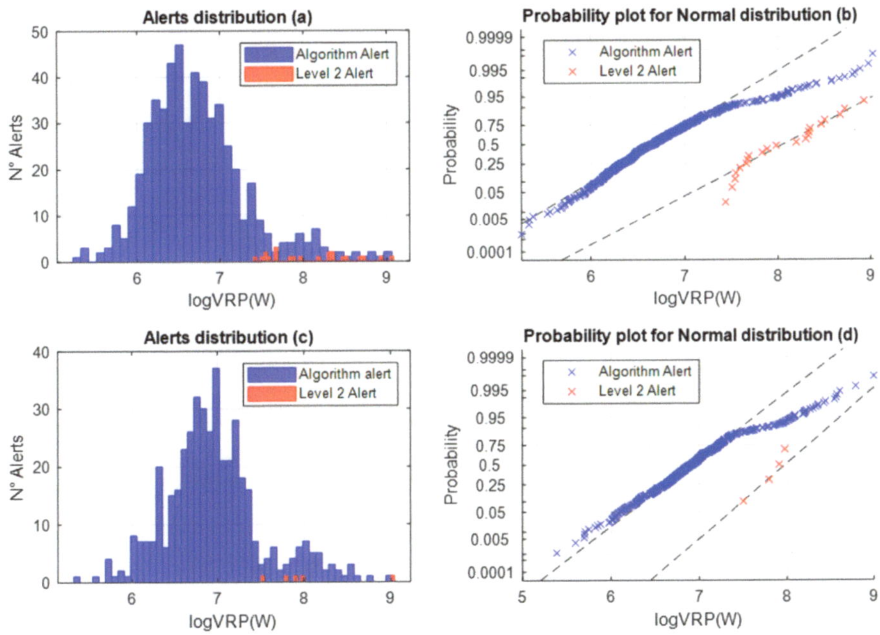

Figure 7. Histograms (**a**,**c**) and probability plots (**b**,**d**) for Stromboli datasets. (**a**,**c**) Histograms display data distribution related to VRP (and FRP) in logarithmic scale; (**a**) blue bars represent the distribution of SLSTR–RSDF algorithm processed data, and red bars represent SLSTR Level 2 product data; (**c**) blue bars represent the distribution of MODIS–RSDF algorithm processed data, and red bars represent MODIS active fire products. (**b**,**d**) Probability plots for normal distribution of RSDF algorithm processed data (blue), and Level 2 product data and MODIS active fire products (red). The dashed grey lines represent the reference lines of the theoretical distributions, and the black dashed line in (**b**) corresponds to the slope change associated with the transition between regimes of background and high thermal activity for Stromboli.

The frequency distribution of alerts from MODIS mirrors the trend observed in the statistical analysis of SLSTR. In Figure 7b, the threshold of 8 MW divides the graph into two distinct zones. On the left, there is a zone indicative of ordinary events, while on the right, a zone emerges with points deviating from the theoretical line, signaling extraordinary events.

Extracting the weekly average VRP [48], as done in the case of Etna, a similar temporal trend is observed in Figure 8, especially for data derived from SLSTR. The continuity in detecting hotspots is also highlighted here, even during periods of mild volcanic activity.

The trends of MODIS and SLSTR, as processed by the RSDF algorithm, exhibit similar patterns in Figure 8, with MODIS showing lower numerical values. The discrepancy in SLSTR and MODIS values can be attributed to the varying revisit times of the sensors and uncertainty conditions during satellite passes.

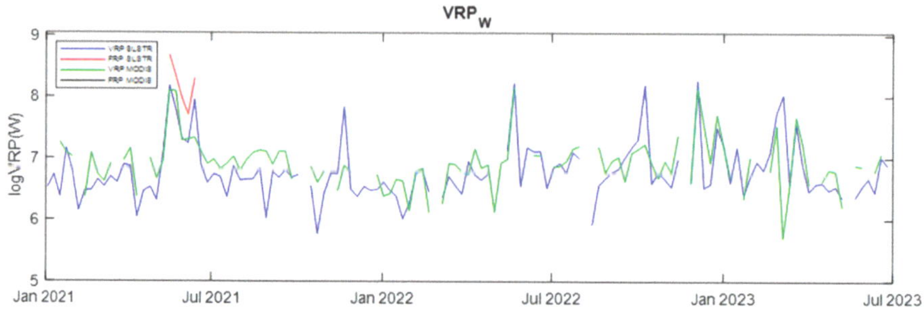

Figure 8. Stacked time series of VRPw (weekly mean) retrieved for SLSTR–RSDF algorithm processed data (blue), SLSTR Level 2 product data (red), MODIS–RSDF algorithm processed data (green), and MODIS active fire products (black) at the Etna volcano, displayed on a logarithmic scale.

4.3. VRE (Volcanic Radiated Energy)

The final volcanic parameter under analysis is the VRE, obtained from the time series of VRP and FRP from both MODIS and SLSTR. Table 3 presents the VRE values (in Joules), allowing for comparisons of eruptions in the different phases identified earlier for both Etna and Stromboli. An initial quantitative comparison between Etna and Stromboli highlights VRE values for Etna that are several orders of magnitude higher than those for Stromboli. This disparity arises from substantial differences in their volcanic activities.

Table 3. Table displaying cumulative total VRE for each Etna phase and Stromboli phase.

	RSDF Algorithm SLSTR	SLSTR Level 2 Product	RSDF Algorithm MODIS	MODIS Level 2 Product
Etna				
15 February 2021–3 April 2021	1.03×10^{10}	2.05×10^{10}	1.21×10^{10}	8.45×10^{6}
17 May 2021–28 September 2021	1.21×10^{10}	1.54×10^{10}	1.58×10^{10}	8.29×10^{7}
12 May 2022–15 June 2022	1.56×10^{10}	2.32×10^{10}	1.08×10^{10}	1.28×10^{7}
27 November 2022–6 February 2023	1.58×10^{10}	1.48×10^{10}	9.46×10^{9}	0
2021–2023	5.94×10^{10}	8.42×10^{10}	5.81×10^{10}	6.86×10^{7}
Stromboli				
13 May 2021–27 May 2021	1.73×10^{9}	2.11×10^{9}	9.02×10^{8}	0
11 May 2022–8 June 2022	2.93×10^{8}	0	2.03×10^{8}	0
27 November 2022–11 December 2022	1.01×10^{9}	0	4.94×10^{8}	0
2021–2023	5.12×10^{9}	2.25×10^{9}	3.01×10^{9}	2.59×10^{8}

Regarding the results of the RSDF algorithm compared to the Level 2 products and MODIS active fire products, it is important to note that during periods between two eruptive activities a constant segment is observed in the Level 2 product, whereas the RSDF algorithm shows a segment that grows linearly before reaching a peak marking the beginning of a new significant volcanic event. For Etna, it is interesting to note that SLSTR VRE values—both for the data processed by the RSDF algorithm and for the Level 2 data—as well as the MODIS data produced by the RSDF algorithm, show similar behavior in quantitative terms. The substantial difference is observed in the graph of the MODIS active fire products, with shows values differing by several orders of magnitude due to the limited amount of data available (Figure 9a,c). In the case of Stromboli, interpreting the comparison is more challenging due to the limited data recorded by the Level 2 products of both sensors. The efficiency of the RSDF algorithm enables tracking the evolution of the volcano's state and distinguishing various significant volcanic events. Separately, in both cases, the results of VRE derived from the MODIS Level 2 product differ significantly from previous values due to lower reliability (Figure 9b,d). The higher VRE values derived from the Level 2 SLSTR products exceed those from RSDF as they are calculated from the VRP time series, highlighting this intensity difference.

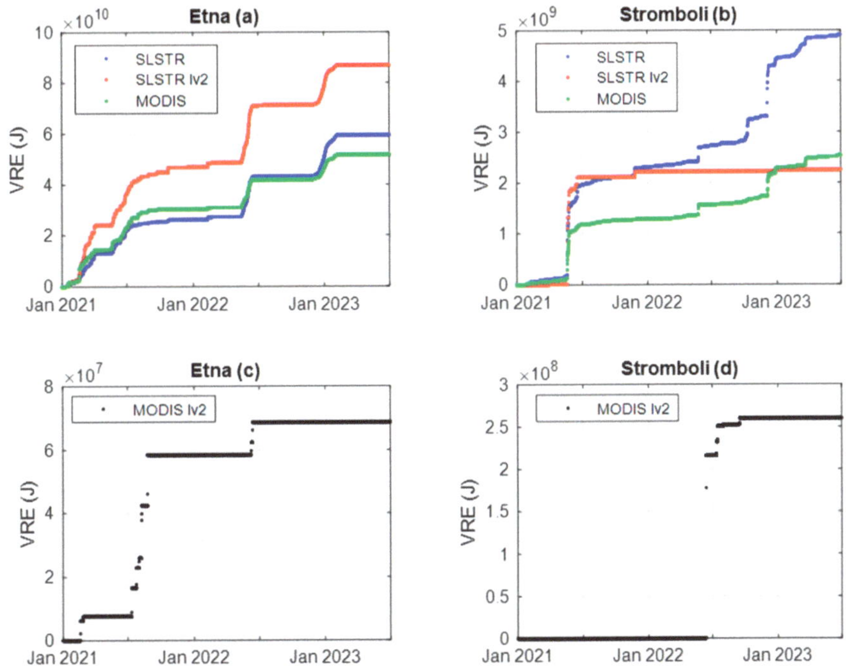

Figure 9. Cumulative Volcanic Radiative Energy (VRE) calculated from VRP (and FRP) using the trapezoidal rule for integration. The blue line represents VRESLSTR, the red dashed line FREMODIS, the green dashed line VREMODIS, and the black dashed line FREMODIS. Panels (**a**,**c**) show data for Etna; panels (**b**,**d**) show data for Stromboli.

5. Discussion

5.1. Performances

The temporal distribution of Etna's VRP demonstrates the RSDF algorithm's remarkable sensitivity in continuous volcanic monitoring over an extended time period. This sensitivity contrasts sharply with the low reliability of observations provided by SLSTR Level 2 products and MODIS active fire products, particularly in the detection of low-intensity thermal anomalies (Figure 2). The consistent results reported above show the algorithm's superior performance in detecting thermal anomalies for the Etna and Stromboli volcanoes when compared to already-processed Level 2 data and MODIS active fire products. Two specific indices, the omissions rate and the false detection rate, were calculated to quantify the algorithm's performance. The omission rate refers to pixels where volcanic activity is present but goes undetected, whereas the false detection rate refers to pixels indicated as true hotspots where no volcanic activity exists. A random set of images was selected from the dataset, and comparisons were conducted on an image by image basis, examining NTI, MIR, and the final mask obtained from the algorithm to identify false or omitted detected pixels. For Etna, using a sample of 200 data points from March 2021 to April 2021 taken from the SLSTR sensor, the RSDF algorithm exhibited a 2.9% omission rate and a 3.6% false detection rate, while the Level 2 product shows a 62% omission rate and a 12% false detection rate. Regarding MODIS data, analyzing a sample of 346 data points from July 2021 to August 2021, the RSDF algorithm showed a 4.5% omission rate and a

4.8% false detection rate, while the MODIS active fire products showed an 89% omission rate and a 10% false detection rate.

In the case of Stromboli, using a sample of 150 data points from June 2021 to August 2022 from the SLSTR sensor dataset, the RSDF algorithm showed a 5.4% omission rate and a 5.1% false detection rate. Conversely, the Level 2 product showed a 90% omission rate and 95% false detection rate. For MODIS data, using a sample of 150 data points from the same period, the RSDF algorithm showed a 5.7% omission rate and a 4.7% false detection rate, while the MODIS active fire products showed a 99% omission rate and a 1% false detection rate (Table 4). The Nalert and Npass parameters were calculated for the entire dataset for both Etna and Stromboli.

Table 4. Comparison between datasets of SLSTR and MODIS from 2021 to 2023 for selected volcanoes.

	False Rate	Omitted Rate	Npass	Nalert (f%)	Mean VRP (MW)	f% Daytime	f% Nightime
Etna							
SLSTR RSDF	3.6%	2.9%	2545	764 (30%)	6.83×10^7	13%	17%
SLSTR Level 2	12%	62%	2545	445 (18%)	5.16×10^7	5%	12%
MODIS RSDF	4.8%	4.5%	5171	858 (17%)	5.44×10^7	10%	7%
MODIS Level 2	10%	89%	5171	17 (0.002%)	3.06×10^7	0.001%	0.001%
Stromboli							
SLSTR RSDF	5.1%	5.4%	2545	605 (24%)	2.16×10^7	17%	7%
SLSTR Level 2	2%	80%	2545	25 (0.01%)	2.36×10^7	0.007%	0.001%
MODIS RSDF	4.7%	5.7%	5171	570 (11%)	6.55×10^6	7%	4%
MODIS Level 2	1%	99%	5171	18 (0.002%)	6.21×10^6	0.002%	0.0%

The anomaly detection frequency is higher for both Etna and Stromboli during daytime as compared to nighttime, primarily due to the influence of solar radiation, which is particularly significant during summer months. The RSDF algorithm demonstrates even greater improvement compared to active fire products for Stromboli. In this case, volcanic products exhibit significant numerical deficiencies, with omission rates of 80% for SLSTR and 99% for MODIS. Generally, a high false rate mainly results from false daytime alerts, often due to solar radiation. However, higher omitted rates tend to balance this out by a reduction in daytime alerts compared to nighttime alerts, given that nighttime thermal gradients are typically higher. SLSTR level 2 products exhibit a significantly higher omitted rate (62%) than false rate (12%), so the overall effect is largely driven by the higher omitted rate. Specifically, SLSTR has a lower overall percentage of Nalert compared to RSDF (18% vs. 30%), with notably higher daytime omissions, having only 5% daytime alerts compared to 13%.

A further classification can be made based on various intensity levels, categorized as low (green), moderate (yellow), high (orange), and extreme (red). Values below the green line belong to the low intensity region, values between the green and the yellow lines belong to the moderate region, values between the yellow and the orange lines are classed as high, and values above the orange line are classified as extreme. The gap between the various levels should be approximately an order of magnitude to distinguish between different volcanic events. These levels are extracted from the datasets using a statistical indicator like the percentile, enabling the comparison of a specific value with the entire matrix. In Figure 10a,b, the radiative power time series is represented in a logarithmic scale, using datasets from the MODIS and SLSTR sensors processed by the RSDF algorithm, for both Etna and Stromboli.

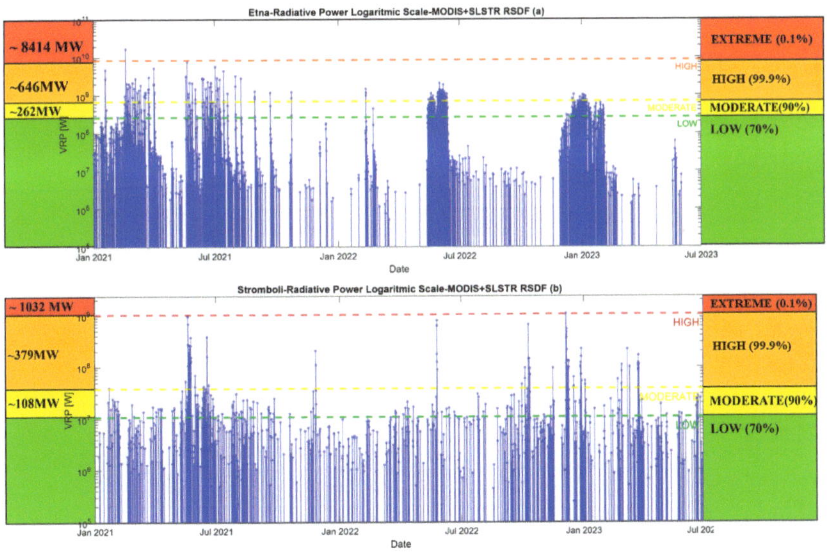

Figure 10. Radiative power time series from SLSTR– and MODIS–RSDF algorithm data with intensity limits categorized as low, moderate, high, and extreme. (**a**) Etna, (**b**) Stromboli.

This classification allows for a clearer understanding of the intensity and frequency of volcanic events, facilitating better risk assessment and decision-making. The use of the RSDF algorithm thus not only enhances detection accuracy but also provides a structured approach to analyzing and categorizing volcanic activities based on their thermal intensity. This data fusion method proves effective for continuous and integrated volcano monitoring in terms of accuracy and the variety of data used, especially for active and complex volcanic regions like Etna and Stromboli.

5.2. Combining SEVIRI, MODIS, SLSTR, and VIIRS

The decision to use several sensors is closely tied to the desire to generate a unique and consistent VRP signal by integrating the findings from the VRP series of individual sensors. This method takes advantage of the various characteristics of the sensors in terms of spatial and temporal resolution. By integrating the results obtained from the RSDF algorithm for VRP, using data from SEVIRI, MODIS, SLSTR, and VIIRS, each with low to medium spatial resolution, it is possible to identify and confirm thermal anomalies across all four sensors. Figure 11a,b shows the time series of above average volcanic activity for Etna, from 1 February 2021 to 30 April 2021, and for Stromboli, from 27 September 2023 to 10 October 2023, respectively.

The dense columns of observable points during significant volcanic events can be attributed to SEVIRI's high temporal resolution, which allows for the monitoring of high-intensity thermal anomalies more frequently than the other sensors.

The adaptability of the RSDF algorithm to different sensor types allows for the derivation of volcanic parameters such as the Time-Averaged Discharge Rate (TADR) and lava flow volume, leveraging the high temporal resolution (every 15 min) of a geostationary sensor like SEVIRI. The volume calculation involves dividing the TADR value by the time difference between consecutive observations and then aggregating it through a cumulative function to yield the total erupted lava volume.

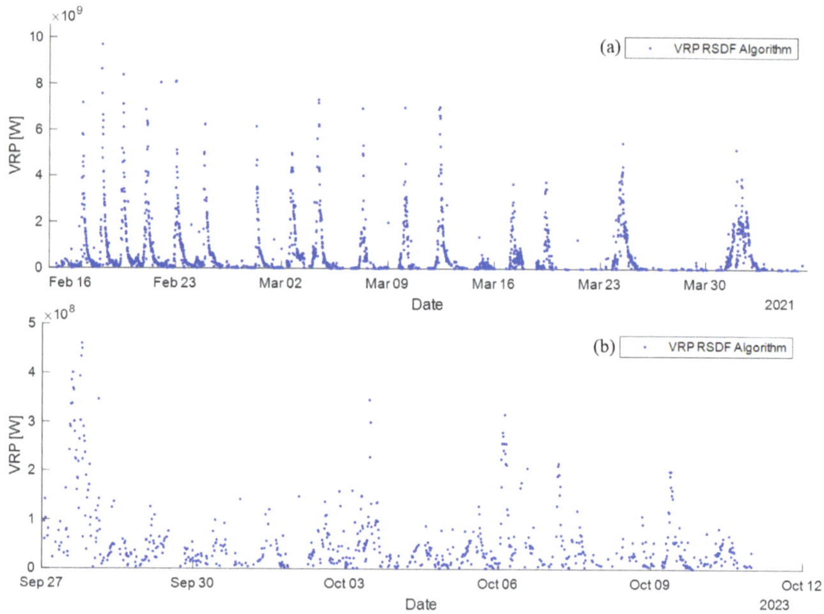

Figure 11. Temporal trend of VRP values derived from the RSDF algorithm for SEVIRI, SLSTR, MODIS, and VIIRS over two periods at Mt. Etna: (**a**) 1 February 2021–30 April 2021, and (**b**) 27 September 2023–10 October 2023.

Cumulative volume curves, derived from various TADR values, exhibit linear growth during the lava flow emplacement, followed by a plateau representing the cooling period between successive flows. For instance, during an effusive event on Mount Etna from 14 May 2022 to 16 June 2022, the volume of erupted lava stabilized at around 680,197 m^3 (Figure 12).

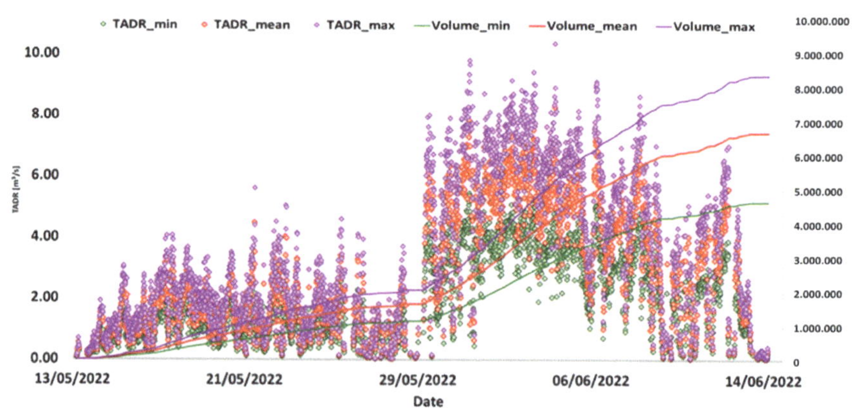

Figure 12. TADR and lava flow volume flux during the effusive event at Etna from 14 May 2022 to 16 June 2022. TADR_max, TADR_mean, and TADR_min are represented by blue, red, and green points, respectively. The total volume_max, volume_mean, and volume_min are represented by blue, red, and green lines, respectively.

To evaluate the adaptability of the RSDF algorithm through the integration of an additional sensor like SEVIRI, we examined a recent effusive eruption on Stromboli from 25 September 2023 to 10 October 2023. This event was used to compute TADR and the volume of lava flows, quantifying the size and magnitude of the volcanic events. The VRP flow analysis identified five distinct volcanic events, each linked to a corresponding lava flow. Notably, the initial event on September 27 at 14:30 UTC significantly contributed to the total erupted volume, constituting 45% of the maximum volume (112.6 m^3/s out of 251.3 m^3/s), the mean volume (56.3 m^3/s out of 125.7 m^3/s), and the minimum volume (37.3 m^3/s out of 83.8 m^3/s) (Figure 13). The limitations requiring greater attention are related to factors of uncertainty that negatively influence the accurate detection of thermal anomalies. Factors such as clouds and satellite viewing angles necessitate careful validation of each individual image. Upcoming developments will include comparing our system with other existing systems to evaluate different data validation methodologies.

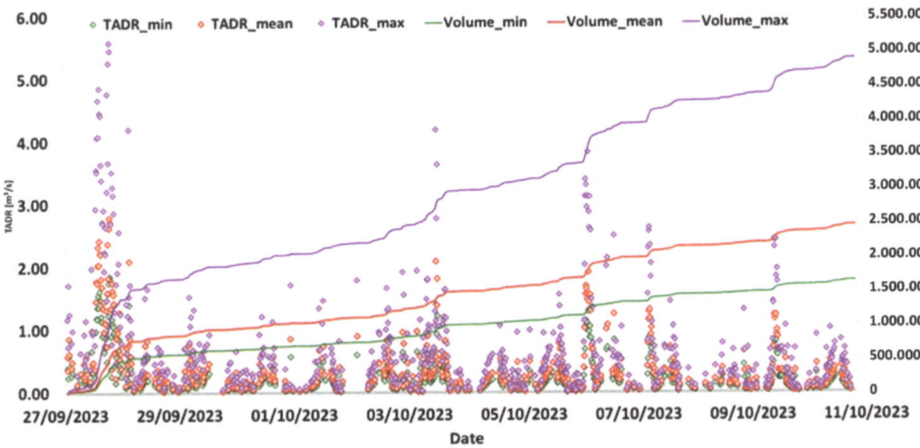

Figure 13. TADR and lava flow volume flux during the effusive event at Stromboli from 27 September 2023 to 10 October 2023. TADR_max, TADR_mean, and TADR_min are represented by blue, red, and green points, respectively. The total volume_max, volume_mean, and volume_min are represented by blue, red, and green lines, respectively.

6. Conclusions

Effective monitoring of active volcanoes requires the use of multiple sensors and accurate hotspot detection algorithms. In this study we demonstrate several key points:

Importance of SLSTR for Volcanic Monitoring: SLSTR, typically not used for volcanic activity monitoring, has proven to be an important sensor due to its ability to detect thermal anomalies effectively.

Enhanced Detection Algorithm: The proposed detection algorithm, which incorporates the Gabor filter on medium spatial resolution satellite sensors, significantly improves sensitivity and reduces false positives. This improvement is crucial for accurate and reliable volcanic monitoring.

Versatility Across Sensors: The detection algorithm's capacity to be applied uniformly across both polar and geostationary sensors showcases its versatility. This uniform application ensures consistent monitoring regardless of the sensor's specific characteristics.

The analysis conducted on the Etna and Stromboli volcanoes from 2021 to 2023 highlighted several improvements. Specifically, the overall accuracy increased significantly, with the omission rate dropping from 75.5% to 3.7% and the false detection rate decreasing from 11% to 4.3%. Additionally, intensity levels based on the logarithmic scale of VRP time series were calculated. These intensity levels are instrumental in pinpointing the volcano's activity status at any given moment.

We underscore the necessity of not only relying on individual sensors but also of integrating information from multiple sensors with different characteristics. By doing so, it is possible to address each sensor's weaknesses and leverage their strengths, depending on their spatial and temporal resolutions. This integrative approach enables the monitoring of hotspots characterized by both high and medium-low intensity, which are particularly challenging for sensors with low spatial resolution.

Finally, the implementation of a multi-sensor approach combined with an enhanced detection algorithm significantly advances the capability to monitor and analyze volcanic activity. This method provides an accurate satellite remote sensing data fusion algorithm for volcano thermal monitoring, essential for timely warnings and forecasting volcanic hazards. The fusion of polar and geostationary data is crucial for advancing volcanic monitoring and ensuring better preparedness for volcanic events. The study demonstrates that leveraging various sensors and sophisticated algorithms is key to improving volcanic monitoring and preparedness. Future developments will involve expanding the number of volcanoes investigated to enable global volcanic monitoring and exploring super resolution techniques.

Author Contributions: Conceptualization, G.S.D.B. and C.C.; methodology, G.S.D.B. and C.C.; software, G.S.D.B., S.C., F.T. and C.C.; validation, S.C. and F.T.; formal analysis, S.C. and F.T.; data curation, G.S.D.B. and C.C.; writing—original draft preparation, G.S.D.B. and C.C.; writing—review and editing, C.D.N., S.C. and F.T.; visualization, G.S.D.B.; supervision, C.D.N.; project administration, C.D.N.; funding acquisition, C.D.N. All authors have read and agreed to the published version of the manuscript.

Funding: This research was funded by ATHOS Research Programme (INGV OB.FU. 0867.010), by the 2019 Strategic Project FIRST—ForecastIng eRuptive activity at Stromboli volcano: timing, eruptive style, size, intensity, and duration—of the INGV Volcanoes Department (Delibera n. 144/2020), and by the Project INGV Pianeta Dinamico VT_ORME 2023–2025 (INGV OB.FU. 1020.010).

Data Availability Statement: All SLSTR, MODIS, VIIRS, and SEVIRI data utilized in this work are available on the Google Earth Engine.

Acknowledgments: This work was developed within the framework of the Laboratory of Technologies for Volcanology (TechnoLab) at the INGV in Catania (Italy). We are grateful to the European Space Agency (ESA), the Italian Space Agency (ASI), and the National Aeronautics and Space Administration (NASA) for satellite data. Note: Google CoLaboratory™ is a trademark of Google LLC—©2018 Google LLC All rights reserved.

Conflicts of Interest: The authors declare no conflicts of interest. The funders had no role in the design of the study; in the collection, analyses, or interpretation of data; in the writing of the manuscript; or in the decision to publish the results.

Abbreviations

FRP	Fire Radiative Power
MIR	Medium Infrared
MODIS	Moderate Resolution Imaging Spectroradiometer
NTI	Normal Thermal Index
NVA	Non-Volcanic Area
RSDF	Remote Sensing Data Fusion
SEVIRI	Spinning Enhanced Visible And Infrared Imager
SLSTR	Sea And Land Surface Temperature Radiometer
SSD	Spatial Standard Deviation
TIR	Thermal Infrared
VA	Volcanic Area
VRE	Volcanic Radiative Energy
VRP	Volcanic Radiative Power

References

1. Corradino, C.; Malaguti, A.B.; Ramsey, M.S.; Del Negro, C. Quantitative Assessment of Volcanic Thermal Activity from Space Using an Isolation Forest Machine Learning Algorithm. *Remote Sens.* **2024**, *16*, 2001. [CrossRef]
2. Torrisi, F.; Corradino, C.; Cariello, S.; Del Negro, C. Enhancing detection of volcanic ash clouds from space with convolutional neural networks. *J. Volcanol. Geotherm. Res.* **2024**, *448*, 108046. [CrossRef]
3. Coppola, D.; Cardone, D.; Laiolo, M.; Aveni, S.; Campus, A.; Massimetti, F. Global radiant flux from active volcanoes: The 2000–2019 MIROVA database. *Front. Earth Sci.* **2023**, *11*, 1240107. [CrossRef]
4. Williams, D.B.; Ramsey, M.S. Infraredspectroscopy of volcanoes: From laboratoryto orbital scale. *Front. Earth Sci.* **2024**, *12*, 1308103. [CrossRef]
5. Mia, M.B.; Fujimitsu, Y.; Nishijima, J. Thermal activity monitoring of an active volcano using Landsat 8/OLI-TIRS sensor images: A case study at the Aso volcanic area in southwest Japan. *Geosciences* **2017**, *7*, 118. [CrossRef]
6. Corradino, C.; Amato, E.; Torrisi, F.; Del Negro, C. Data-Driven Random Forest Models for Detecting Volcanic Hot Spots in Sentinel-2 MSI Images. *Remote Sens.* **2022**, *14*, 4370. [CrossRef]
7. Marchese, F.; Genzano, N.; Nolde, M.; Falconieri, A.; Pergola, N.; Plank, S. Mapping and characterizing the Kīlauea (Hawai'i) lava lake through Sentinel-2 MSI and Landsat-8 OLI observations of December 2020–February 2021. *Environ. Model. Softw.* **2022**, *148*, 105273. [CrossRef]
8. Del Negro, C.; Cappello, A.; Neri, M.; Bilotta, G.; Herault, A.; Ganci, G. Lava flow hazards at Mount Etna: Constraints imposed by eruptive history and numerical simulations. *Sci. Rep.* **2013**, *3*, 3493. [CrossRef]
9. Torrisi, F.; Amato, E.; Corradino, C.; Mangiagli, S.; Del Negro, C. Characterization of Volcanic Cloud Components Using Machine Learning Techniques and SEVIRI Infrared Images. *Sensors* **2022**, *22*, 7712. [CrossRef]
10. Ramsey, M.S.; Harris, A.J.L. Volcanology 2020: How will thermal remote sensing of volcanic surface activity evolve over the next decade? *J. Volcanol. Geotherm. Res.* **2013**, *249*, 217–233. [CrossRef]
11. Steffke, A.M.; Harris, A.J.L. A review of algorithms for detecting volcanic hot spots in satellite infrared data. *Bull. Volcanol.* **2011**, *73*, 1109–1137. [CrossRef]
12. Tramutoli, V.; Cuomo, V.; Filizzola, C.; Pergola, N.; Pietrapertosa, C. Assessing the potential of thermal infrared satellite surveys for monitoring seismically active areas: The case of Kocaeli (İzmit) earthquake, August 17, 1999. *Remote Sens. Environ.* **2005**, *96*, 409–426. [CrossRef]
13. Wright, R.; Flynn, L.; Garbeil, H.; Harris, A.; Pilger, E. MODVOLC: Near-real-time thermal monitoring of global volcanism. *J. Volcanol. Geotherm Res.* **2004**, *135*, 29–49. [CrossRef]
14. Higgins, J.; Harris, A. VAST: A program to locate and analyse volcanic thermal anomalies automatically from remotely sensed data. *Comput. Geosci.* **1997**, *23*, 627–645. [CrossRef]
15. Tramutoli, V. Robust AVHRR techniques (RAT) for environmental monitoring: Theory and applications. *Proc. SPIE Earth Surf. Remote Sens. II* **1998**, *3496*, 101–113. [CrossRef]
16. Wright, R.; Flynn, L.; Garbeil, H.; Harris, A.; Pilge, E. Automated volcanic eruption detection using MODIS. *Remote Sens. Environ.* **2002**, *82*, 135–155. [CrossRef]
17. Dehn, J.; Dean, K.; Engle, K. Thermal Monitoring of North Pacific Volcanoes from Space. *Geology* **2000**, *28*, 755–758. [CrossRef]
18. Genzano, N.; Pergola, N.; Marchese, F. A Google Earth Engine Tool to Investigate, Map and Monitor Volcanic Thermal Anomalies at Global Scale by Means of Mid-High Spatial Resolution Satellite Data. *Remote Sens.* **2020**, *12*, 3232. [CrossRef]
19. Falconieri, A.; Genzano, N.; Mazzeo, G.; Pergola, N.; Marchese, F. First Implementation of a Normalized Hotspot Index on Himawari-8 and GOES-R Data for the Active Volcanoes Monitoring: Results and Future Developments. *Remote Sens.* **2022**, *14*, 5481. [CrossRef]
20. Ganci, G.; Vicari, A.; Del Negro, C. The HOTSAT Volcano Monitoring System Based on a Combined Use of SEVIRI and MODIS Multispectral Data. *Ann. Geophys.* **2011**, *54*, 5338. [CrossRef]
21. Ganci, G.; Cappello, A.; Bilotta, G.; Del Negro, C. How the variety of satellite remote sensing data over volcanoes can assist hazard monitoring efforts: The 2011 eruption of Nabro volcano. *Remote Sens. Environ.* **2020**, *236*, 111426. [CrossRef]
22. Corradino, C.; Ganci, G.; Bilotta, G.; Cappello, A.; Del Negro, C.; Fortuna, L. Smart Decision Support Systems for Volcanic Applications. *Energies* **2019**, *12*, 1216. [CrossRef]
23. Coppola, D.; Laiolo, M.; Cigolini, C.; Massimetti, F.; Delle Donne, D.; Ripepe, M.; Arias, H.; Barsotti, S.; Parra, C.B.; Centeno, R.G.; et al. Thermal remote sensing for global volcano monitoring experiences from the Mirova system. *Front. Earth Sci.* **2020**, *7*, 362. [CrossRef]
24. Gouhier, M.; Guéhenneux, Y.; Labazuy, P.; Cacault, P.; Decriem, J.; Rivet, S. HOTVOLC: A web-based monitoring system for volcanic hot spots. *Geol. Soc. Lond. Spec. Publ.* **2016**, *426*, 223–241. [CrossRef]
25. Torrisi, F.; Amato, E.; Corradino, C.; Del Negro, C. The FastVRP automatic platform for the thermal monitoring of volcanicactivity using VIIRS and SLSTR sensors: FastFRP to monitor volcanic radiative power. *Ann. Geophys.* **2023**, *65*, 1.
26. Wooster, M. Fire radiative energy for quantitative study of biomass burning derivation from the bird experimental satellite and comparison to MODIS fire products. *Remote Sens Environ.* **2003**, *86*, 83–107. [CrossRef]

27. Xu, W.; Wooster, M.J. Sentinel-3 SLSTR active fire (AF) detection and FRP daytime product-Algorithm description and globalintercomparison to MODIS, VIIRS and landsat AF data. *Sci. Remote Sens.* **2023**, *7*, 100087. [CrossRef]
28. Coppola, D.; Laiolo, M.; Cigolini, C.; Delle Donne, D.; Ripepe, M. Enhanced volcanic hot-spot detection using MODIS IR data: Results from the MIROVA system. *Geol. Soc. Spec. Publ.* **2016**, *426*, 181–205. [CrossRef]
29. Campus, A.; Aveni, S.; Laiolo, M.; Coppola, D. Thermal unrest at La Fossa (Vulcano Island, Italy): The 2021–2023 VIIRS 375 m MIROVA-processed dataset. *Bull. Volcanol.* **2024**, *86*, 25. [CrossRef]
30. Cariello, S.; Corradino, C.; Torrisi, F.; Del Negro, C. Cascading Machine Learning to Monitor Volcanic Thermal Activity Using Orbital Infrared Data: From Detection to Quantitative Evaluation. *Remote Sens.* **2024**, *16*, 171. [CrossRef]
31. Ramsey, M.; Corradino, C.; Thompson, J.; Leggett, T.N. Statistical retrieval of volcanic activity in long time series orbital data: Implications for forecasting future activity. *Remote Sens. Environ.* **2023**, *295*, 113704. [CrossRef]
32. Corradino, C.; Ramsey, M.S.; Pailot-Bonnetat, S.; Harris, A.J.L.; Del Negro, C. Detection of Subtle Thermal Anomalies: Deep Learning Applied to the ASTER Global Volcano Dataset. *IEEE Trans. Geosci. Remote Sens.* **2023**, *61*, 5000715. [CrossRef]
33. De Beni, E.; Proietti, C.; Scollo, S.; Cantarero, M.; Mereu, L.; Romeo, F.; Pioli, L.; Sciotto, M.; Alparone, S. A Hidden Eruption: The 21 May 2023 Paroxysm of the Etna Volcano (Italy). *Remote Sens.* **2024**, *16*, 1555. [CrossRef]
34. Calvari, S.; Nunnari, G. Statistical Insights on the Eruptive Activity at Stromboli Volcano (Italy) Recorded from 1879 to 2023. *Remote Sens.* **2023**, *15*, 4822. [CrossRef]
35. Wooster, M.J.; Xu, W.; Nightingale, T. Sentinel-3 SLSTR Active Fire Detection and FRP Product: Pre-Launch Algorithm Development and Performance Evaluation Using MODIS and ASTER Datasets. *Remote Sens. Environ.* **2012**, *120*, 236–254. [CrossRef]
36. Justice, C.O.; Giglio, L.; Korontzi, S.; Owens, J.; Morisette, J.T.; Roy, D.; Descloitres, J.; Alleaume, S.; Petitcolin, F.; Kaufman, Y. The MODIS Fire Products. *Remote Sens. Environ.* **2002**, *83*, 244–262. [CrossRef]
37. Del Negro, C.; Cappello, A.; Ganci, G. Quantifying lava flow hazards in response to effusive eruption. *Geol. Soc. Am. Bull.* **2016**, *128*, 752–763. [CrossRef]
38. Del Negro, C.; Cappello, A.; Bilotta, G.; Ganci, G.; Hérault, A.; Zago, V. Living at the Edge of an Active Volcano: Risk from Lava Flows on Mt Etna. *GSA Bull.* **2020**, *132*, 1615–1625. [CrossRef]
39. Calvari, S.; Nunnari, G. Comparison between automated and manual detection of lava fountains from fixed monitoring thermalcameras at Etna volcano, Italy. *Remote Sens.* **2022**, *14*, 2392. [CrossRef]
40. Marchese, F.; Filizzola, C.; Lacava, T.; Falconieri, A.; Faruolo, M.; Genzano, N.; Mazzeo, G.; Pietrapertosa, C.; Pergola, N.; Tramutoli, V.; et al. Mt. Etna paroxysms of February–April 2021 monitored and quantified through a multi-platform satellite observing system. *Remote Sens.* **2021**, *13*, 3074. [CrossRef]
41. Guerrieri, L.; Corradini, S.; Theys, N.; Stelitano, D.; Merucci, L. Volcanic Clouds Characterization of the 2020–2022 Sequence of Mt. Etna Lava Fountains Using MSG-SEVIRI and Products' Cross-Comparison. *Remote Sens.* **2023**, *15*, 2055. [CrossRef]
42. Métrich, N.; Bertagnini, A.; Pistolesi, M. Paroxysms at Stromboli Volcano (Italy): Source, Genesis and Dynamics. *Front. Earth Sci.* **2021**, *9*, 593339. [CrossRef]
43. Corradino, C.; Amato, E.; Torrisi, F.; Calvari, S.; Del Negro, C. Classifying Major Explosions and Paroxysms at Stromboli Volcano (Italy) from Space. *Remote Sens.* **2021**, *13*, 4080. [CrossRef]
44. Plank, S.; Marchese, F.; Filizzola, C.; Pergola, N.; Neri, M.; Nolde, M.; Martinis, S. The July/August 2019 lava flows at the Sciara Del Fuoco, Stromboli—Analysis from multi-sensor infrared satellite imagery. *Remote Sens.* **2019**, *11*, 2879. [CrossRef]
45. Giudicepietro, F.; Calvari, S.; De Cesare, W.; Di Lieto, B.; Di Traglia, F.; Esposito, A.M.; Orazi, M.; Romano, P.; Tramelli, A.; Nolesini, T.; et al. Seismic and thermal precursors of crater collapses and overflows at Stromboli volcano. *Sci. Rep.* **2023**, *13*, 11115. [CrossRef]
46. Calvari, S.; Di Traglia, F.; Ganci, G.; Bruno, V.; Ciancitto, F.; Di Lieto, B.; Gambino, S.; Garcia, A.; Giudicepietro, F.; Inguaggiato, S.; et al. Multi-parametric study of an eruptive phase comprising unrest, major explosions, crater failure, pyroclastic density currents and lava flows: Stromboli volcano, 1 December 2020–30 June 2021. *Front. Earth Sci.* **2022**, *10*, 899635. [CrossRef]
47. EUMETSAT. Available online: https://data.eumetsat.int/ (accessed on 30 November 2023).
48. LAADS-DAAC. Available online: https://ladsweb.modaps.eosdis.nasa.gov/ (accessed on 18 January 2024).
49. COPERNICUS. Available online: https://sentinels.copernicus.eu/documents/247904/0/SLSTR_Level-2_Fire_ATBD.pdf/ (accessed on 1 November 2023).
50. Giglio, L.; Descloitres, J.; Justice, C.O.; Kaufman, Y. An enhanced contextual fire detection algorithm for MODIS. *Remote Sens. Environ.* **2003**, *87*, 273–282. [CrossRef]
51. Giglio, L.; Schroeder, W.; Justice, C.O. The Collection 6 MODIS active fire detection algorithm and fire products. *Remote Sens. Environ.* **2016**, *178*, 31–41. [CrossRef]
52. Coppola, D.; Laiolo, M.; Piscopo, D.; Cigolini, C. Rheological control on the radiant density of active lava flows and domes. *J. Volcan. Geotherm. Res.* **2013**, *249*, 39–48. [CrossRef]
53. Ganci, G.; Vicari, A.; Cappello, A.; Del Negro, C. An emergent strategy for volcano hazard assessment: From thermal satellite monitoring to lava flow modeling. *Remote Sens. Environ.* **2012**, *119*, 197–207. [CrossRef]
54. Plank, S.; Massimetti, F.; Soldati, A.; Hess, K.U.; Nolde, M.; Martinis, S.; Dingwell, D.B. Estimates of lava discharge rate of 2018 Kīlauea Volcano, Hawaiʻi eruption using multi-sensor satellite and laboratory measurements. *Int. J. Remote Sens.* **2021**, *42*, 1492–1511. [CrossRef]

55. Harris, A.J.L.; Steffke, A.; Calvari, S.; Spampinato, L. Thirty years of satellite-derived lava discharge rates at Etna: Implications for steady volumetric output. *J. Geophys. Res.–Solid Earth* **2011**, *116*, B8. [CrossRef]
56. Campus, A.; Laiolo, M.; Massimetti, F.; Coppola, D. The Transition from MODIS to VIIRS for Global Volcano Thermal Monitoring. *Sensors* **2022**, *22*, 1713. [CrossRef] [PubMed]

remote sensing

MDPI

Article

Refined Intelligent Landslide Identification Based on Multi-Source Information Fusion

Xiao Wang [1,2], Di Wang [3], Chenghao Liu [4], Mengmeng Zhang [4], Luting Xu [1], Tiegang Sun [5], Weile Li [6], Sizhi Cheng [7] and Jianhui Dong [1,*]

1 School of Architecture and Civil Engineering, Chengdu University, Chengdu 610106, China; wangxiao@cdu.edu.cn (X.W.); xuluting@cdu.edu.cn (L.X.)
2 Key Laboratory of Earth Exploration and Information Techniques, Ministry of Education, Chengdu University of Technology, Chengdu 610059, China
3 The State Key Laboratory of Information Engineering in Surveying, Mapping and Remote Sensing, Wuhan University, Wuhan 430079, China; wd10486@whu.edu.cn
4 College of Earth Sciences, Chengdu University of Technology, Chengdu 610059, China; 2023020190@stu.cdut.edu.cn (C.L.); zhangmengmeng@stu.cdut.edu.cn (M.Z.)
5 China Building Materials Southwest Survey and Design Co., Ltd., Chengdu 610052, China; stg@cnbm.com.cn
6 State Key Laboratory of Geohazard Prevention and Geoenvironment Protection, Chengdu University of Technology, Chengdu 610059, China; liweile08@mail.cdut.edu.cn
7 Sichuan Earthquake Agency, Chengdu 610041, China; csz7895@scdzj.gov.cn
* Correspondence: dongjianhui@cdu.edu.cn

Abstract: Landslides are most severe in the mountainous regions of southwestern China. While landslide identification provides a foundation for disaster prevention operations, methods for utilizing multi-source data and deep learning techniques to improve the efficiency and accuracy of landslide identification in complex environments are still a focus of research and a difficult issue in landslide research. In this study, we address the above problems and construct a landslide identification model based on the shifted window (Swin) transformer. We chose Ya'an, which has a complex terrain and experiences frequent landslides, as the study area. Our model, which fuses features from different remote sensing data sources and introduces a loss function that better learns the boundary information of the target, is compared with the pyramid scene parsing network (PSPNet), the unified perception parsing network (UPerNet), and DeepLab_V3+ models in order to explore the learning potential of the model and test the models' resilience in an open-source landslide database. The results show that in the Ya'an landslide database, compared with the above benchmark networks (UPerNet, PSPNet, and DeepLab_v3+), the Swin Transformer-based optimization model improves overall accuracies by 1.7%, 2.1%, and 1.5%, respectively; the F1_score is improved by 14.5%, 16.2%, and 12.4%; and the intersection over union (IoU) is improved by 16.9%, 18.5%, and 14.6%, respectively. The performance of the optimized model is excellent.

Keywords: landslide intelligent identification; Swin Transformer; multi-source information; PSPNet; UPerNet; DeepLab_V3+

Citation: Wang, X.; Wang, D.; Liu, C.; Zhang, M.; Xu, L.; Sun, T.; Li, W.; Cheng, S.; Dong, J. Refined Intelligent Landslide Identification Based on Multi-Source Information Fusion. *Remote Sens.* **2024**, *16*, 3119. https://doi.org/10.3390/rs16173119

Academic Editors: Fumio Yamazaki and Michele Saroli

Received: 19 June 2024
Revised: 17 August 2024
Accepted: 22 August 2024
Published: 23 August 2024

1. Introduction

Landslides are among the most impactful types of geological disasters worldwide, posing serious threats to the lives and property of people everywhere due to their suddenness and uncertainty. In particular, China is significantly affected by geological disasters [1]. According to the China Statistical Yearbook (https://www.stats.gov.cn/sj/ndsj/2023/indexch.htm, accessed on 11 May 2024) released by the National Bureau of Statistics (NBS), a total of 5659 geological disasters occurred in 2022, of which 3919 were landslides, accounting for 69.2% of these disasters [2]. In this paper, the term "landslide" refers specifically to landslides, where rock and soil bodies slide along a certain sliding surface as an entire piece of mass under the influence of natural or anthropogenic factors. The economic losses caused by the landslides that occurred in 2022 amounted to CNY 1502.81 million. In order to reduce the hazards posed by landslide disasters, governmental disaster reduction departments have organized landslide censuses and detailed investigations at different scales and with different accuracies, and they have identified numerous potential landslide disaster sites, thereby providing an important foundation for landslide disaster reduction efforts [3]. To date, landslide surveys and detailed investigations have mainly been conducted utilizing remote sensing interpretation and field trips. The identification of landslides through fieldwork is not only time-consuming and laborious, but it is also limited by the vegetation on the surface of the slope and the professional ability of the inspectors [4].

Remote sensing, with its outstanding advantages of large-scale synchronous observation and cost-effectiveness, is being increasingly used for the investigation of geological disasters. Although the identification of landslides using remote sensing interpretation methods has greatly improved in terms of efficiency, existing remote sensing-based landslide interpretation methods are primarily visual interpretation methods (i.e., involving the interpretation of visual aerial and satellite remote sensing imagery and manual visual interpretation with the help of some auxiliary equipment); however, these methods not only have a low interpretation efficiency but also a high rate of omission. This is also one of the reasons why many landslide hazards still exist, even though many landslide censuses and detailed investigations have been conducted [5]. In this context, the use of rapidly developing image-processing technology in the construction of an automatic landslide identification model based on remote sensing imagery in order to improve the efficiency and accuracy of landslide identification can be considered a more effective method for landslide identification, providing a basis for landslide disaster mitigation [6]. Landslide identification based on remote sensing images has a long history, where most sample objects selected for landslide identification in previous studies were seismic-type landslides; in this context, the vegetation overlying the landslide mass is destroyed, and the spectral characteristics of the ground surface significantly differ from those of normal ground surfaces. These landslides are relatively easier to recognize, and computer recognition technology for this type of landslide based on remote sensing imagery is relatively mature [7]. However, the spectral characteristics of most landslides are not significant in that the spectral characteristics and topographic features should be integrated to identify landslides at different stages of development. Landslides change the original surface structure and destroy the original topography, drainage, and vegetation conditions. These features are very prominent in remote sensing images, providing automatic identification possibilities of landslides based on remote sensing. Mapping the process of the visual interpretation of landslides by professionals to the computational space is key to solving the automatic landslide identification problem; in particular, it is necessary to incorporate other feature factors besides landslide images in the process of landslide identification.

In recent years, deep learning technology has realized remarkable achievements in remote sensing image-processing applications [8]. Deep learning, as a derivative field of artificial intelligence, involves the use of deeper neural networks. A core goal of such approaches is to present image data as multi-level associative structures, which can be represented as multi-scale semantic feature associations, gradually linking local features to global features and thus realizing a high-level feature representation of low-level features. In semantic segmentation, high-level feature expressions can represent semantic information that guides the segmentation process [9]. Deep learning does not require the artificial extraction of features because—through the model's own learning ability—the essential features and intrinsic laws in image data can be gradually extracted, providing good adaptability. After iteratively adjusting the model parameters, the model is able to better learn and construct high-level feature structures of the image from the training image samples in order to obtain optimal segmentation performance results. As a result, a semantic segmentation model that can be automated and adapted to a given task is obtained. In the field of semantic segmentation, many techniques have been found to further enhance the expression of advanced semantic information, thus improving the segmentation accuracy of models [10–12]. The transformer approach has achieved remarkable results in the translation task. Connecting the encoder and decoder through an attention mechanism, it is able to efficiently direct attention to the important positions in the input sequence, thus achieving better performance than some CNN networks. The first transformer to be applied to computer vision tasks was the vision transformer (ViT), and numerous experiments have shown that this constitutive converter structure achieves feature extraction results comparable to those of CNNs [13,14]. ViT uses the same image facet size at each stage in the process of extracting image features. This processing method ignores feature information at different scales in high-resolution remote sensing images. The Swin Transformer algorithm was optimized on the basis of ViT, incorporating a new pyramid-stage processing method and a sliding window to realize efficient transformer computation.

For this study, we chose Ya'an City, Sichuan Province, as the study area and constructed a landslide sample dataset for the study area from 2021 to 2023. As landslide hazards account for a relatively small proportion of the study area, in order to address the problem of low recognition accuracy for a few target classes, we effectively improved the landslide hazard recognition accuracy by fusing the features of multi-source data and introducing the boundary loss function.

The main contributions of this study are as follows:

(1) We constructed a sample library for landslide recognition in Ya'an City with multi-source features.
(2) We analyzed the learning process of the deep learning model on multi-source feature samples using a heat map.
(3) We constructed a Swin Transformer-based model based on landslide recognition and introduced a boundary loss function to address the problem of low recognition accuracy for a few target classes. We compared our model with classical networks, and it was observed that, after multi-source feature fusion, the model effectively solved the abovementioned problems and improved landslide recognition accuracies.
(4) We explored a potential application of the Swin Transformer-based model in landslide detection. Based on the Bijie landslide dataset, the expressiveness of the Swin Transformer-based model based on the boundary constraint function and multi-source feature fusion was explored in order to validate its effectiveness and explore its generalization ability.

2. Materials

2.1. Study Area

Ya'an City is located in the western part of Sichuan Province (1019°56′–102°28′E, 28°51′–30°56′N). As Ya'an is located in an area characterized by mountainous terrain, the terrain is complicated by a great variation in altitude, resulting in an evident gradual decrease in temperature with increasing altitude. It has a high annual precipitation and heavy rainfall in the summer; hence, it is named "Rainy City." The crust of this region is in active motion, and the fracture zones are widely distributed. These active fracture zones not only affect the stability of the regional crust but may also trigger geological disasters, such as landslides. The city has jurisdiction over Yucheng, Meishan, Xingjing, Hanyuan, Shimian, Tianquan, Lushan, and Songxing, which include two districts and six counties (Figure 1) [15].

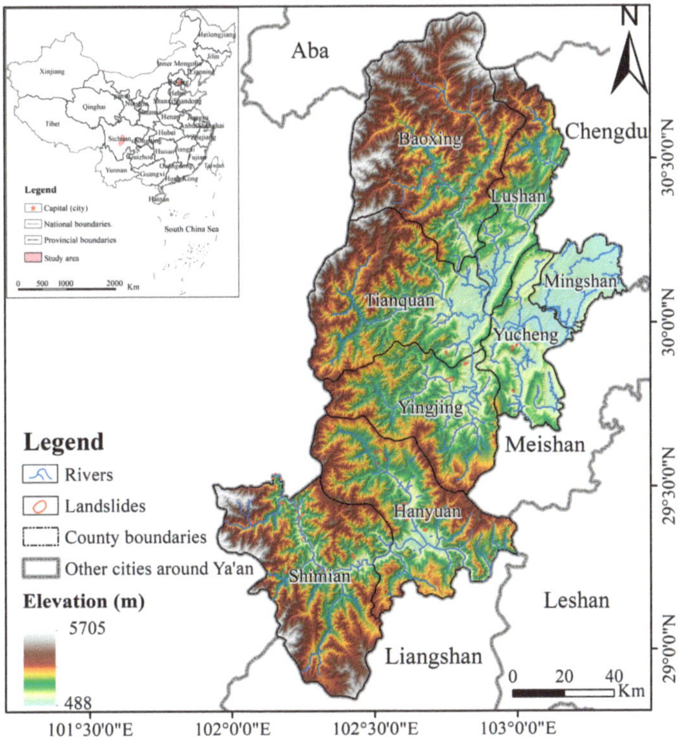

Figure 1. Distribution of landslides in Ya'an (2021–2023).

2.2. *Landslide Identification Database*

2.2.1. Visual Interpretation of Landslides

For this study, a detailed landslide hazard interpretation was completed in Ya'an based on field landslide site survey data for 2021–2023 from the Ya'an Bureau of Natural Resources and Planning, as well as high-resolution Google Earth imagery. The steps were as follows: 1. We overlaid the known landslide sites with Google Earth images of Ya'an, and the location and extent of the landslides were initially determined; 2. Staff experienced in landslide interpretation carefully outlined the landslide boundaries according to the texture and color characteristics of each landslide on the images, and some landslides that could not be identified in the images were removed. The detailed landslide interpretation results are shown in Figure 2.

Some of the landslide field verification images are shown in Figure 3.

2.2.2. Database Production for Multi-Source Data

Optical satellite imagery and a digital elevation model (DEM) were used to compile a multi-source database for landslide identification. The historical optical satellite images were provided by Google Earth. The images displayed on Google Earth integrate various data such as aerial photos, satellite images, remote sensing, and a geographic information system (GIS), which are spliced together according to the geographic coordinates and certain mathematical rules, resulting in rich imagery, comprehensive information, and a balance between data accuracy and display realism. The optical images used in this study were captured in 2023, and they have excellent image quality, no cloud cover, and a ground resolution of 2.15 m. In the model training and recognition phase, DEM data were added as the first dimension of the data input, in addition to the red, green, and blue (RGB) satellite image information used to provide auxiliary data for establishing a high-precision landslide intelligent recognition model and to further improve the recognition accuracy of the network. DEM data were obtained from the 1:10,000 DEM of Ya'an City with an accuracy of 5 m, and the range was consistent with that of the images. All data were mapped to the WGS_1984_UTM_Zone_48N coordinate system.

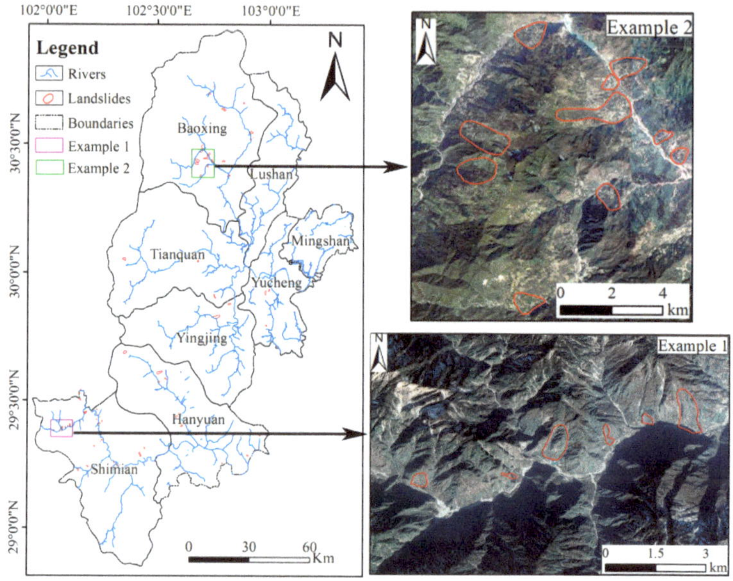

Figure 2. Interpretation of landslides in Ya'an (2021–2023).

Figure 3. Field validation of landslides (2021–2023).

As the size and extent of each landslide in the images differed, the edges of each landslide were extended in all directions when the landslide samples were cropped, and a variety of backgrounds were added to the landslide samples using the outward cropping operation, which helped the network model to learn the unique features of the landslides themselves without interference from the backgrounds. In this study, a total of 108 landslide hazards were identified, and 306 positive samples of landslide instances were produced by combining historical images. Examples of landslide instances are shown in Figure 4.

Figure 4. Images of some landslides in Ya'an (2021–2023).

Sample points identified as non-landslide features were manually selected as negative samples on the satellite images, including non-landslide samples such as mountains, villages, roads, rivers, forests, and farmland. A total of 306 non-landslide samples were used in this study.

Applying a deep learning model for landslide hazard identification requires the creation of a dataset that is suitable with respect to the network structure, potentially requiring the dimensionality of the dataset to be increased in geospatial terms, where each sample is a multi-dimensional matrix containing the information of neighboring pixels. Considering the influence of neighboring pixels on the target elements, the spatial information of landslides is extracted by expanding a single pixel into a two-dimensional pixel matrix consisting of the location and points surrounding it. In order to improve the performance of the network and reduce the risk of overfitting, it was necessary to set the features and labels according to the sample's information during the construction of the dataset. The dimensions of the pixel matrix, as well as the number of landslide influence factors, together constitute the features of the samples. Whether or not a landslide had occurred at the location corresponding to the pixel matrix was used as a condition to label the samples, with 1 indicating that a landslide had occurred and 0 indicating that a landslide had not occurred. The pixel values of all input feature layers also needed to be normalized within the range of 0–1. The database production process is depicted in Figure 5.

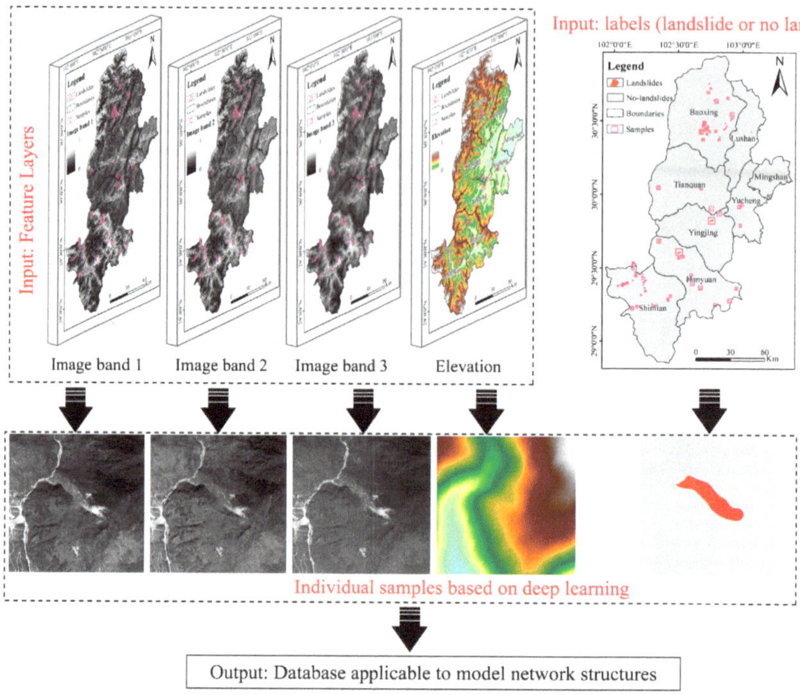

Figure 5. Landslide identification database construction process.

3. Methods

3.1. Swin Transformer

The vision transformer (ViT), as a model representative of the application of transformers in image processing, has achieved better results than traditional CNNs in the field of image recognition, which proves that the transformer is able to replace CNNs in image processing to a certain extent; thus, it has excellent application prospects [16,17]. The shifted window (Swin) transformer modifies the original multi-head self-attention (MSA) structure of the transformer and introduces a window-based MSA (W-MSA) structure, which decomposes the image into multiple non-overlapping windows and calculates the attention in each window separately [18,19], significantly reducing the computation load compared to calculating the attention of the entire picture, as is carried out in ViT. In addition, in order to address the problem of features not being transferred between different windows, the Swin transformer also includes a shifted window-based MSA (SW-MSA) structure, which enhances the recognition accuracy of the network by transferring feature information from different windows through shifting their positions [20]. The Swin transformer model structure and some hyperparameters are shown in Figure 6.

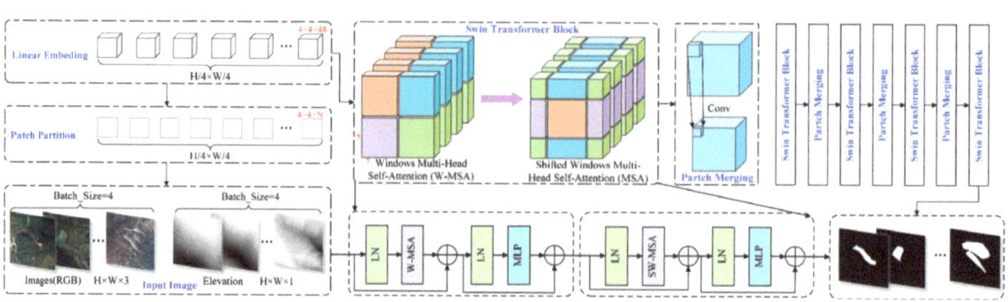

Figure 6. Swin transformer network architecture.

Similarly to the ViT, after entering the Swin transformer, the images are first processed using the patch partition structure for chunking and flattening in the channel dimension. After this, these feature maps pass through four stages to extract the feature information [21]. In the first stage, each block is linearly transformed and then fed into two consecutive Swin transformer blocks [22]. In the subsequent stages, in order to further extract the features and reduce the size of the feature map, the feature map is down-sampled utilizing patch merging, and it is then inputted into the next Swin transformer block [23]. The patch-merging structure divides each neighboring pixel of 2 × 2 size into a patch, and the pixels at the same position in each patch are then divided into two patch-merging blocks and spliced in the depth direction. Next, the feature map is fed into the layer normalization (LayerNorm) layer for normalization; finally, it is downscaled in the depth direction using a fully connected layer. After down-sampling utilizing a patch-merging structure, the width and height of the feature map are reduced to half of the previous values, and the depth is doubled compared to its previous value [24].

The Swin transformer block is the core component of the Swin transformer, and it is composed of two consecutive transformer modules in series, forming the basis upon which the Swin transformer achieves efficient feature extraction and classification [25]. Compared with the traditional transformer module, it replaces the MSA part with W-MSA and SW-MSA structures [26]. Assuming that the size of the input feature map is H × W and the window size of the Swin transformer is set to M × M, when the feature map enters the Swin transformer block, it is segmented by W-MSA into HM × WM windows with a size of M × M; following this, self-attention computation is carried out for each window. Next, the feature map is passed through the multi-layer perceptron (MLP) module, and residual connections are introduced [27]. This structure significantly reduces the computational effort compared to the traditional transformer module; however, as self-attention computation is only performed within the windows, no information can be transferred from window to window [28]. Therefore, the feature maps are obtained after the W-MSA structure passes through the SW-MSA structure again, which shifts the position of the window by M/2 pixels in both the horizontal and vertical directions to complete the transfer of feature information between the different windows [29]. In order to improve robustness and generalization abilities through training, a LayerNorm layer is included between each structure for normalization; finally, the feature extraction results are output through the MLP module.

3.2. Loss Functions

3.2.1. Binary Cross-Entropy (BCE)

The loss function measures the predicted values against the true values: the smaller the model loss, the better the generalization performance of the model [30]. The ultimate goal of model training is to achieve minimization of the loss function under conditional constraints [31]. When considering different tasks, the selected loss function also differs [32]. As its mathematical form is simpler to calculate, the cross-entropy loss is often used as the loss function for training models [33]. The binary cross-entropy (BCE) loss function is based on the cross-entropy loss function but adds proportional weights (0.5) to the number of positive and negative samples. The cross-entropy loss function is used in binary classification problems and is particularly suitable when there is a probabilistic output and only two types of labels (usually labeled 0 and 1) [34]. In machine learning and deep learning, when we train a model to predict the probability of an event occurring, the BCE loss helps to measure the gap between the probability distribution predicted by the model and the actual observed labels, providing a basis for optimizing the model [35]. The BCE loss function can be expressed as follows [36]:

$$L_{BCE} = -\frac{1}{N}\sum_{i=1}^{N} [y_i \log(p_i) + (1 - y_i) \log(1 - p_i)] \quad (1)$$

where N is the number of image pixels, and $y_i \in \{0, 1\}$ [33] is the label category of pixel i in the image, where $y_i = 1$ denotes landslide slopes and $y_i = 0$ denotes non-landslide slopes. The probability of predicting a pixel i as having label y_i is denoted as $p_i \in [0, 1]$.

3.2.2. Boundary Loss Function

There is usually a severe category imbalance problem in landslide image recognition; that is, the size of the target foreground region (the extent of the landslide) is often several orders of magnitude smaller than that of the background region (other features). This is due to the fact that landslides are relatively rare in nature, compared to other landforms.

The cross-entropy loss function, which is commonly used in the field of landslide identification, has a well-known drawback in highly unbalanced cases; namely, it assumes that all of the samples and categories are of equal importance, which usually leads to training instability and causes the decision boundaries to be biased towards a high number of categories [37]. A common strategy for the category imbalance problem is to down-sample a high number of categories to re-balance the prior distribution of the categories; however, this strategy limits the use of training images. Another strategy is weighting: assigning greater weights to categories with a small number of samples and smaller weights to categories with a large number of samples. Although this method is effective for some imbalance problems, difficulties are still faced when dealing with extremely imbalanced data. The cross-entropy gradient computed over a few pixels usually contains noise, and assigning larger weights to a few categories further increases this noise, thus leading to unstable training [38]. Considering these problems associated with the BCE loss, researchers have proposed a boundary-based loss function, which takes the form of a distance metric on the contour space rather than the region space. Instead of integrating over regions, the boundary loss computes integrals over the boundaries between regions, thus alleviating the problems associated with region loss in highly unbalanced segmentation problems [39].

In order to construct the boundary loss, we first calculate the closest distance from the landslide area of the real label (foreground) to the background d_F^i (Euclidean distance between the pixel points). We also calculate the closest distance from the background of the real label to the landslide area (foreground) d_B^j. The expression is as follows [40]:

$$d_F^i = \min_{j \in B}\left(\sqrt{(x_i - x_j)^2 + (y_i - y_j)^2}\right), (i \in F) \tag{2}$$

$$d_B^j = \min_{j \in F}\left(\sqrt{(x_j - x_i)^2 + (y_j - y_i)^2}\right), (i \in B) \tag{3}$$

where F denotes the set of foreground pixel points, B is the set of background pixel points, and x and y are the horizontal and vertical coordinates where the pixel points are located, respectively. The weight function for the boundary penalty is then obtained by combining d_F^i and d_F^i:

$$W_{boundary}^k = \begin{cases} 1 - d_F^k, & (k \in F) \\ d_B^k, & (k \in B) \end{cases} \tag{4}$$

When the pixel belongs to the background and is far away from the boundary, it will be assigned a larger penalty weight. When the pixel belongs to the foreground and is the boundary, its penalty weight is 0. When the pixel belongs to the foreground and is not the boundary, its weight is negative. Finally, the boundary penalty loss function is defined as follows [41]:

$$L_{boundary} = \frac{1}{N}\sum_{k}^{N} p_k W_{boundary}^k \tag{5}$$

where p_k is the prediction probability of pixel k and N is the number of pixel points in the entire image. In the process of gradient descent, if a pixel point is predicted to be outside the range of the landslide region, a positive gradient will be obtained, and, therefore, the prediction probability will be reduced. In contrast, a negative gradient inside the landslide region will increase the prediction probability during gradient descent.

3.3. Precision Evaluation Indicator

In landslide identification experiments, selecting appropriate evaluation indices is crucial to better reflect the model's real performance. For all of the experiments conducted in this study, precision (P), recall (R), F1-score (F1), intersection over union (IoU), and overall accuracy (OA) were used as evaluation indices to measure the performance of the model [42]. For the evaluation metrics, the higher the P, the more accurate the detected pixels, and a higher recall rate indicates that fewer wrong pixels were identified by the model. The F1 score takes into account both the precision and recall rate of the classification model, providing a balanced evaluation of the model's performance. IoU is used to measure the correlation between the real and the predicted values, and the OA reflects the overall precision of the prediction results, which was used as an evaluation index to measure the model performance [43,44]. Both indices range from 0 to 1, with higher values indicating better prediction results. The index formulas are as follows:

$$P = \frac{TP}{TP + FP} \tag{6}$$

$$R = \frac{TP}{TP + FN} \tag{7}$$

$$F_1 = \frac{2 \times P \times R}{P + R} \tag{8}$$

$$IoU = \frac{TP}{TP + FP + FN} \tag{9}$$

$$OA = \frac{TP + TN}{TP + FP + TN + FN} \tag{10}$$

where *TP* (true positive) denotes the true-positive class, *FP* (false positive) is the false-positive class, *FN* (false negative) is the false-negative class, and *TN* (true negative) is the true-negative class.

3.4. Experimental Environment Settings

The proposed semantic segmentation model was implemented using the Pytorch deep learning framework, leveraging its end-to-end machine learning capabilities and using Python as the development language. Its ability to carry out inverse auto-derivation provides a great advantage in the computational process of deep learning tasks. Essential functionalities for building a neural network architecture, such as basic convolution, activation, and pooling, are provided by the torch.nn library. In addition, a graphics processing unit (GPU) was introduced to improve the efficiency of the code. The utilized experimental environment and configuration are described in Table 1.

Table 1. Hardware and software details.

Hardware and Software	Parameters
CPU	13th Gen Intel(R) Core(TM) i7-13700KF
GPU	NVIDIA GeForce RTX 4090(NVIDIA Corporation, located in Santa Clara, CA, USA)
Operating Memory	64 GB
Total Video Memory	24 GB
Operating System	Windows 11
Python	Python 3.7.16

Table 1. *Cont.*

Hardware and Software	Parameters
IDE	PyCharm 2023.1 (Professional Edition) PyCharm
CUDA	CUDA 11.1
CUDNN	CUDNN 8.0.1
Deep Learning Architecture	PyTorch 1.8.1

4. Experimental Analysis

4.1. Training Details

4.1.1. Model Training Strategies

The Swin transformer-based model was implemented using the open-source deep learning library Pytorch, and for our experiment, the crop size was set to 256. First, the input source images were randomly cropped to ensure a consistent resolution of 256 × 256 for all images processed using the model. The initial loss function was set as the cross-entropy loss, and the initial learning rate was set as 0.00013. The SGD+Momentum stochastic optimization algorithm was used to train the model; the momentum of the optimization algorithm was 0.9, and the weight decay factor was 0.0005. Each training round consisted of 100 rounds, and the batch size (batch size) was 4. After the completion of each round of training, the validation dataset was used to evaluate the performance of the model at that stage, and the optimal model was then preserved for comparison purposes. Finally, a model with the best segmentation effect was obtained.

Due to the deeper layers of the network model proposed in this study and the up-sampling process, the entire training process is highly time-consuming. Therefore, in the training of deep learning models, researchers generally use the pre-training method, which involves initializing the parameters of the training model using model parameters with better convergence. In this study, we used the method illustrated in Figure 6 to carry out the model-training process. First, we used the proposed backbone network to carry out training on an easy-to-classify dataset (ImageNet) and saved the trained parameters; then, we carried out training again on the collected landslide dataset. The model was initialized to use the pre-training parameters of the backbone network, and the constant updating of the loss value and the weights helped the model adapt to the landslide dataset, reducing the time required for model training while greatly improving classification accuracy during feature extraction. The pre-training process of the model is depicted in Figure 7.

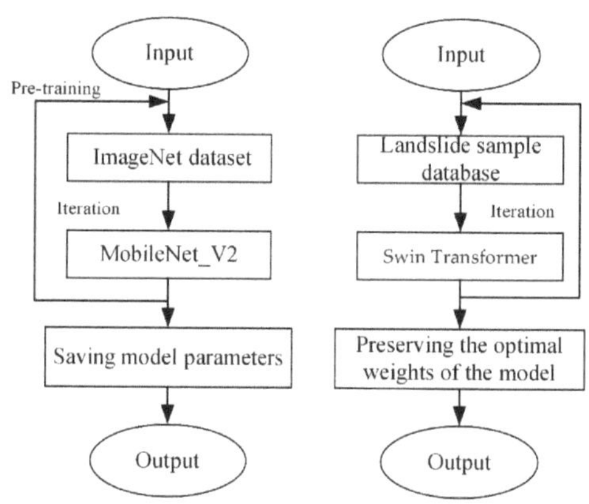

Figure 7. Model pre-training process.

4.1.2. Online Data Enhancement

The learning ability of a neural network mainly depends on the number of data samples, which affects the generalization ability and robustness of the model. When the amount of data is limited, the network is prone to overfitting the data. Furthermore, labeling additional data is both time-consuming and labor-intensive [45]. To address these issues, when faced with limited image data, enhancing the dataset to expand the sample size is necessary. Data enhancement can be divided into offline data enhancement and online data enhancement. Offline data enhancement refers to the preprocessing of data before data training, and it is commonly performed using methods such as scaling and cropping. In contrast, online data enhancement refers to the enhancement of data before each iteration in the training process in order to enlarge the scale of the dataset and improve the generalization ability and stability of the model [46]. The main steps of online enhancement are as follows [47]: (1) the model starts training; (2) a small batch of data is extracted from the original dataset according to specific or random rules; (3) the small batch of original data is sent to the data intensifier, which then carries out geometric, color, and pixel transformations according to specific rules; (4) the data enhancer inputs the enhanced small batch of data into the model for training; and (5) steps (2) to (4) are repeated until the training is completed. Figure 8 presents a schematic diagram of the workflow of the online data enhancement method.

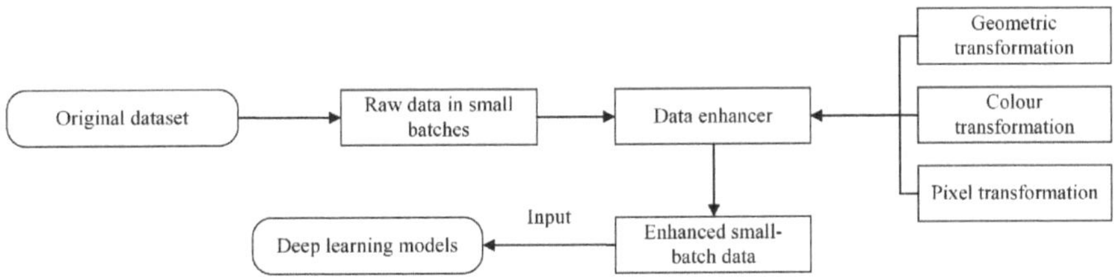

Figure 8. Online data enhancement module operation flow.

In this study, data enhancement was carried out by means of a two-dimensional online data enhancement module—including geometric shape and image color transformations—in order to improve the training effect of the model. Particular attention should be paid to the fact that as the elevation data are formatted as a single-band image, the color change has no meaning; thus, only geometric transformation is performed on these data.

Geometric Shape Transformation

The geometric transformation of image data generally includes flipping, rotating, cropping, deforming, scaling, and other operations [48]. After a large number of experiments, for image recognition or classification, the flipping and rotating transformations of the image were found to effectively enhance the data. Flipping is a method similar to mirror folding; in contrast, rotating refers to clockwise or counterclockwise rotation. It should be noted that it is best to carry out rotations by an interval of 90° to 180°; otherwise, there will be inconsistency in the image sample's scale. In addition, the enhanced images obtained via flipping and rotating differ from the original image only in terms of the direction and angle, while the image size remains the same [49,50]. In this study, we chose to use random rotation and flipping methods to carry out the data enhancement of landslide sample images. When geometric transformations were performed on the training images, the labels were also transformed accordingly; otherwise, the labels would not correspond to the transformed images. Examples of the geometrically transformed images are shown in Figure 9.

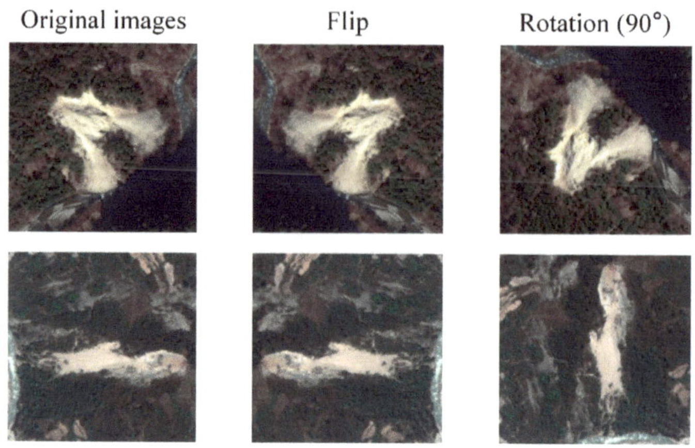

Figure 9. Example of geometry transformation of some landslide samples in Ya'an County.

Image Color Transformation

The color transformation process does not directly change the information in the image sample itself: it only selects a portion of the original image according to certain rules or re-distributes the pixels of the original image according to a certain ratio.

In simple terms, the different colors in an image represent different pixels. Therefore, color transformation is conducted to enhance an image by changing the information (pixels) in the image itself. The general methods of color transformation include contrast, brightness, color enhancement, and image-sharpening transformations [51]. Among them, brightness transformations involve adjusting the brightness and darkness of the original image, contrast transformations involve transformation by adjusting the brightness ratio between the brightest and darkest parts of the image, and image sharpening involves making the image clearer by compensating for the color at the edges of the image. An example of partial color transformation of a landslide image is shown in Figure 10.

Figure 10. Example of color transformation of some landslide samples in Ya'an County.

4.2. Analysis of Experimental Results

4.2.1. Comparison of Different Feature Extraction Networks

For this experiment, the landslide dataset was divided into training, validation, and testing sets. The model performances of the Swin transformer-based network and classical feature extraction networks—UPerNet, pyramid scene parsing network (PSPNet), and

DeepLab_v3+—were compared. Figure 10 presents graphs showing loss and accuracy changes during the training process for the above models. As observed in Figure 11, UPerNet, Swin transformer, and DeepLab_v3+ fit faster during the training process, and the Swin transformer-based model exhibited the highest training accuracy. Meanwhile, PSPNet had a slower fit and the lowest training accuracy. The results indicate that the Swin transformer-based model reached the convergence state earlier, exhibited more stable and effective training performances, did not overfit, and had a stronger generalization ability compared to the other models.

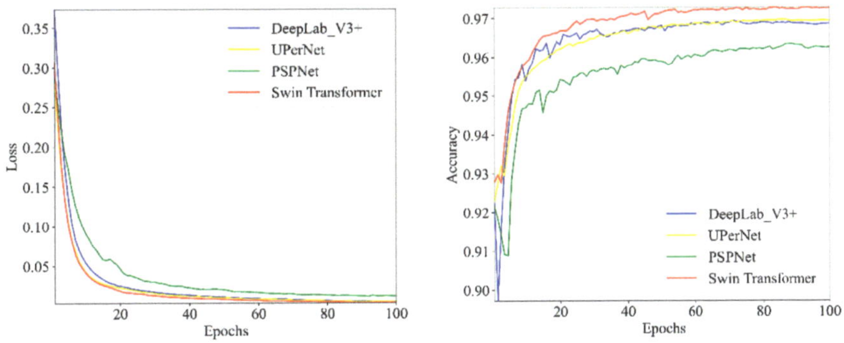

Figure 11. Training curves for different models.

Table 2 presents the results of the comparison between the different models in the landslide testing set.

Table 2. Comparison of the segmentation performances of the different models on the testing set.

Models	OA	F1-Score	IoU	PA	Recall
UPerNet	0.942	0.611	0.439	0.769	0.506
PSPNet	0.938	0.594	0.423	0.663	0.539
DeepLab_v3+	0.944	0.632	0.462	0.713	0.567
Swin Transformer	0.945	0.693	0.531	0.629	0.772

When evaluating the prediction performance of different models, several indicators need to be considered to make a judgment. From the above results, it can be observed that the Swin transformer-based model outperformed the other models in terms of the OA, F1-score, IoU, and R metrics. The OA of the Swin transformer-based model was 0.3%, 0.7%, and 0.1% higher than those of UPerNet, PSPNet, and DeepLab_v3+, respectively. The F1-scores of UPerNet, PSPNet, and DeepLab_v3+ were 61.1%, 59.4%, and 63.2%, respectively, while the Swin transformer-based model outperformed the other three models with a score of 69.3%. The Swin transformer-based model's IoU was also better than those of the UPerNet, PSPNet, and DeepLab_v3+ in several aspects, including OA and R, with OA being 9.2%, 10.8%, and 6.9% higher and R being 26.6%, 23.3%, and 20.5% higher than those of the original UPerNet, PSPNet, and DeepLab_v3+, respectively. However, the Swin transformer-based model had a lower P than the other models. The higher P of UPerNet means that it is more accurate in predicting instances with positive samples, but this may come at the expense of the R value. In many cases, we want the model to have both a high P and R. However, the two are often contradictory, and increasing one of these metrics may lead to a decrease in the other. Therefore, the F1-score—as the reconciled average of the P and R—is often used to comprehensively evaluate the performance of a model. As mentioned above, the F1-score of the Swin transformer-based model was much higher than that of the other three models. According to the above analysis, the Swin transformer-based model performed better than the other three models in terms of several recognition accuracy metrics; thus, it is more suitable for landslide hazard recognition tasks.

The Swin transformer-based model obtained better results in terms of the recognition of landslide disasters in Ya'an City. We randomly selected the results of image recognition using the above models in the testing dataset for comparison, as shown in Figure 12a–f, which show the original image, the labeled image, and the prediction results obtained with PSPNet, UPerNet, DeepLab_v3+, and the Swin transformer-based model, respectively.

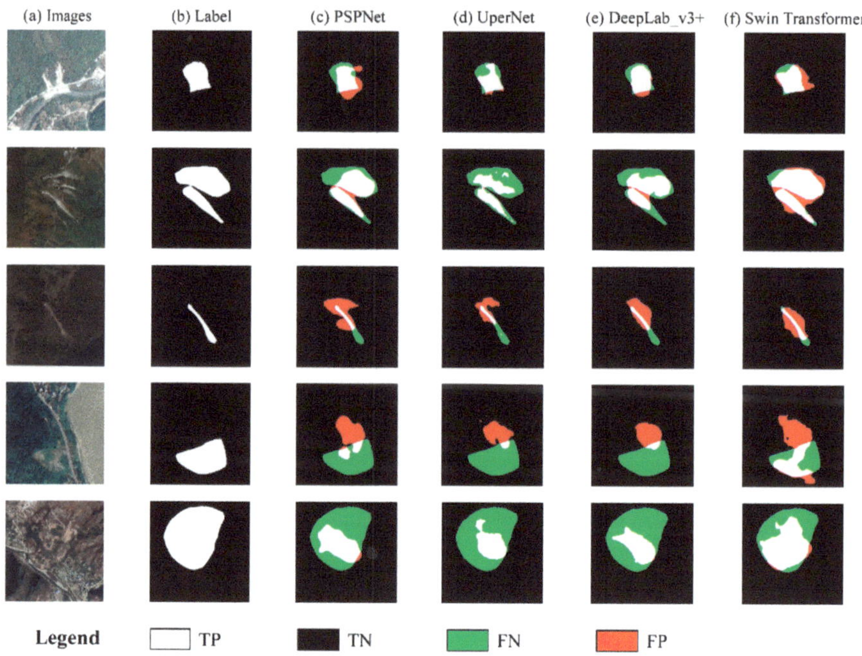

Figure 12. Comparison of different models' prediction results: (**a**) Images, (**b**) Labels, (**c**) Predicted results of PSPNet, (**d**) Predicted results of UPerNet, (**e**) Predicted results of DeepLab_V3+, (**f**) Predicted results of Swin Transformer.

As observed in Figure 12c–f, the Swin transformer-based model obtained the best results in terms of recognizing landslides. Even for the last row, which shows ancient landslide images, it was still the best model. PSPNet captures contextual information at different scales using the pyramid pooling module (PPM), but its core is still based on the architecture of a CNN, which may have some limitations in processing global information. Landslide phenomena are diverse—that is, they have different shapes, sizes, and contexts—and UPerNet may encounter difficulties in processing such complex image data, namely, misidentifying non-landslide areas as landslide areas or failing to recognize landslide areas in the image. DeepLab_v3+ is based on the encoder–decoder architecture, enhancing the encoder's ability to extract semantic information through the use of an ASPP module. However, compared with the Swin transformer, it may be slightly insufficient in processing high-resolution images and capturing complex contextual relationships. In summary, the reasons that the Swin transformer-based model outperformed UPerNet, PSPNet, and DeepLab_v3+ in terms of landslide recognition accuracy are mainly its unique model design, adaptability to visual tasks, and high performance, as demonstrated in several experiments. These advantages enable the Swin transformer-based model to capture semantic information in an image more accurately when dealing with complex visual tasks such as landslide recognition, thus achieving higher recognition accuracy.

4.2.2. Network Comparison Experiments after Adding DEM Features

In complex tasks such as landslide identification, non-landslide image samples often present textures and shapes that are similar to those of landslide areas, and these interfering factors greatly increase the risk of misclassification when relying only on optical images for landslide identification [52]. However, the rich topographic information contained in digital elevation models, such as elevation changes, can provide valuable information that is complementary to optical imagery, which can help the network model to more accurately recognize landslide areas. In order to verify the practical value of using DEMs in landslide identification tasks, we explored improvements in the landslide identification performance of the network model after adding DEM data through comparative experiments. As shown in Table 3, when we trained and tested the network containing the DEM data as the fourth channel of the optical image co-input, the four network models were significantly improved with respect to the testing set in terms of all four main metrics. Among them, the OA, F1-score, IoU, and R of UPerNet were improved by 0.7%, 6.3%, 7%, and 11.8%, respectively; the F1-score, IoU, and R of PSPNet were improved by 2.8%, 2.9%, and 6.2%, respectively; the OA, F1-score, IoU, P, and R of DeepLab_v3+ were improved by 0.6%, 2.4%, 0.6%, and 9.9%, respectively; the OA, F1-score, IoU, and P of the Swin transformer-based model were improved by 1.2%, 5.2%, and 6.2%, respectively. As shown in Table 3, when we input the DEM data as the fourth channel of the optical image into the network for training and testing, based on the testing set, the four main indices of the four network models were significantly improved.

Table 3. Comparison of model segmentation performance of fused DEM and RGB.

Models	Input Samples	OA	F1-Score	IoU	PA	Recall
UPerNet	RGB	0.942	0.611	0.439	0.769	0.506
	RGB+DEM	0.949	0.674	0.509	0.733	0.624
PSPNet	RGB	0.938	0.594	0.423	0.663	0.539
	RGB+DEM	0.938	0.622	0.452	0.646	0.601
DeepLab_v3+	RGB	0.944	0.632	0.462	0.713	0.567
	RGB+DEM	0.941	0.656	0.489	0.718	0.666
Swin Transformer	RGB	0.945	0.693	0.531	0.629	0.772
	RGB+DEM	0.957	0.747	0.596	0.755	0.739

First, DEM data provide rich topographic information, which plays a crucial role in landslide identification. As a surface deformation phenomenon, the occurrence of a landslide is often closely related to topographic conditions [53]. By fusing DEM data, the deep learning model is able to learn more spatial features related to landslides and, thus, can more accurately identify landslide areas. Second, the DEM data, as a stable information source, are not affected by factors such as light conditions and vegetation cover and can provide more reliable and stable inputs for the model. In optical remote sensing imagery, changes in the lighting conditions may lead to the degradation of the image quality, affecting the recognition performance of the model. The interference of vegetation cover may also make it difficult to accurately recognize landslide areas. By incorporating DEM data, the model is able to maintain a high recognition accuracy under different environmental conditions, improving the robustness and generalization ability of the model. In addition, DEM data can provide additional contextual information to the model, which allows it to better understand the relationship between the landslide area and the surrounding environment. In landslide identification tasks, contextual information often has a significant impact on the model's judgment. By fusing DEM data, the model is able to gain a more comprehensive understanding of topographic features, surrounding geomorphology, and possible geologic conditions of the landslide area in order to more accurately determine whether an area is a landslide area or not. The experimental results provided in Table 3 indicate that the UPerNet, PSPNet, DeepLabv3+, and Swin transformer-based deep learning models obtained significantly better values for all metrics (e.g., accuracy, recall, F1-score) in the landslide identification task when DEM data were used as an additional input channel in combination with optical remote sensing images. This fully demonstrates the important role and value of DEM data in the landslide recognition task. This is mainly due to the rich terrain information provided by DEM data, which is characterized by a stable information source, additional contextual information, and complementarity with other input information.

4.2.3. Optimization Experiments Using Boundary Loss Functions

In a landslide dataset, the landslide samples (positive classes) tend to make up only a small portion of the entire dataset, while the non-landslide samples (negative classes) make up the vast majority. This imbalance causes the model to be more likely to favor the majority class (i.e., non-landslide samples) during the training process, as the majority class contains a large number of samples, and it is easier for the model to reduce the loss of the majority class during the optimization process. This can have a significant impact on the model's learning ability, resulting in a decrease in the overall recognition accuracy of the model. Although the prediction accuracy of the Swin transformer-based model was better compared to classical networks, this model still has some room for improvement. When dealing with extremely unbalanced classification tasks such as the considered landslide dataset, the introduction of the boundary loss function (BLF) can improve the performance of the Swin transformer, which relies only on the binary cross-entropy function. The changes in the specific evaluation indices are shown in Table 4. When the Swin transformer-based model was trained using optical images and the boundary loss function was added, the OA, F1-score, IoU, and R of the model improved by 0.6%, 0.5%, 0.5%, and 8.2%, respectively. When the optical images and DEM data were fused using the Swin transformer and the boundary loss function was added, the OA, F1-score, IoU, and R of the model improved by 0.2%, 0.9%, 1.2%, and 2.2%, respectively.

Table 4. Comparison of model predictions after adding the boundary loss function.

Input Samples	Models	OA	F1-Score	IoU	PA	Recall
RGB	Swin Transformer (BCE)	0.945	0.693	0.531	0.629	0.772
	Swin Transformer (BCE+Boundary)	0.951	0.698	0.536	0.711	0.684
RGB+DEM	Swin Transformer (BCE)	0.957	0.747	0.596	0.755	0.739
	Swin Transformer (BCE+Boundary)	0.959	0.756	0.608	0.753	0.761

Compared with the Swin transformer-based model using only optical images as input, the accuracy evaluation indices were significantly improved when fusing elevation information and introducing the boundary loss function; in particular, the OA, F1-score, IoU, and P of the model improved by 1.4%, 6.3%, 7.7%, and 12.4%, respectively. In order to visually compare the degree of optimization of the prediction results with the introduction of the boundary loss function, the recognition image results with respect to the testing dataset were selected. The results of the comparison are shown in Figure 13a–d, which show the original image, the labeled image, the Swin transformer-based model trained on the optical images (this model uses the binary cross-entropy function), and the Swin transformer-based model trained on optical images and DEM data (this model uses the binary cross-entropy function and boundary loss function).

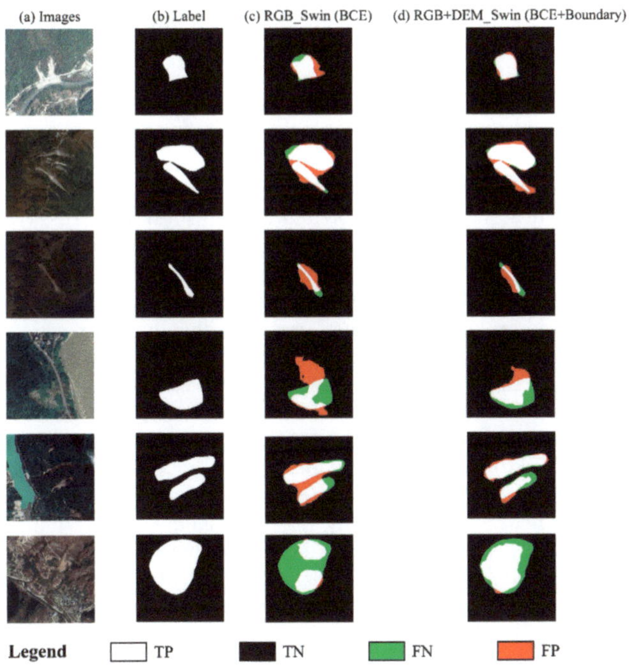

Figure 13. Comparison of model predictions before and after adding the boundary loss function.

When dealing with landslide datasets, particularly when the landslide areas are closely adjacent and difficult to distinguish, traditional identification models are often challenged and may incorrectly identify two landslides that are in close proximity to each other as one large landslide. Before the introduction of the boundary loss function, it was difficult for the model to accurately capture subtle differences between landslides due to the unbalanced nature of the dataset and the fuzzy boundaries between the landslide regions. A consequence of this was that neighboring landslides that were very close to one another could be mistakenly recognized as the same landslide during the classification process and, thus, were merged into one landslide during identification.

However, by introducing a boundary loss function, the model is able to focus more attention on the samples near the classification boundary, especially regions located between neighboring landslides. The design of the boundary loss function allows the model to give special consideration to these boundary samples during the optimization process, and it attempts to create a clearer demarcation between them. This strategy allows the model to more accurately identify the boundaries between neighboring landslides, thus avoiding incorrect merging into one large landslide. Landslide samples usually account for a relatively small proportion of the dataset; this results in the phenomenon in which traditional models become easily biased towards the non-landslide samples (i.e., the majority class) during the training process, thus failing to accurately identify landslide samples (i.e., the minority class). The introduction of the boundary loss function not only helps the model pay more attention to the samples near the classification boundary—especially the landslide and non-landslide samples that are difficult to distinguish—but also improves the sensitivity of the model to landslide samples. This strategy not only helps improve the model's recognition accuracy relative to the landslide samples, but, more importantly, it can also influence the model in carrying out more refined recognition during the classification process. By suppressing the risk of misclassifying samples and optimizing the model's decision-making at the classification boundary, the boundary loss function helps achieve the precise classification of landslide samples, thus improving the overall landslide detection and identification accuracy and providing more reliable technical support for the early warning and prevention of geological disasters.

5. Discussion

5.1. Interpretation of the Models' Visualization Results

In order to obtain the part of the image that contributes the most to the output category and improve the accuracy of the feature matching during image retrieval, we introduce the heat map for intuition and visualization purposes. We mapped the extracted feature data to the corresponding locations on the map to form a preliminary spatial distribution map. Different colors are assigned to each region of the image based on the values of the feature data (predicted probability values for the testing set) in order to create a heatmap. In the heatmap, red indicates high-risk areas, while blue indicates low-risk areas. In this paper, we use heatmaps to compare the intelligent identification results of the optimal Swin transformer-based model to the benchmark network without fused DEM features, which can visualize the prediction ability of the Swin transformer-based model. As shown in Figure 14, the feature fusion enhancement allowed the Swin transformer-based model to more accurately identify landslide areas during prediction. The areas with very high and very low landslide risks are shown in dark red and dark blue, respectively, on the heat map; specifically, the dark red areas represent areas with a high probability of landslide identification, while the dark blue areas represent areas with a low probability of landslide identification. This process of feature fusion directly results in a reduction in the transition colors, as the model may have learned to predict landslide areas more directly and accurately.

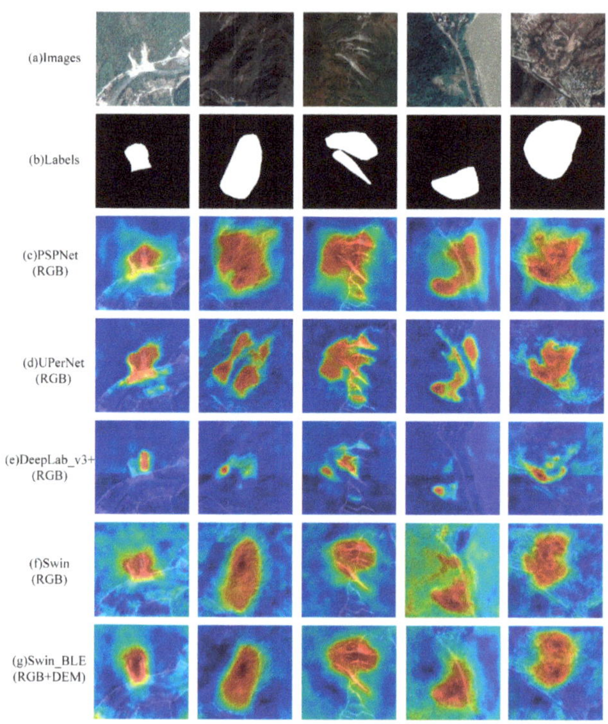

Figure 14. Comparison of heat maps with different input features: (**a**) Images, (**b**) Labels, (**c**) Predicted results of PSPNet (RGB), (**d**) Predicted results of UPerNet (RGB), (**e**) Predicted results of DeepLab_V3+ (RGB), (**f**) Predicted results of Swin Transformer (RGB), (**g**) Predicted results of Swin Transformer combined with boundary loss function (RGB+DEM).

Optical images and DEM data each have unique information characteristics. Optical images mainly record the spectral information and texture characteristics of the ground surface, which can reflect intuitive information, such as the vegetation cover and soil type. Meanwhile, DEM data provide three-dimensional information about the terrain, which is crucial for understanding the occurrence mechanism of geohazards. Therefore, the fusion of these two data features can fully leverage the complementarity between them, enabling the model to capture more comprehensive and accurate landslide-related features. The Swin transformer, as a deep learning model based on the Transformer architecture, has a powerful feature extraction capability. In the feature extraction stage, the model is able to gradually fuse the features of the optical image and DEM through multi-layer convolution and pooling operations. This fusion process not only superimposes the features of the two datasets, but also further enhances the key features related to landslide occurrence through operations such as non-linear transformation and feature selection. The effects of this feature enhancement allowed the model to identify landslide areas more accurately during prediction and to refine the prediction results.

5.2. Model Resilience Analysis Based on a Publicly Available Landslide Dataset

In the field of landslide hazard prediction and prevention, the adaptability of a model is one of the most important indicators of its performance. In order to fully evaluate the effectiveness of our selected and modified models for practical applications, we conducted a test on a publicly available landslide dataset [54] of Bijie City, Guizhou Province, China.

Bijie City is located in the northwestern part of Guizhou Province, and it is characterized by complex geological conditions and frequent landslide disasters. The considered dataset comprises data related to 770 landslide events in Bijie City, including labeled information such as the extent and size of the landslides; high-resolution optical satellite imagery, which visualizes the textural characteristics of the landslides; and elevation data, which provide key information for understanding the topographic and geomorphic conditions under which the landslides occurred. Examples of the landslides are shown in Figure 15.

Figure 15. Partial landslides in the Bijie dataset.

During the experiment, four benchmark networks (UPerNet, PSPNet, DeepLab_v3+, and the Swin transformer-based model), as mentioned above, were selected for testing. According to the pre-determined experimental scheme, the data in the Bijie landslide dataset were randomly divided into training, validation, and test sets, and data enhancement was performed on the training set samples. To ensure the fairness and accuracy of the experiments, these networks were not re-trained, but the optimal weights for each previously trained model were directly used. The initial performance evaluation of the models loaded with optimal weights was performed using the validation set's data in order to ensure that the best model was used. The results of the model testing are detailed in Table 5.

Table 5. Performance comparison of different models on the Bijou test set.

Models	Input Samples	OA	F1-Score	IoU	PA	Recall
UPerNet	RGB	0.961	0.815	0.688	0.769	0.867
	RGB+DEM	0.963	0.816	0.689	0.790	0.844
PSPNet	RGB	0.958	0.792	0.655	0.780	0.804
	RGB+DEM	0.962	0.817	0.691	0.779	0.859
DeepLab_v3+	RGB	0.966	0.823	0.700	0.846	0.802
	RGB+DEM	0.966	0.831	0.710	0.824	0.837
Swin Transformer	RGB	0.965	0.836	0.718	0.782	0.897
	RGB+DEM	0.970	0.856	0.748	0.821	0.894

From the above table, it can be observed that the model accuracy performed better overall after loading the optimal weights from previous training. This is because the pre-training weights were obtained through previous training on a dataset; thus, the model had already learned a rich feature representation. When these weights were used to initialize a new task, they were able to take advantage of the already learned knowledge, exhibiting high initial landslide recognition accuracy with respect to Bijie data. Meanwhile, the effectiveness of the pre-training weights depends to a large extent on the relevance of the pre-training task to the target task. As all landslides were targeted in this study, the improvement introduced by previous pre-training weights was more significant. After adding DEM data, the overall performance of the models showed an upward trend, which is sufficient to conclude that the DEM data provide more dimensional information for the model, thus helping the model to understand landslide phenomena more comprehensively.

The Swin transformer-based model effectively handles high-resolution images through its hierarchical structure, windowed self-attention mechanism, and shifted window technique, and it achieved excellent landslide recognition performance with respect to the Bijie dataset. In order to improve the recognition accuracy of the model, it was chosen to utilize the boundary constraint function such that the final prediction results were more in line with a priori knowledge. The obtained evaluation indices are provided in Table 6.

Table 6. Comparison of the Bijou test set after adding the boundary loss function.

Input Samples	Models	OA	F1-Score	IoU	PA	Recall
RGB	Swin Transformer (BCE)	0.965	0.836	0.718	0.782	0.897
	Swin Transformer (BCE+Boundary)	0.970	0.855	0.746	0.816	0.898
RGB+DEM	Swin Transformer (BCE)	0.970	0.856	0.748	0.821	0.894
	Swin Transformer (BCE+Boundary)	0.973	0.868	0.766	0.843	0.895

The experimental results show that all accuracy indices of the model were improved after adding the boundary constraint function, which further verifies the effectiveness of the boundary constraint function in improving the performance of the model. Compared to the Swin transformer-based model with only optical images as input, the accuracy evaluation index of the Swin transformer-based model was significantly improved after fusing the elevation information and introducing the boundary loss function. The OA, F1 score, IoU, and PA of the model were improved by 0.8%, 3.2%, 4.8%, and 6%, respectively. In order to visually compare the optimization degree of the prediction results after the introduction of the boundary loss function, the results for certain images in the Bijie test dataset were selected, as shown in Figure 16. Figure 16a–d show the original image, labeled image, the Swin transformer-based model's prediction results when trained on optical images (using the binary cross-entropy function), and the prediction results when the model was trained on optical images and DEM data (using the binary cross-entropy function and a boundary loss function). When the elevation information was combined with the boundary constraint function, they complemented each other and jointly enhanced the recognition effect of the model. Elevation information can be combined with other image features (e.g., color, texture) to form a richer and more accurate feature representation. This multi-dimensional feature representation helps the model better capture the essential features of the target object, thus improving the recognition effect. The boundary constraint function, on the other hand, further improves the recognition accuracy of the model by introducing prior knowledge and optimizing the decision boundary. This combination makes the model more accurate and efficient in dealing with targets with significant elevation features.

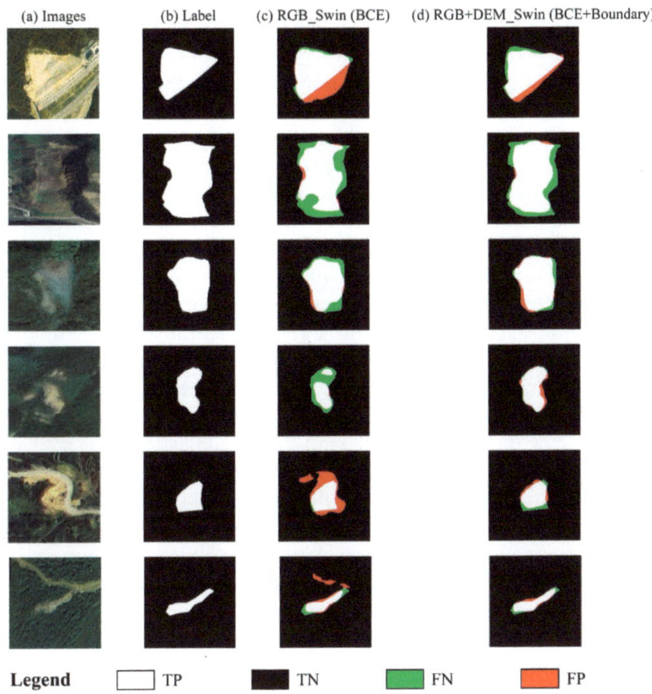

Figure 16. Effectiveness of landslide identification in Bijie: (**a**) Images, (**b**) Labels, (**c**) Predicted results of Swin Transformer (RGB), (**d**) Swin Transformer combined with boundary loss function (RGB+DEM).

5.3. Comparison of Model Recognition Performance for Different Datasets

In the field of landslide identification, the diversity and complexity of datasets pose a serious challenge to model performance [55,56]. In this study, four deep learning models—Swin transformer, UPerNet, PSPNet, and DeepLab_v3+—were applied for landslide recognition experiments for the Ya'an and Bijie datasets, respectively. With respect to the two different datasets, the Swin transformer model superimposed on the boundary loss function exhibited optimal recognition performance, while the UPerNet, PSPNet, and DeepLab_v3+ models performed slightly worse. Both the Ya'an dataset and the Bijie dataset contain elevation data and optical imagery, which provide rich information on topographic landforms and texture features for landslide identification. Elevation data [57] provide information on the elevation of the terrain, which helps identify changes in the terrain, such as ground subsidence or uplift caused by landslides, while optical imagery [58] provides information on the ground cover and texture, which helps identify features such as vegetation damage, cracks, etc., in landslide areas. Moreover, the landslide types in both datasets cover both recent landslides and deformed slopes, which further increases the complexity of the identification task. Recent landslides are usually characterized by more obvious surface deformation and cracks [59], while deformed slopes [60] may show minor ground surface deformations or changes in vegetation cover, and high-precision feature extraction and recognition capabilities for accurate judgment are needed.

The optimized Swin transformer model exhibited the best recognition results relative to both datasets, and this is mainly due to its powerful feature extraction capability and adaptability to complex scenes. The model is able to effectively capture key features of the landslide region, such as topographic changes, crack distribution, and differences in vegetation cover, through the self-attention mechanism and window delineation strategy. Simultaneously, the Swin transformer is able to take full advantage of the complementary information of these two data sources when processing complex datasets containing both DEM data and optical imagery to achieve more accurate landslide identification. The boundary loss function plays an important role in landslide identification tasks. The boundaries between landslide and non-landslide areas are often irregular, and they are affected by a variety of factors, such as topography, vegetation, and rainfall. This makes it challenging for the model to determine the boundary. The boundary loss function, however, optimizes the decision boundary of the model by increasing the focus on the samples near the classification boundary. When the model's prediction of the boundary samples does not match the true labels, the boundary loss function exhibits a large penalty, prompting the model to adjust its parameters to better fit the true classification boundary. This mechanism helps improve the model's ability to recognize boundary samples, resulting in more accurate results for landslide identification.

In contrast, the UPerNet, PSPNet, and DeepLab_v3+ models show some limitations when dealing with these complex scenarios. Although these models have achieved good results in their respective areas of expertise, there is still more room for performance improvement when facing the specific task of landslide identification. This may be related to the strategies of these models in feature extraction, contextual information fusion, and multi-scale processing. We can make improvements with respect to the following: first, the model's adaptive training can be strengthened relative to complex scenes, and the model's sensitivity to key features, such as terrain changes and crack distribution, can be improved; second, the model's contextual information fusion strategy should be modified, and complementary information relative to DEM data and optical imagery should be made use of; third, a multi-scale feature extraction and processing mechanism should be introduced in order to improve the model's recognition of landslide targets at different scales.

6. Conclusions

This study focused on Ya'an City, a landslide-prone area in the Sichuan Basin. We collected geological, environmental, and optical imagery data. Using remote sensing and AI, we addressed the challenge of landslide identification, a small-sample learning problem. We employed a Swin transformer-based model to enhance recognition accuracy by integrating optical images and DEM data. The introduction of a boundary constraint function further refined the results compared to classical networks (PSPNet, UPerNet, and DeepLab_V3+). The key findings include the following.

The performance of the original Swin transformer-based model, relying solely on the binary cross-entropy function, was significantly enhanced by incorporating the boundary loss function. Compared to only using optical images, the model's accuracy evaluation indices, including OA, F1-score, IoU, and P, were improved by 1.4%, 6.3%, 7.7%, and 12.4%, respectively, when integrating DEM information and using the boundary loss function.

Testing using the landslide dataset of Bijie City, Guizhou Province, to verify the model's resilience revealed that adding pre-training weights enhanced the performance. With the inclusion of elevation information and the boundary constraint function, the Swin transformer-based model's OA, F1 score, IoU, and PA improved by 0.8%, 3.2%, 4.8%, and 6% respectively, compared to only using optical images. This demonstrated that combining elevation information and the boundary constraint function further boosted the model's recognition ability.

However, this study has limitations. Firstly, the model's generalization ability may be impacted by the study area and sample size constraints. To validate its applicability, future tests in diverse geological and environmental conditions are necessary. Secondly, while the Swin transformer exhibited remarkable results with respect to landslide identification, its computational complexity is high, and the model size is large, posing challenges for real-time performance and computational resources in practical applications. Exploring lightweight and efficient model architectures is a crucial direction for future research.

Author Contributions: Conceptualization, X.W., S.C. and D.W.; methodology, X.W., W.L. and J.D.; software, D.W. and L.X.; validation, T.S. and J.D.; formal analysis, C.L.; investigation, X.W.; resources, D.W.; data curation, X.W.; writing—original draft preparation, X.W. and M.Z.; writing—review and editing, M.Z.; visualization X.W.; supervision, X.W.; project administration, X.W.; funding acquisition, L.X. All authors have read and agreed to the published version of the manuscript.

Funding: This research was funded by the National Natural Science Foundation of China, grant number 52208006, and the Technology Innovation Center for Geological Disaster Prevention and Ecological Restoration in Western China, MNR (Chengdu University of Technology), grant number TICGP2023K002.

Data Availability Statement: The original contributions presented in the study are included in the article, further inquiries can be directed to the corresponding author.

Acknowledgments: We thank LetPub (www.letpub.com) for its linguistic assistance during the preparation of this manuscript.

Conflicts of Interest: Author Tiegang Sun was employed by the company China Building Materials Southwest Survey and Design Co., Ltd. The remaining authors declare that the research was conducted in the absence of any commercial or financial relationships that could be construed as a potential conflict of interest.

References

1. Zhao, Z.; Lan, H.; Li, L.; Strom, A. Landslide Spatial Prediction Using Cluster Analysis. *Gondwana Res.* **2024**, *130*, 291–307. [CrossRef]
2. Chen, X.; Chen, W. GIS-Based Landslide Susceptibility Assessment Using Optimized Hybrid Machine Learning Methods. *Catena* **2021**, *196*, 104833. [CrossRef]
3. Bui, D.T.; Tsangaratos, P.; Nguyen, V.-T.; Van Liem, N.; Trinh, P.T. Comparing the Prediction Performance of a Deep Learning Neural Network Model with Conventional Machine Learning Models in Landslide Susceptibility Assessment. *Catena* **2020**, *188*, 104426. [CrossRef]
4. Yang, C.; Liu, L.-L.; Huang, F.; Huang, L.; Wang, X.-M. Machine Learning-Based Landslide Susceptibility Assessment with Optimized Ratio of Landslide to Non-Landslide Samples. *Gondwana Res.* **2023**, *123*, 198–216. [CrossRef]
5. Cheng, G.; Wang, Z.; Huang, C.; Yang, Y.; Hu, J.; Yan, X.; Tan, Y.; Liao, L.; Zhou, X.; Li, Y.; et al. Advances in Deep Learning Recognition of Landslides Based on Remote Sensing Images. *Remote Sens.* **2024**, *16*, 1787. [CrossRef]
6. Xu, Y.; Ouyang, C.; Xu, Q.; Wang, D.; Zhao, B.; Luo, Y. CAS Landslide Dataset: A Large-Scale and Multisensor Dataset for Deep Learning-Based Landslide Detection. *Sci. Data* **2024**, *11*, 12. [CrossRef]
7. Ju, Y.; Xu, Q.; Jin, S.; Li, W.; Su, Y.; Dong, X.; Guo, Q. Loess Landslide Detection Using Object Detection Algorithms in Northwest China. *Remote Sens.* **2022**, *14*, 1182. [CrossRef]
8. Ma, L.; Liu, Y.; Zhang, X.; Ye, Y.; Yin, G.; Johnson, B.A. Deep Learning in Remote Sensing Applications: A Meta-Analysis and Review. *ISPRS J. Photogramm. Remote Sens.* **2019**, *152*, 166–177. [CrossRef]
9. Audebert, N.; Le Saux, B.; Lefèvre, S. Semantic Segmentation of Earth Observation Data Using Multimodal and Multi-Scale Deep Networks. In Proceedings of the Asian Conference on Computer Vision, Taipei, Taiwan, 20–24 November 2016; Springer: Berlin/Heidelberg, Germany, 2016; pp. 180–196.
10. Tao, A.; Sapra, K.; Catanzaro, B. Hierarchical Multi-Scale Attention for Semantic Segmentation. *arXiv* **2020**, arXiv:2005.10821.
11. Hoyer, L.; Dai, D.; Van Gool, L. Daformer: Improving Network Architectures and Training Strategies for Domain-Adaptive Semantic Segmentation. In Proceedings of the IEEE/CVF Conference on Computer Vision and Pattern Recognition, New Orleans, LA, USA, 18–24 June 2022; pp. 9924–9935.
12. He, H.; Cai, J.; Pan, Z.; Liu, J.; Zhang, J.; Tao, D.; Zhuang, B. Dynamic Focus-Aware Positional Queries for Semantic Segmentation. In Proceedings of the IEEE/CVF Conference on Computer Vision and Pattern Recognition, Vancouver, BC, Canada, 17–24 June 2023; pp. 11299–11308.
13. Nguyen, T.; Nguyen, L.; Tran, P.; Nguyen, H. Improving Transformer-Based Neural Machine Translation with Prior Alignments. *Complexity* **2021**, *2021*, 5515407. [CrossRef]

14. Liu, Z.; Lin, Y.; Cao, Y.; Hu, H.; Wei, Y.; Zhang, Z.; Lin, S.; Guo, B. Swin Transformer: Hierarchical Vision Transformer Using Shifted Windows. In Proceedings of the IEEE/CVF International Conference on Computer Vision, Montreal, BC, Canada, 11–17 October 2021; pp. 10012–10022.
15. Pourghasemi, H.R.; Sadhasivam, N.; Amiri, M.; Eskandari, S.; Santosh, M. Landslide Susceptibility Assessment and Mapping Using State-of-the Art Machine Learning Techniques. *Nat. Hazards* **2021**, *108*, 1291–1316. [CrossRef]
16. Lee, S.I.; Koo, K.; Lee, J.H.; Lee, G.; Jeong, S.; O, S.; Kim, H. Vision Transformer Models for Mobile/Edge Devices: A Survey. *Multimed Syst.* **2024**, *30*, 109. [CrossRef]
17. Liu, Y.; Nand, P.; Hossain, M.A.; Nguyen, M.; Yan, W.Q. Sign Language Recognition from Digital Videos Using Feature Pyramid Network with Detection Transformer. *Multimed. Tools Appl.* **2023**, *82*, 21673–21685. [CrossRef]
18. Dong, Z.; Wang, Q.; Zhu, P. Multi-Head Second-Order Pooling for Graph Transformer Networks. *Pattern Recognit. Lett.* **2023**, *167*, 53–59. [CrossRef]
19. Nie, J.; Xie, J.; Sun, H. Remote Sensing Image Dehazing via a Local Context-Enriched Transformer. *Remote Sens.* **2024**, *16*, 1422. [CrossRef]
20. Pacal, I.; Alaftekin, M.; Zengul, F.D. Enhancing Skin Cancer Diagnosis Using Swin Transformer with Hybrid Shifted Window-Based Multi-Head Self-Attention and SwiGLU-Based MLP. *J. Imaging Inform. Med.* **2024**, 1–19. [CrossRef]
21. Kim, H.; Yim, C. Swin Transformer Fusion Network for Image Quality Assessment. *IEEE Access* **2024**, *12*, 57741–57754. [CrossRef]
22. Zhong, F.; He, K.; Ji, M.; Chen, J.; Gao, T.; Li, S.; Zhang, J.; Li, C. Optimizing Vitiligo Diagnosis with ResNet and Swin Transformer Deep Learning Models: A Study on Performance and Interpretability. *Sci. Rep.* **2024**, *14*, 9127. [CrossRef]
23. Liu, E.; He, B.; Zhu, D.; Chen, Y.; Xu, Z. Tiny Polyp Detection from Endoscopic Video Frames Using Vision Transformers. *Pattern Anal. Appl.* **2024**, *27*, 38. [CrossRef]
24. Ramkumar, K.; Medeiros, E.P.; Dong, A.; de Albuquerque, V.H.C.; Hassan, M.R.; Hassan, M.M. A Novel Deep Learning Framework Based Swin Transformer for Dermal Cancer Cell Classification. *Eng. Appl. Artif. Intell.* **2024**, *133*, 108097. [CrossRef]
25. Pacal, I. A Novel Swin Transformer Approach Utilizing Residual Multi-Layer Perceptron for Diagnosing Brain Tumors in MRI Images. *Int. J. Mach. Learn. Cybern.* **2024**, *15*, 3579–3597. [CrossRef]
26. Dai, Y.; Liu, F.; Chen, W.; Liu, Y.; Shi, L.; Liu, S.; Zhou, Y. Swin MAE: Masked Autoencoders for Small Datasets. *Comput. Biol. Med.* **2023**, *161*, 107037.
27. Masood, A.; Naseem, U.; Kim, J. Multi-Level Swin Transformer Enabled Automatic Segmentation and Classification of Breast Metastases. In Proceedings of the 2023 45th Annual International Conference of the IEEE Engineering in Medicine & Biology Society (EMBC), Sydney Australia, 24–27 July 2023; IEEE: Piscataway, NJ, USA, 2023; pp. 1–4.
28. Guo, Z.; He, Z.; Lyu, L.; Mao, A.; Huang, E.; Liu, K. Automatic Detection of Feral Pigeons in Urban Environments Using Deep Learning. *Animals* **2024**, *14*, 159. [CrossRef]
29. Gao, L.; Zhang, J.; Yang, C.; Zhou, Y. Cas-VSwin Transformer: A Variant Swin Transformer for Surface-Defect Detection. *Comput. Ind.* **2022**, *140*, 103689. [CrossRef]
30. Yuan, W.; Xu, W. Neighborloss: A Loss Function Considering Spatial Correlation for Semantic Segmentation of Remote Sensing Image. *IEEE Access* **2021**, *9*, 75641–75649. [CrossRef]
31. Yeung, M.; Sala, E.; Schönlieb, C.-B.; Rundo, L. Unified Focal Loss: Generalising Dice and Cross Entropy-Based Losses to Handle Class Imbalanced Medical Image Segmentation. *Comput. Med. Imaging Graph.* **2022**, *95*, 102026. [CrossRef]
32. Guo, Q.; Wang, C.; Xiao, D.; Huang, Q. A Novel Multi-Label Pest Image Classifier Using the Modified Swin Transformer and Soft Binary Cross Entropy Loss. *Eng. Appl. Artif. Intell.* **2023**, *126*, 107060. [CrossRef]
33. Agarwal, N.; Balasubramanian, V.N.; Jawahar, C.V. Improving Multiclass Classification by Deep Networks Using DAGSVM and Triplet Loss. *Pattern Recognit. Lett.* **2018**, *112*, 184–190. [CrossRef]
34. Xiang, S.; Liang, Q.; Hu, Y.; Tang, P.; Coppola, G.; Zhang, D.; Sun, W. AMC-Net: Asymmetric and Multi-Scale Convolutional Neural Network for Multi-Label HPA Classification. *Comput. Methods Programs Biomed.* **2019**, *178*, 275–287. [CrossRef] [PubMed]
35. Zhang, H.; Qian, Z.; Tan, Y.; Xie, Y.; Li, M. Investigation of Pavement Crack Detection Based on Deep Learning Method Using Weakly Supervised Instance Segmentation Framework. *Constr. Build. Mater.* **2022**, *358*, 129117. [CrossRef]
36. Zhang, M.; Wang, H.; Wang, L.; Saif, A.; Wassan, S. CIDN: A Context Interactive Deep Network with Edge-Aware for X-Ray Angiography Images Segmentation. *Alex. Eng. J.* **2024**, *87*, 201–212. [CrossRef]
37. Pawara, P.; Okafor, E.; Groefsema, M.; He, S.; Schomaker, L.R.B.; Wiering, M.A. One-vs-One Classification for Deep Neural Networks. *Pattern Recognit.* **2020**, *108*, 107528. [CrossRef]
38. Wang, P.; Chung, A.C.S. Relax and Focus on Brain Tumor Segmentation. *Med. Image Anal.* **2022**, *75*, 102259. [CrossRef]
39. Wang, G.; Wang, F.; Zhou, H.; Lin, H. Fire in Focus: Advancing Wildfire Image Segmentation by Focusing on Fire Edges. *Forests* **2024**, *15*, 217. [CrossRef]
40. Ma, J.; Liang, P.; Yu, W.; Chen, C.; Guo, X.; Wu, J.; Jiang, J. Infrared and Visible Image Fusion via Detail Preserving Adversarial Learning. *Inf. Fusion* **2020**, *54*, 85–98. [CrossRef]
41. Zhai, J.; Mu, C.; Hou, Y.; Wang, J.; Wang, Y.; Chi, H. A Dual Attention Encoding Network Using Gradient Profile Loss for Oil Spill Detection Based on SAR Images. *Entropy* **2022**, *24*, 1453. [CrossRef] [PubMed]
42. Li, H.; He, Y.; Xu, Q.; Deng, J.; Li, W.; Wei, Y. Detection and Segmentation of Loess Landslides via Satellite Images: A Two-Phase Framework. *Landslides* **2022**, *19*, 673–686. [CrossRef]

43. Zhou, Y.; Wang, H.; Yang, R.; Yao, G.; Xu, Q.; Zhang, X. A Novel Weakly Supervised Remote Sensing Landslide Semantic Segmentation Method: Combining CAM and CycleGAN Algorithms. *Remote Sens.* **2022**, *14*, 3650. [CrossRef]
44. Feng, X.; Du, J.; Wu, M.; Chai, B.; Miao, F.; Wang, Y. Potential of Synthetic Images in Landslide Segmentation in Data-Poor Scenario: A Framework Combining GAN and Transformer Models. *Landslides* **2024**, *21*, 2211–2226. [CrossRef]
45. Lan, H.; Liu, X.; Li, L.; Li, Q.; Tian, N.; Peng, J. Remote Sensing Precursors Analysis for Giant Landslides. *Remote Sens.* **2022**, *14*, 4399. [CrossRef]
46. Grigoryan, A.M.; Agaian, S.S. Monotonic Sequences for Image Enhancement and Segmentation. *Digit. Signal Process.* **2015**, *41*, 70–89. [CrossRef]
47. Liang, L.; Zhang, Z.-M. Structure-Aware Enhancement of Imaging Mass Spectrometry Data for Semantic Segmentation. *Chemom. Intell. Lab. Syst.* **2017**, *171*, 259–265. [CrossRef]
48. Domokos, C.; Kato, Z. Parametric Estimation of Affine Deformations of Planar Shapes. *Pattern Recognit.* **2010**, *43*, 569–578. [CrossRef]
49. Qin, Z.; Chen, Q.; Ding, Y.; Zhuang, T.; Qin, Z.; Choo, K.-K.R. Segmentation Mask and Feature Similarity Loss Guided GAN for Object-Oriented Image-to-Image Translation. *Inf. Process. Manag.* **2022**, *59*, 102926. [CrossRef]
50. Schmitter, D.; Unser, M. Shape Projectors for Landmark-Based Spline Curves. *IEEE Signal Process. Lett.* **2017**, *24*, 1517–1521. [CrossRef]
51. Mehrish, A.; Subramanyam, A.V.; Emmanuel, S. Sensor Pattern Noise Estimation Using Probabilistically Estimated RAW Values. *IEEE Signal Process. Lett.* **2016**, *23*, 693–697. [CrossRef]
52. Yang, Y.; Mei, G. Deep Transfer Learning Approach for Identifying Slope Surface Cracks. *Appl. Sci.* **2021**, *11*, 11193. [CrossRef]
53. Li, D.; Tang, X.; Tu, Z.; Fang, C.; Ju, Y. Automatic Detection of Forested Landslides: A Case Study in Jiuzhaigou County, China. *Remote Sens.* **2023**, *15*, 3850. [CrossRef]
54. Ji, S.; Yu, D.; Shen, C.; Li, W.; Xu, Q. Landslide Detection from an Open Satellite Imagery and Digital Elevation Model Dataset Using Attention Boosted Convolutional Neural Networks. *Landslides* **2020**, *17*, 1337–1352. [CrossRef]
55. Ma, Z.; Mei, G. Deep Learning for Geological Hazards Analysis: Data, Models, Applications, and Opportunities. *Earth-Sci. Rev.* **2021**, *223*, 103858. [CrossRef]
56. Merghadi, A.; Yunus, A.P.; Dou, J.; Whiteley, J.; ThaiPham, B.; Bui, D.T.; Avtar, R.; Abderrahmane, B. Machine Learning Methods for Landslide Susceptibility Studies: A Comparative Overview of Algorithm Performance. *Earth-Sci. Rev.* **2020**, *207*, 103225. [CrossRef]
57. Becek, K.; Ibrahim, K.; Bayik, C.; Abdikan, S.; Kutoglu, H.S.; Glabicki, D.; Blachowski, J. Identifying Land Subsidence Using Global Digital Elevation Models. *IEEE J. Sel. Top. Appl. Earth Obs. Remote Sens.* **2021**, *14*, 8989–8998. [CrossRef]
58. Zhong, C.; Liu, Y.; Gao, P.; Chen, W.; Li, H.; Hou, Y.; Nuremanguli, T.; Ma, H. Landslide Mapping with Remote Sensing: Challenges and Opportunities. *Int. J. Remote Sens.* **2020**, *41*, 1555–1581. [CrossRef]
59. Lian, X.-G.; Li, Z.-J.; Yuan, H.-Y.; Liu, J.-B.; Zhang, Y.-J.; Liu, X.-Y.; Wu, Y.-R. Rapid Identification of Landslide, Collapse and Crack Based on Low-Altitude Remote Sensing Image of UAV. *J. Mt. Sci.* **2020**, *17*, 2915–2928. [CrossRef]
60. Wasowski, J.; Bovenga, F. Investigating Landslides and Unstable Slopes with Satellite Multi Temporal Interferometry: Current Issues and Future Perspectives. *Eng. Geol.* **2014**, *174*, 103–138. [CrossRef]

remote sensing

MDPI

Article

InSAR-Driven Dynamic Landslide Hazard Mapping in Highly Vegetated Area

Liangxuan Yan [1,2], Qianjin Xiong [3], Deying Li [1,*], Enok Cheon [2], Xiangjie She [1] and Shuo Yang [1]

[1] Faculty of Engineering, China University of Geosciences, Wuhan 430074, China; yanliangxuan@cug.edu.cn (L.Y.); shexiangjie@cug.edu.cn (X.S.); yang_shuo@cug.edu.cn (S.Y.)
[2] Norwegian Geotechnical Institute, 0484 Oslo, Norway; enok.cheon@ngi.no
[3] Hubei Geological Survey, Wuhan 430034, China; 1202011450@cug.edu.cn
* Correspondence: lideying@cug.edu.cn

Abstract: Landslide hazard mapping is important to urban construction and landslide risk management. Dynamic landslide hazard mapping considers landslide deformation with changes in the environment. It can show more details of the landslide process state. Landslides in highly vegetated areas are difficult to observe directly, which makes landslide hazard mapping much more challenging. The application of multi-InSAR opens new ideas for dynamic landslide hazard mapping. Specifically, landslide susceptibility mapping reflects the spatial probability of landslides. For rainfall-induced landslides, the scale exceedance probability reflects the temporal probability. Based on the coupling of them, dynamic landslide hazard mapping further considers the landslide deformation intensity at different times. Zigui, a highly vegetation-covered area, was taken as the study area. The landslide displacement monitoring effect of different band SAR datasets (ALOS-2, Sentinel-1A) and different interpretation methods (D-InSAR, PS-InSAR, SBAS-InSAR) were studied to explore a combined application method. The deformation interpreted by SBAS-InSAR was taken as the main part, PS-InSAR data were used in towns and villages, and D-InSAR was used for the rest. Based on the preliminary evaluation and the displacement interpreted by fusion InSAR, the dynamic landslide hazard mappings of the study area from 2019 to 2021 were finished. Compared with the preliminary evaluation, the dynamic mapping approach was more focused and accurate in predicting the deformation of landslides. The false positives in very-high-hazard zones were reduced by 97.8%, 60.4%, and 89.3%. Dynamic landslide hazard mapping can summarize the development of and change in landslides very well, especially in highly vegetated areas. Additionally, it can provide trend prediction for landslide early warning and provide a reference for landslide risk management.

Keywords: landslide hazard mapping; dynamic assessment; InSAR; highly vegetated coverage

Citation: Yan, L.; Xiong, Q.; Li, D.; Cheon, E.; She, X.; Yang, S. InSAR-Driven Dynamic Landslide Hazard Mapping in Highly Vegetated Area. *Remote Sens.* **2024**, *16*, 3229. https://doi.org/10.3390/rs16173229

Academic Editor: Fumio Yamazaki

Received: 18 July 2024
Revised: 16 August 2024
Accepted: 29 August 2024
Published: 31 August 2024

1. Introduction

Landslides are common natural disasters that cause thousands of casualties and huge property losses every year [1]. Preventing and reducing landslide risk effectively is challenging work [2]. Landslide hazard mapping (LHM), which provides a reference for urban planning, is essential for building site selection and emergency evacuation [3].

LHM is a comprehensive evaluation of the spatial probability, temporal probability, and scale probability of landslides based on susceptibility assessment [4]. Early on, there was no unified terminology for LHM, and there were many discussions on the content of evaluation. Then, LHM was further developed in the direction of quantification and systematization [5,6]. Glade et al. [7] and Fell et al. [8] established a landslide risk framework and clarified the concept and process of LHM. The spatial probability of landslides is described by susceptibility assessment based on landslide events [9]. The temporal probability of landslides is analyzed by physical models [10] or statistical models [11,12]. LHM is usually combined with GIS and has been rapidly developed and applied worldwide [13,14].

However, the evolution of landslides is a dynamic process [15]. At different times, due to changes in triggers, such as rainfall [16], earthquakes [17], and reservoir water level fluctuations [18], landslide hazards also change accordingly. Generally, this is positively correlated to rainfall, the earthquake magnitude, and the magnitude and speed of the water level changes. Static LHM cannot reflect such changes. Therefore, dynamic triggers are introduced to characterize the changes in landslide hazards. Guzzetti et al. [19] and Lee et al. [20] calculated the landslide hazard under different extreme rainfalls through the rainfall return period. Kirschbaum et al. [21] and Gariano et al. [22] used dynamic rainfall in different periods as an evaluation factor to carry out dynamic LHM.

The landslide deformation velocity can also be used directly as a time probability factor for dynamic LHM. InSAR has good regional deformation recognition capabilities and can monitor the deformation state of regional landslides in real time [23]. It can provide deformation for dynamic LHM and has been widely used [24,25]. Synthetic aperture radar (SAR) is an all-weather active sensor. Interferometric SAR (InSAR) can implement the high-precision observation of the surface [26]. D-InSAR [27], PS-InSAR [28], and SBAS-InSAR [29] have been widely used in surface deformation detection and landslide monitoring [30–34].

In areas with high vegetation, different radar datasets and solution methods greatly differ in their accuracy of deformation interpretation. Although SAR is a remote sensing radar, it is not affected by weather or light [35]. However, the absorption of radar waves by plant leaves will affect the accuracy of interpretation. Liang et al. [36] and Zhang et al. [37] explored the identification and monitoring of landslides in areas with different densities of vegetation coverage using radar remote sensing in various bands, which shows the anti-interference ability of the C band to vegetation coverage. Walker et al. [38], Oveisgharan et al. [39], and Maghsoudi et al. [40] studied the detection of the terrain, vegetation canopy, and deformation in forest areas using different InSAR interpretation methods. A large number of landslide identification and monitoring studies have shown that landslide deformation monitoring in areas with high vegetation coverage is still quite challenging. Many mountainous areas have lush vegetation and are also prone to landslides. An important task in landslide monitoring and early warning in these areas is to identify surface deformation. This study explores the applicability of different SAR data sources and different interpretation methods in such areas and how to obtain the best possible identification results.

Zigui, a county in the Three Gorges Reservoir Area, was chosen as the study area for dynamic LHM. It has hundreds of landslides and is covered by forest. A preliminary LHM was obtained through landslide susceptibility mapping (LSM) and rainfall return period analysis. A multi-InSAR interpretation method was explored and summarized to calculate the landslide deformation velocity. It was coupled with the preliminary LHM. The dynamic LHM considering landslide deformation provides a more accurate and practical reference for landslide management and provides trend prediction for landslide early warning.

2. Methods

Landslide hazard refers to the possibility of landslides occurring on a certain scale in a certain area within a certain period. It can be evaluated by analyzing the probability of occurrence.

According to the risk framework [7,41], the spatial probability of a landslide is based on susceptibility mapping. In the study area, rainfall is the main factor leading to landslides. The rainfall data are statistically used to calculate the extreme rainfall that can trigger a landslide of a specific scale. The extreme rainfall return period is analyzed to obtain the time probability of landslides. The evaluation formula of a regional landslide hazard (H) is as follows:

$$H = P(S) \times P(N_L) \tag{1}$$

where $P(S)$ is the landslide spatial probability, and $P(N_L)$ is the landslide time probability.

The evolution of landslides is a dynamic process. The risk will also change at different periods due to changes in triggers. The dynamic assessment of landslide hazards needs to capture the impact of these changes on landslides. InSAR can monitor the regional landslide deformation and provide the deformation data (D) for dynamic LHM. The improved formula is as follows:

$$H = P(S) \times P(N_L) \times D \tag{2}$$

The workflow of dynamic LHM includes four steps, as shown in Figure 1. (1) Based on remote sensing identification, field investigation, and landslide cataloging, analyze the landslide characteristics and triggering factors. The spatial probability of landslides is calculated through LSM. (2) Analyze the landslide time probability in different rainfall return periods. Define the scenarios and calculate the preliminary landslide hazard. (3) Calculate the surface deformation velocity in areas with high vegetation coverage by multi-InSAR. (4) According to Equation (2), construct a hazard dynamic evaluation matrix considering the landslide deformation to carry out dynamic LHM.

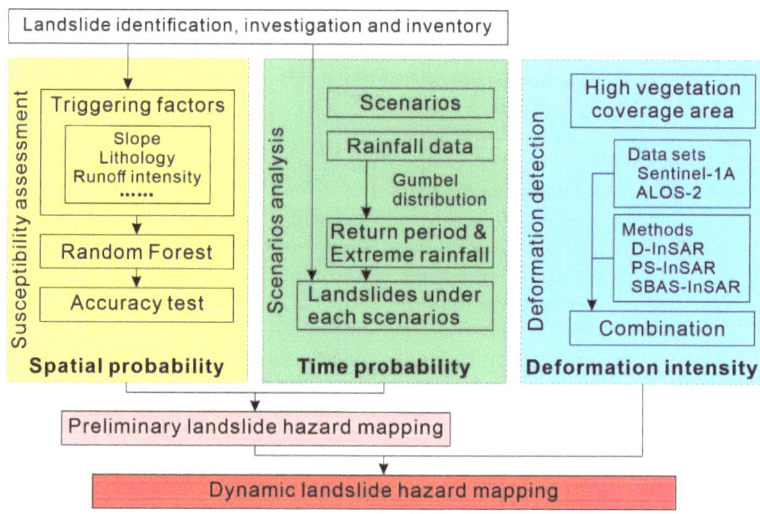

Figure 1. Dynamic LHM workflow.

2.1. Landslide Hazard Mapping

2.1.1. Landslide Spatial Probability Based on LSM

LSM is an intuitive expression of the spatial probability of landslides. Its essence is to use Engineering Geological Analogies (EGAs) to infer the landslide probability based on landslides that have already occurred. LSM methods are divided into three categories: heuristic models, statistical models, and physical models [42,43]. Random forest (RF) was used for the LSM in this study. RF is a commonly used statistical method with good accuracy and generalization abilities [44]. It takes the decision tree model as the basic unit. A "forest" supervised learning algorithm is obtained by integrating many unrelated decision tree models. The result is obtained by calculating the average or voting. During the decision tree building process, features that reduce information uncertainty most are selected to divide the dataset [45]. The evaluation criteria for decision tree node splitting generally include the Information Gain Ratio (IGR) [46] and the Gini Index [47]. To carry out LSM, the regional landslide characteristics need to be analyzed. The factors for landslide susceptibility evaluation can be chosen according to the landslide characteristics, including the geology, topography, hydrology, etc. The landslide data and evaluation factors data are unified and standardized. Landslide points and non-landslide points are randomly selected as positive and negative samples for model training. The model training accuracy is evaluated and debugged to achieve the expected accuracy. The optimized model is used to predict the landslide susceptibility probability in the entire study area. The susceptibility probability is classified, and LSM can be obtained by GIS visualization.

2.1.2. Landslide Time Probability and Preliminary LHM

The landslide time probability can be calculated by its scale exceedance probability. For landslides with detailed records, it is assumed that the time of landslide occurrence follows a Poisson distribution. The time probability ($P(N_L)$) of at least a landslide in each evaluation unit within a certain recurrence period, such as 5 years, 10 years, 20 years, or 50 years, can be calculated. For rainfall-induced landslides, the Gumbel Model [48] can be used to solve the exceedance probability ($P(N_L)$) of landslides under specific extreme rainfall conditions, such as 5-year, 10-year, 20-year, and 50-year rainstorm conditions. The preliminary LHM can be calculated according to Equation (1) and divided into four levels (H4: very high hazard; H3: high hazard; H2: moderate hazard; H1: low hazard) according to the local standard [49].

2.2. Multi-InSAR Displacement Detection

Synthetic aperture radar (SAR) moves a small-aperture antenna in a direction. The data obtained at different locations are comprehensively processed to form an equivalent large-aperture radar, thereby improving the azimuth resolution and detection accuracy [50].

InSAR uses two SAR images of the same area to obtain the phase information of each pixel through interference processing. Thus, the positional relationship between ground objects and satellites can be obtained. Differential interferometry InSAR (D-InSAR) removes the terrain phase (φ_t), atmospheric phase (φ_a), and noise phase (φ_n) from the initial interference phase (φ) to obtain the information phase (φ_d) (Equation (3)):

$$\varphi = \varphi_d + \varphi_t + \varphi_f + \varphi_a + \varphi_n \tag{3}$$

PS-InSAR takes $N + 1$ SAR images covering the same area, selects one of them as the main image, and registers the rest of them to this image to generate a time series differential interferogram. The points that maintain high coherence will be selected to establish and solve the deformation inversion model. It can effectively separate other error phase components and obtain the rate of surface deformation.

Unlike PS-InSAR, SBAS-InSAR takes the time baseline and space baseline as a condition, with less than a certain threshold, and all data are combined by interference pairs. For the No. j scene differential interference pattern ($T_b > T_a$) generated from image T_a and main image T_b, after spatial unwrapping, the interference phase ($\delta\varphi_j$) at any point can be expressed as follows:

$$\delta\varphi_j = \varphi_{T_b} - \varphi_{T_a} \approx \frac{4\pi}{\lambda}(d_{T_b} - d_{T_a}) \tag{4}$$

where φ_{T_a} and φ_{T_b} are the phase differences at T_a and T_b, respectively. $j \in [1, \cdots, N]$. d_{T_b} and d_{T_a} are the accumulated deformation amounts in the line-of-sight direction at T_a and T_b time relative to the initial time ($d_{T_0} = 0$), respectively.

Assuming that the deformation in the time interval between T_a and T_b is linear, the surface deformation is piecewise linear during the entire period, and then the deformation phase value of the No. j scene interference pattern is as follows:

$$\delta\varphi_j = \sum_{k=T_{a,j}+1}^{T_{b,j}} (T_k - T_{k-1})v_k \tag{5}$$

The integral of the speed in each period is superimposed over the time interval of the master–slave image. All unwrapped differential interference phases are combined into a matrix:

$$\delta\varphi = \boldsymbol{Bv} \tag{6}$$

$\boldsymbol{Bv}$ is an $M \times N$ matrix. The generalized inverse matrix of $\boldsymbol{B}$ can be obtained through the singular value decomposition method. The final deformation rate or cumulative displacement can be calculated.

2.3. Dynamic LHM Based on Slope Deformation

Based on the preliminary LHM, the dynamic LHM is calculated through the landslide deformation velocity. Rainfall is one of the main factors inducing landslides. Surface deformation is a direct manifestation of landslide failure. LHM based on the rainfall return period assumes that rainfall has the same impact on landslides, and InSAR monitoring results represent the status of landslides. Introducing the deformation into hazard assessment can correct errors and improve the accuracy and timeliness of evaluation results. The deformation obtained by multi-InSAR is divided into four levels according to the velocity and reginal deformation characteristics [51,52]: fast (V4), moderate (V3), slow (V2), and very slow (V1). The velocity classification standard is determined according to the maximum deformation velocity in the study area. The preliminary LHM is superimposed with the slope deformation level to establish a dynamic LHM matrix [53], as shown in Table 1.

Table 1. Dynamic LHM matrix.

	V4	V3	V2	V1
H4	VHH	VHH	HH	MH
H3	VHH	HH	MH	MH
H2	VHH	HH	MH	LH
H1	HH	MH	LH	LH

VHH: very high hazard (red); HH: high hazard (orange); MH: moderate hazard (yellow); LH: low hazard (green).

3. Database

3.1. Geological Background of Zigui

Zigui is in central China, with an area of 485.5 km^2, and it is 30 km away from the Three Gorges Dam (Figure 2). It has a subtropical continental monsoon climate, with an average annual rainfall of 1336.1 mm. The Yangtze River traverses 64 km of the county. It has 135 first-level tributaries, forming a complex and tortuous reservoir bank.

Zigui belongs to the folded mountains of western Hubei, with an average altitude of over 800 m. It has towering peaks, deep valleys, and a relative height difference of 500 m–1300 m. The terrain slope in the area changes greatly. The Yangtze River gorge area and the transition zone often form steep cliffs. The strata exposure from the Proterozoic to the Quaternary. They spread from old to young, from east to west. They are mainly constructed of terrigenous clastic rocks and carbonate rocks. Magmatic rocks are distributed in the east, and the lithology is mainly neutral and acidic rocks.

Zigui is located at the composite site of the northern end of the Hubei uplift belt of the Neocathaysian tectonic system and the Huaiyang Mountain-shaped tectonic system, which has complex structures. The N-W trending structures developed in the Pre-Sinian metamorphic rock series are composed of several folds and faults accompanied by magmatic activity. The W-E trending structures are distributed in the south and are dominated by folds composed of sedimentary cover. The tectonic of Zigui is stable. Since the Cenozoic, there have been intermittent uplifts and local faulting activities. These faults produced differential activities in the Quaternary, which promoted landslides.

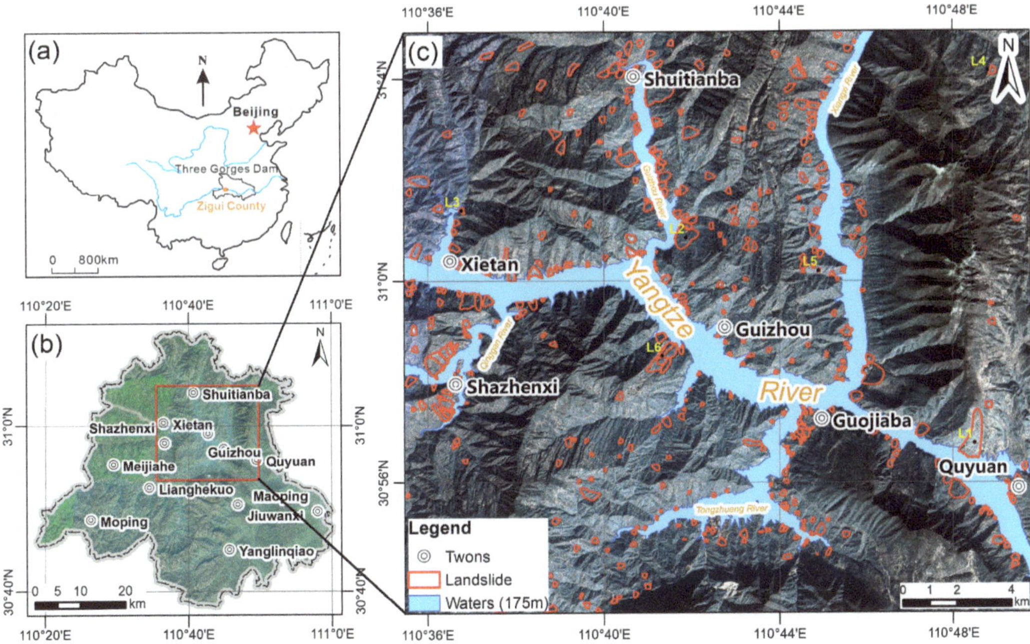

Figure 2. (**a**) Zigui County in China; (**b**) study area in Zigui County; (**c**) landslide map of study area.

3.2. Dataset

Sentinel-1 and ALOS-2 radar remote sensing datasets were collected for this study. Sentinel-1 carries a C-band synthetic aperture radar with a wavelength of 5.5 cm and a resolution of 5 m × 20 m. A total of 81 scenes of Sentinel-1A ascending orbit data from January 2019 to December 2021 were used to carry out PS-InSAR and SBAS-InSAR monitoring. ALOS-2 was launched on 24 May 2014 and is equipped with the PALSAR-2 sensor. The operating band is the L-band with a wavelength of 25 cm and a resolution of 10 m × 10 m. A total of two scenes of ascending orbit data from 20 May 2019 and 15 July 2019 were obtained for D-InSAR. According to the characteristics of electromagnetic waves, the L-band ALOS-2 image with a longer wavelength has better penetration into vegetation. This gives it a unique advantage in areas with high vegetation coverage. The Copernicus DEM GLO-30 (https://doi.org/10.5270/ESA-c5d3d65, accessed on 28 August 2024), archived on 8 April 2021, was collected. Since the Three Gorge Reservoir Area has been impounded since 2003 and was first impounded to 175 m in 2010, the DEM after 2011 is more suitable for this study. Dataset information is shown in Table 2. All images fully cover the study area.

Table 2. Dataset information.

Type	Name	Resolution/m	Phase	Coverage
SAR	ALOS-2	10	20 May 2019, 15 July 2019	2 scenes, covering 485.5 km^2
	Sentinel-1A	5 × 20	1 January 2019~ 28 December 2021	81 scenes, covering 485.5 km^2
DEM	COP GLO-30	30	2011~2015	Covering 485.5 km^2

3.3. Landslide Inventory

Based on historical landslide data and field investigation, a total of 413 landslides were confirmed in the study area. A detailed landslide inventory recorded the types, time of failure, location, deformation, elements at risk, and other information. The locations and boundaries of these landslides are marked in Figure 1. Most of these landslides were accumulation landslides. They are elongated, tongue-shaped, or semicircular. A total of 104 of them had a volume of more than 10^6 m^3. The largest one was the Xintan landslide (L1 in Figure 2), with a volume of 3×10^7 m^3, destroyed on 12 June 1985. A total of 28 landslides were unstable, with a total volume of 7.32×10^7 m^3. Most landslides are distributed along the reservoir bank and are closely related to rainfall and reservoir level changes. A total of 198 are on high-steep slopes far from the fluctuation zone. The high vegetation in the study area brings challenges to landslide deformation detection.

4. Results

4.1. LSM in Zigui

The joint action of the geological environment and external triggers causes landslides. The former includes topography, lithology, structures, etc. The latter includes the hydrological environment and others. According to the survey results, the following 12 factors were extracted as indicators for the LSM of the study area: Elevation, Slope, Aspect, Terrain Roughness Index (TRI), Curvature, Plane Curvature, Section Curvature, Lithology, Distance to Fault, Topographic Wetness Index (TWI), Stream Power Index (SPI), and Distance to River. These factors are shown in Figure 3. After collinearity analysis and information gain calculation, the Curvature, TWI, and SPI factors were eliminated, and the remaining nine indicators were used for LSM.

The information value of each factor on the occurrence of landslides was calculated, and the IGR was further calculated, as shown in Figure 4. The IGR reflected the contribution of the factor to the LSM. The factor with the highest IGR was Elevation, reaching 0.0172. The lowest one was the SPI, which is slightly higher than 0.0001. To screen important evaluation factors and improve the accuracy of the LSM, the factors with the lowest IGRs were eliminated to verify the model accuracy. If all factors were retained, the model accuracy was 0.868. The model accuracy was improved to 0.872 when the SPI was eliminated. The model accuracy was further improved to 0.875 when the SPI and Curvature were eliminated. When the SPI, Curvature, and TWI were eliminated, the model accuracy was improved to 0.877, which was the highest among all six models. If we continue to eliminate the Plane Curvature and Profile Curvature, the model accuracy decreases instead. Therefore, the best model is to eliminate the three SPI, TWI, and Curvature factors and retain the remaining nine factors, as shown in Table 3.

Table 3. LSM modeling accuracy statistics table.

Elimination Factors	RF Model Accuracy
None	0.868
SPI	0.872
SPI, Curvature	0.875
SPI, Curvature, TWI	0.877
SPI, Curvature, TWI, Plane Curvature	0.873
SPI, Curvature, TWI, Plane Curvature, Profile Curvature	0.872

All evaluation factors and landslide datasets were resampled to a grid resolution of 25 m × 25 m. Grids without landslides were defined as non-landslide grids. There were 27,785 grids of landslides in the study area. A total of 27,785 non-landslide grids were randomly selected. They were combined with all 27,785 landslide grids to form training samples and verification data. The RF model was constructed in SPSS Pro. The training sample and validation data size ratio was 1:1. All factors were standardized by Z-Score. The optimal hyperparameters of the RF in this study were as follows: bootstrp: true; oob_score: no; max_features: auto; max_depth: 100; min_sanples_leaf: 1; min_samples_split: 2; max_leaf_nodes: 10; min_impurity_decrease: 0; criterion: Gini; min_sample_leaf: 0.

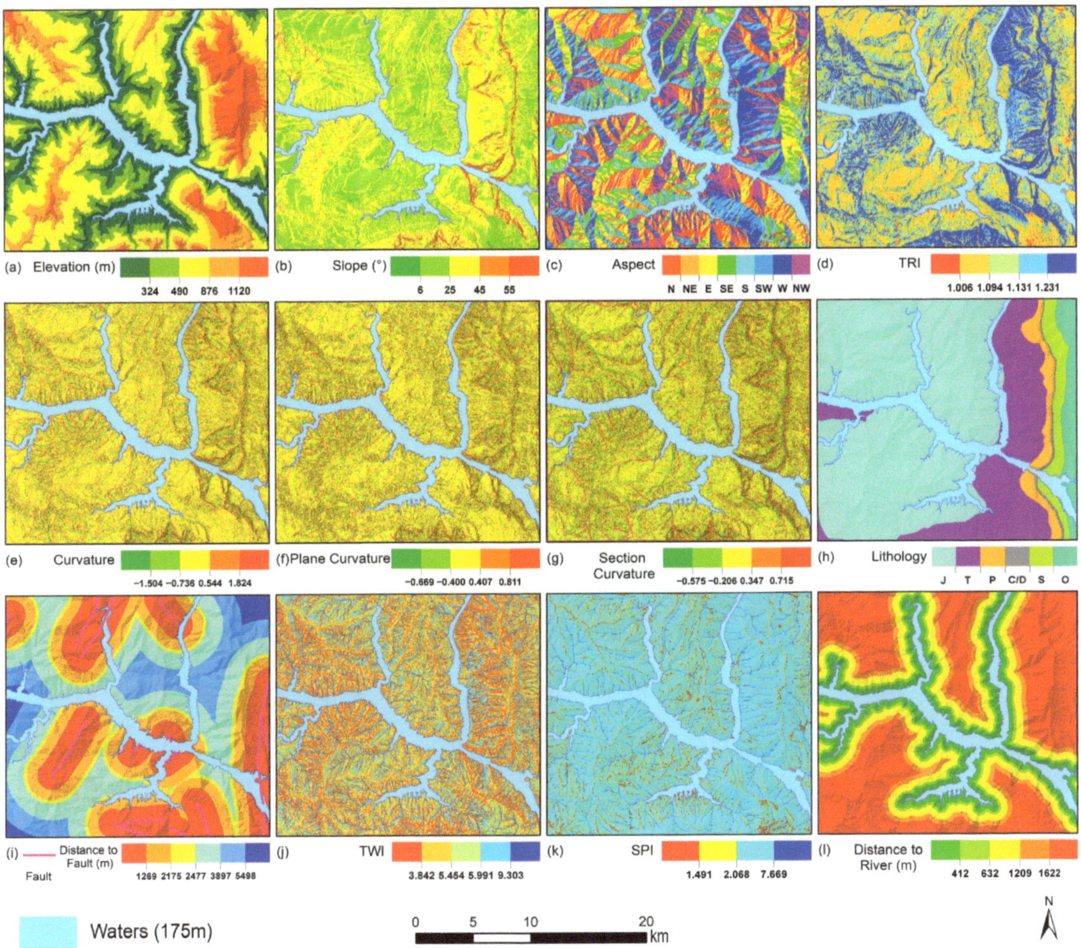

Figure 3. Evaluation factors of LSM: (**a**) Elevation; (**b**) Slope; (**c**) Aspect; (**d**) TRI; (**e**) Curvature; (**f**) Plane Curvature; (**g**) Section Curvature; (**h**) Lithology; (**i**) Distance to Fault; (**j**) TWI; (**k**) SPI; (**l**) Distance to River.

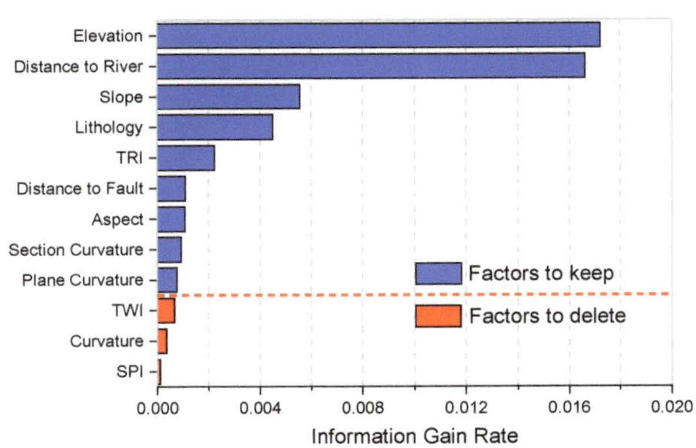

Figure 4. IGR of LSM factors.

The study area was divided into low-susceptibility (LS) zones, moderate-susceptibility (MS) zones, high-susceptibility (HS) zones, and very-high-susceptibility (VHS) zones. They were divided according to the probability of landslides as [0, 0.4), [0.4, 0.6), [0.6, 0.8), and [0.8, 1] [45], as shown in Figure 5. The VHS zones were distributed along both banks of the Yangtze River trunk and tributaries, while the MS and LS zones were mainly in high-altitude areas and far away from the river. ROC analysis was used for the accuracy evaluation [54], with AUC = 0.877. The frequency ratio method was used to analyze the prediction accuracy, as shown in Table 4. In the LS zones, the frequency ratio was 0.084. In the HS zones, the frequency ratio was 1.257, reaching 3.489 in the VHS zones. The landslide frequency ratio increased significantly with increasing susceptibility. This showed that the RF performed well in predicting landslides in the study area.

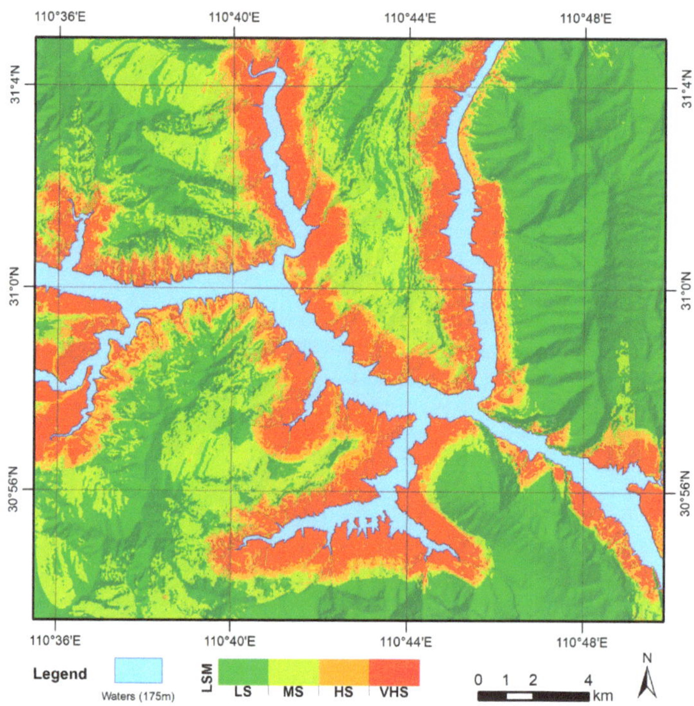

Figure 5. LSM of the study area.

Table 4. Frequency ratio analysis of LSM.

Landslide Susceptibility Level	LS	MS	HS	VHS
Landslide grids	1123	4256	4181	18,225
Zone grids	333,901	148,659	83,469	131,069
Landslide percentage	4.04%	15.32%	15.05%	65.59%
Zone percentage	47.90%	21.33%	11.97%	18.80%
Frequency ratio	0.084	0.718	1.257	3.489

4.2. Scenarios and Preliminary LHM

Statistics on the correlation between landslides and rainfall in the study area show that landslides have the greatest correlation with 3-day rain. Based on the Gumbel distribution model and daily rainfall data from 1980 to 2020, the extreme values of the 3-day rainfall of different return periods in the study area were predicted (Table 5).

Table 5. Extreme rainfall under different return periods in the study area and scenarios.

Scenarios	A	B	C	D
Return period	5 years	10 years	20 years	50 years
3-day rainfall/mm	147.16	171.15	197.14	223.96
Probability	0.2	0.1	0.05	0.02

Landslide information, such as the failure time, rainfall intensity, and duration, was extracted from the landslide inventory. The scale of landslides was calculated by comparing the extreme rainfall in each return period. Then, it was compared with the area of the landslide susceptibility zoning. The spatial probabilities of landslides in different return periods and different susceptibility zoning can be calculated. The preliminary LHM was based on the spatial probability, multiplied by the time probability of the return period. The time probability is the reciprocal of the return period. The spatial probability in different return periods is shown in Table 6.

Table 6. Preliminary LHM based on rainfall return period.

LSM	Calculated Item	Scenarios			
		A	B	C	D
VHS	Zone grids	131,069			
	Landslide grids	142	1743	6871	18225
	Spatial probability (%)	0.11	1.33	5.24	13.90
	Preliminary hazard ($\times10^{-3}$)	0.20	1.30	2.60	2.80
HS	Zone grids	83,469			
	Landslide grids	47	861	1932	4181
	Spatial probability (%)	0.06	1.03	2.31	5.01
	Preliminary hazard ($\times10^{-3}$)	0.10	1.00	1.20	1.00
MS	Zone grids	148,659			
	Landslide grids	18	364	1944	4256
	Spatial probability (%)	0.01	0. 24	1.31	2.86
	Preliminary hazard ($\times10^{-3}$)	0.00	0.20	0.70	0.60
LS	Zone grids	333,901			
	Landslide grids	0	547	547	1123
	Spatial probability (%)	0.00	0.16	0.16	0.34
	Preliminary hazard ($\times10^{-3}$)	0.00	0.20	0.10	0.10

The probability of extreme rainfall-inducing landslides under each scenario was calculated. It was combined with the LSM to realize the preliminary LHM of the study area (Figure 6). As the rainfall intensity increased, the landslide hazard level increased. The high-hazard zoning in each scenario was distributed within about 1 km on both banks of the Yangtze River trunk and tributaries. Among them, the very-high-hazard zonings were mainly located in the following regions: (1) the Xintan landslide (L1 in Figure 2) and surrounding area in the east of the study area; (2) the middle section of the Xiangxi River; (3) near the Kaziwan landslide (L2 in Figure 2) on the left bank of the Guizhou River; (4) the southeast slope on the left bank of the Qinggan River.

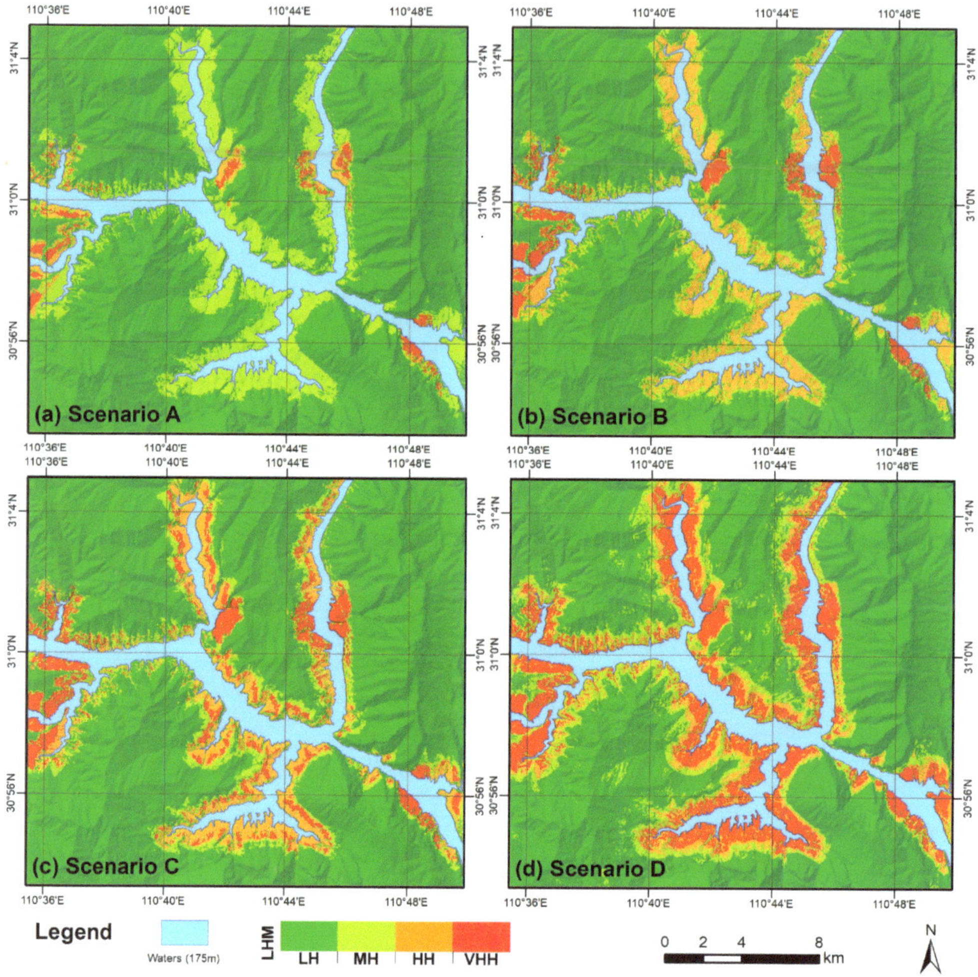

Figure 6. Preliminary LHM of the study area in different scenarios (**a**) Scenario A; (**b**) Scenario B; (**c**) Scenario C; (**d**) Scenario D.

4.3. Slope Deformation in High-Vegetation-Coverage Area

A total of 81 Sentinel-1A ascending orbit images in the study area from 2019 to 2021 were used to interpret the landslide deformation. In 2019, two additional ALOS ascending orbit images were added for D-InSAR interpretation to fill in the gaps in the data, further improving the accuracy of the surface deformation. The visibility of the image sets was 83%, which met the evaluation needs (Figure 7). In this study, a multi-InSAR interpretation fusion method was adopted via SARscape with the high vegetation coverage. The interpretation results of the SBAS-InSAR were taken as the main body, and the PS-InSAR interpretation results were superimposed onto the urban area. The D-InSAR interpretation results were used to supplement the rest of the area in 2019. For the fused deformation map, the Radial Basis Functions were selected for spatial interpolation and divided into four levels [51,52]: V1 (0~10 mm/year), V2 (10~20 mm/year), V3 (20~30 mm/year), and V4 (>30 mm/year). The surface deformation velocity map of the study area was formed with annual time intervals, as shown in Figure 8.

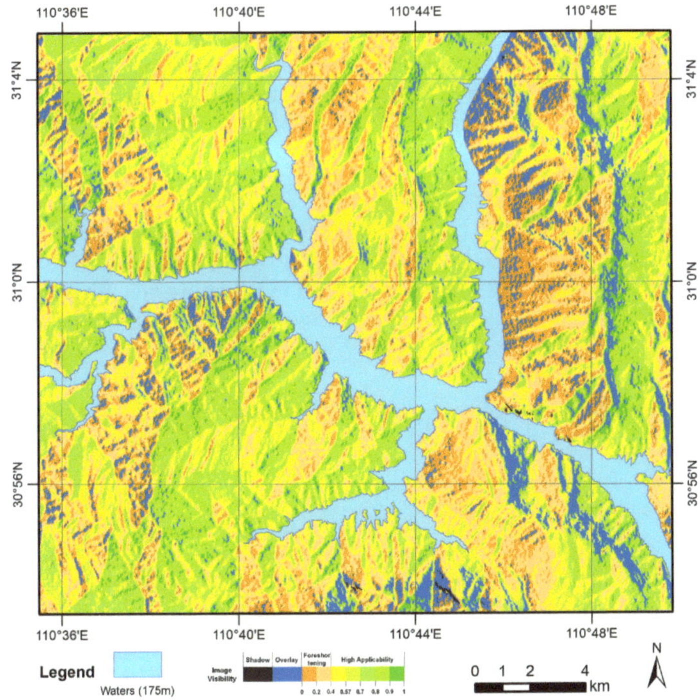

Figure 7. Visibility graph of the SAR datasets.

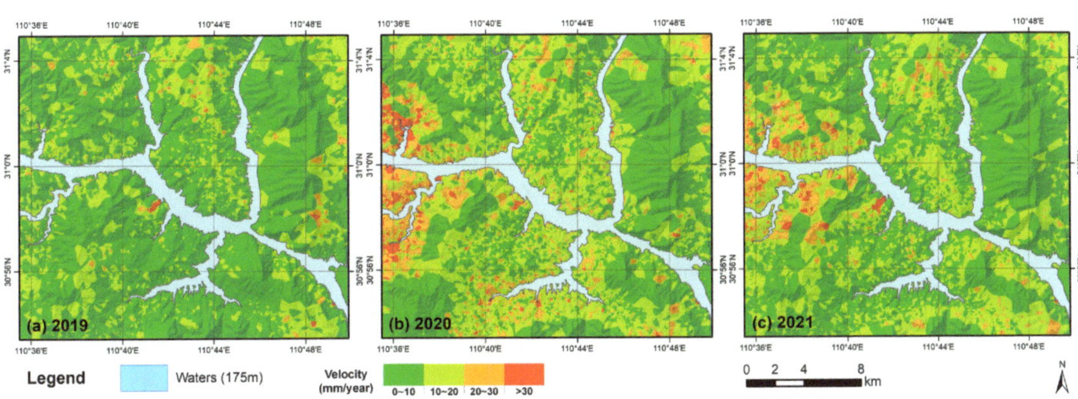

Figure 8. Deformation maps of the study area from 2019 to 2021 by multi-InSAR: (**a**) 2019; (**b**) 2020; (**c**) 2021.

The slope deformation in the study area was negligible in 2019. In 2020, the deformation velocity increased significantly, reaching the highest level in three years. The areas with deformation velocities above 20 mm/year were mainly concentrated in the western and central–southern parts of the study area. In the western part of the study area, the north bank of the Yangtze River is undergoing a highway reconstruction and expansion project, which results in a large deformation variable monitored by InSAR. In 2021, the areas with large slope deformations were the western regions in the study area; the deformations were smaller in 2020. According to the field verification survey, 70.6% of the InSAR interpretation results were consistent with the field deformation evidence. Thanks to the slow deformation of the unstable slopes in the study area, a high accuracy was obtained in the deformation identification of InSAR. When the surface deformation exceeds $1/2\lambda$ in a revisiting period, the excessive deformation rate will make phase unwrapping difficult, resulting in errors. Most of the erroneous identifications came from the rapid changes in the surface caused by engineering construction. Eliminating these errors can improve landslide monitoring and early warning [55]. They were corrected in the dynamic LHM.

4.4. Dynamic LHM of Zigui

The dynamic LHM of the study area (Figure 9) can be calculated by matrix overlaying the deformation map from 2019 to 2021 and the preliminary LHM (Scenario B) according to Table 1. The area of VHH in the study area was smallest in 2019 and largest in 2020. According to the landslide inventory, the number of landslides in 2020 was the highest, and it was the lowest in 2019, which was consistent with the evaluation results.

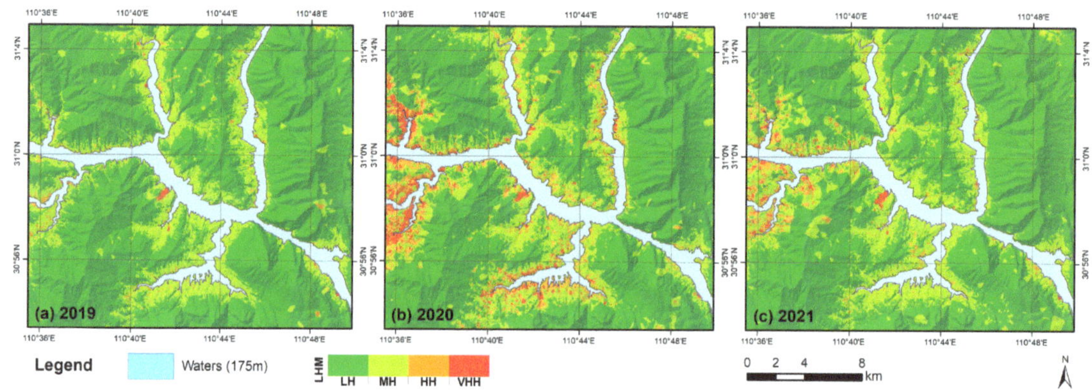

Figure 9. Dynamic LHM in the study area from 2019 to 2021: (**a**) 2019; (**b**) 2020; (**c**) 2021.

5. Discussion

5.1. Accuracy Correction of Hazard Assessment Considering Deformation

In the preliminary LHM (Scenario B), there were 40,990 grids in H4, accounting for 5.88% of the entire study area. However, there were some false-positive errors in the evaluation results. For example, the Xintan landslide in the southeastern part of the study area showed very-high-hazard (H4) classification. With years of continuous engineering management, GPS monitoring data in recent years showed that it was generally stable. Therefore, the hazard level in this area should be reduced. After introducing slope deformation for dynamic LHM, this type of error has been effectively corrected. The Xintan landslide appeared in the LH or MH state. As shown in Table 7, in the dynamic LHM from 2019 to 2021, the numbers of grids in VHH were 934, 16,255, and 4373, accounting for 0.13%, 2.33%, and 0.63% of the study area, respectively. Compared with the preliminary LHM, the areas of VHH decrease, reaching 97.8%, 60.4%, and 89.3%, respectively. The areas of high-hazard zones in these 3 years were reduced by 87.9%, 26.6%, and 62.1%. This significant reduction trend also continued in HH. Dynamic LHM can significantly reduce the scope of high-level landslide hazard zones, greatly reducing unnecessary response and panic.

The Kamenziwan (L3 in Figure 2) landslide in 2019, the Tanjiawan (L4 in Figure 2) landslide in 2020, and the Xiaoyantou (L5 in Figure 2) landslide in 2021 (Figure 10a,b) were all in HH or VHH of the corresponding years. The Lianhuatuo (L6 in Figure 2) landslide in the central part of the study area was in VHH for three years from 2019 to 2021. InSAR monitoring showed that the average displacement velocity of the landslide during these three years exceeded 40 mm/year. InSAR monitoring indicates that the Xiaoyantou landslide experienced major deformations in 2020 and 2021 (Figure 10c–e). In the dynamic

LHM, it was in MH in 2019, upgraded to HH in 2020, and became VHH in 2021. The dynamic LHM demonstrates great evaluation accuracy. Therefore, the dynamic LHM of regional landslides based on InSAR is a summary of the landslide hazard in this period and a prediction of the landslide hazard for the future.

Table 7. Comparison of LHM.

Landslide Hazard Level	Preliminary LHM		Dynamic LHM					
	Scenario B		2019		2020		2021	
	Grids	%	Grids	%	Grids	%	Grids	%
LH	530,585	76.11	546,267	78.36	481,268	69.04	503,123	72.17
MH	57,380	8.23	141,698	20.33	149,580	21.46	163,792	23.50
HH	68,143	9.78	8199	1.18	49,995	7.17	25,810	3.70
VHH	40,990	5.88	934	0.13	16,255	2.33	4373	0.63

Figure 10. Photos of Xiaoyantou landslide: (**a**) landslide image on 12 November 2020, by GaoFeng-1; (**b**) landslide failure image on 27 November 2021, by GaoFeng-2; (**c**) landslide failure photo on August 20, 2021; (**d**) scarp on landslide crown; (**e**) cracks on landslide right edge.

In terms of the accuracy of hazard assessment, dynamic LHM also showed significant superiority. In 2019, 2020, and 2021, 8, 78 and 38 landslides deformed, respectively. Based on LHM and dynamic LHM, the landslide prediction accuracy in each year can be obtained (Table 8). According to Table 8, in 2019, the accuracies of the two were the same. But in 2020 and 2021, the accuracy of the dynamic LHM improved by 47.5 and 44.3 percentage points, respectively.

The comparison showed that after introducing the deformation interpreted by InSAR for LHM, the area of VHH zoning was reduced, and the false-positive error was also reduced. Dynamic LHM also improved the prediction accuracy. It generally shows an excellent summary and prediction performance and can provide references for landslide early warning and risk management in the study area.

Table 8. Comparison of prediction accuracy between LHM and dynamic LHM.

Years		Total Landslides	Forecast Number	Accuracy
2019	LHM	8	7	87.5%
	Dynamic LHM	8	7	87.5%
2020	LHM	78	68	87.2%
	Dynamic LHM	78	31	39.7%
2021	LHM	38	27	71.1%
	Dynamic LHM	38	14	36.8%

5.2. *Different InSAR in High-Vegetation-Coverage Area*

In view of the high vegetation coverage in the study area, this research used SAR data of different resolutions and bands (Sentinel-1A, ALOS-2) combined with different methods, such as SBAS-InSAR, PS-InSAR, and D-InSAR.

Because there are only two images of ALOS-2 of the study area, only two scenes in 2019 can be used for D-InSAR analysis. The Sentinel-1A datasets had 81 scenes, and two sets of SBAS-InSAR detection and two sets of PS-InSAR detection were performed. During the interpretation process, the time baseline was set to 120 d, the spatial baseline was set to 2%, the coherence coefficient of the PS-InSAR was 0.6, the coherence coefficient of the SBAS-InSAR was 0.1, and the other parameters were set to the default values. In the first group of SBAS-InSAR monitoring, 2 images could not be connected to generate image pairs because of the 2% spatial baseline, and 32 images remained after elimination. After filtering the interferograms, the numbers of connected image pairs of the two sets of SBAS-InSAR were 64 and 314, respectively. The connection of four sets of time series InSAR image pairs is shown in Figure 11.

Five sets of InSAR deformation recognition were conducted in total. There were large differences in the number of deformation points or deformation areas. However, the deformation results were highly consistent. The longer the radar wavelength, the better it can penetrate vegetation. ALOS-2 is in the L-band. Compared with the C-band of Sentinel-1A, ALOS-2 can obtain more deformation information in high-vegetation-coverage areas. The ALOS-2 resolution is 10 m, and the Sentinel-1A resolution is 15 m. Therefore, ALOS-2 data can obtain more information in the same area. D-InSAR obtained 352,586 deformation points carried out by ALOS-2 data, with an area of 277.61km^2, accounting for 57.18% of the study area. This was much better than the time series InSAR carried out with Sentinel-1A data. This indicates that ALOS-2 performs better in extracting deformation information.

Regarding Sentinel-1A time series measurement, the deformation points calculated by PS-InSAR (34 scenes) were more than those calculated by PS-InSAR (81 scenes). The reason is that the PS-InSAR (34 scenes) had many isolated points with low coherence (0.6), which suggests low reliability. This showed that the interpretation results were positively correlated with the amount of data.

This was also true for the SBAS-InSAR. The more data the SBAS-InSAR used, the more connected image pairs it generated, the more deformation points it could ultimately obtain, and the larger the deformation area. However, even for SBAS-InSAR using 81 scenes of Sentinel-1A data, the generated deformation area was only 68.63km^2, accounting for only 14.14% of the study area. The deformation areas in the other three sets are all less than 10%. The comparison of the five sets of InSAR is shown in Table 8.

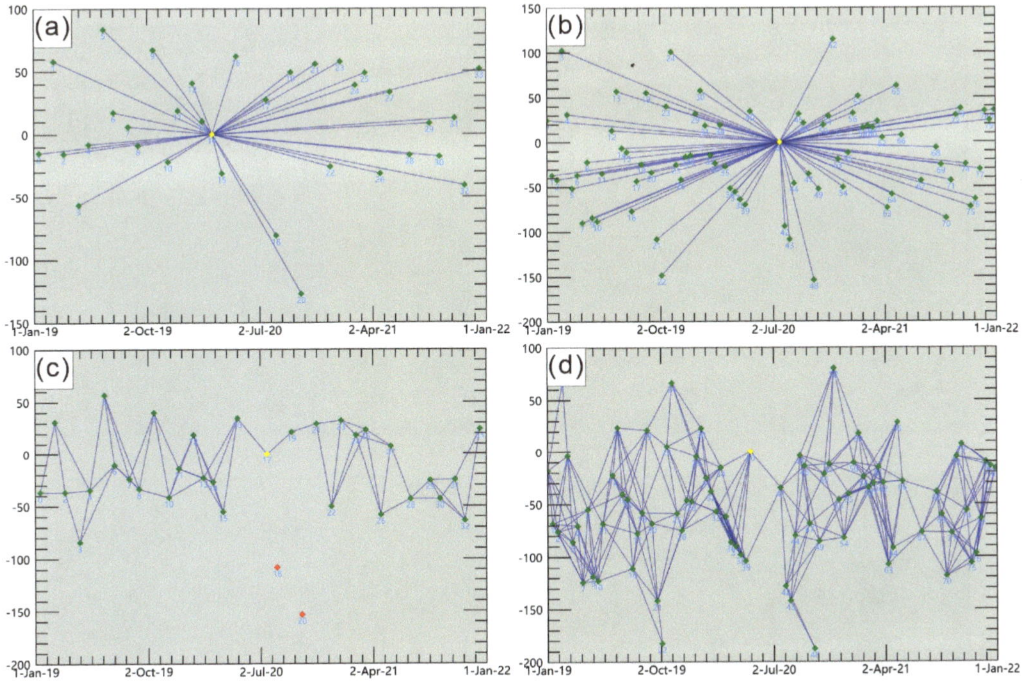

Figure 11. Time series InSAR spatiotemporal connection diagram: (**a**) PS-InSAR (34 scenes); (**b**) PS-InSAR (81 scenes); (**c**) SBAS-InSAR (32 scenes); (**d**) SBAS-InSAR (81 scenes).

5.2.1. Different InSAR Interpretation Methods in High-Vegetation-Coverage Area

As shown in Table 9 and Figure 12, the average deformations of the four sets of Sentinel-1A InSAR were close. They could confirm each other in most areas and had the same deformation trends. However, the two sets of PS-InSAR showed abnormal deformation in dense vegetation areas. The SBAS-InSAR incorrectly generated deformation points in waters. Taken together, PS-InSAR is more accurate in open urban areas. And SBAS-InSAR is better in areas with high vegetation coverage.

Table 9. Information and results of five sets of InSAR.

Algorithm	D-InSAR	PS-InSAR	PS-InSAR	SBAS-InSAR	SBAS-InSAR
Datasets	ALOS-2 Ascending orbit (2 scenes)	Sentinel-1A Ascending orbit (34 scenes)	Sentinel-1A Ascending orbit (81 scenes)	Sentinel-1A Ascending orbit (32 scenes)	Sentinel-1A Ascending orbit (81 scenes)
Connected image pairs	1	33	80	64	314
Total time span	56 days	1068 days	1092 days	1068 days	1092 days
Output coherence threshold	0.3	0.6	0.6	0.1	0.1
Average coherence	0.34	0.71	0.72	0.25	0.3
Deformation points	1,352,586	107,565	92,556	74,495	173,093
Deformation area/km^2	277.61	46.14	28.61	28.73	68.63
Area proportion	57.18%	9.50%	5.89%	5.92%	14.14%
Average deformation	0.07mm	−0.98 mm/year	−0.68 mm/year	1.09 mm/year	0.93 mm/year
Maximum	61.97mm	32.01 mm/year	45.89 mm/year	47.49 mm/year	52.73 mm/year
Minimum	−66.03mm	−43.36 mm/year	−57.02 mm/year	−54.21 mm/year	−63.53 mm/year

D-InSAR monitoring time interval was short without annual average deformation velocity.

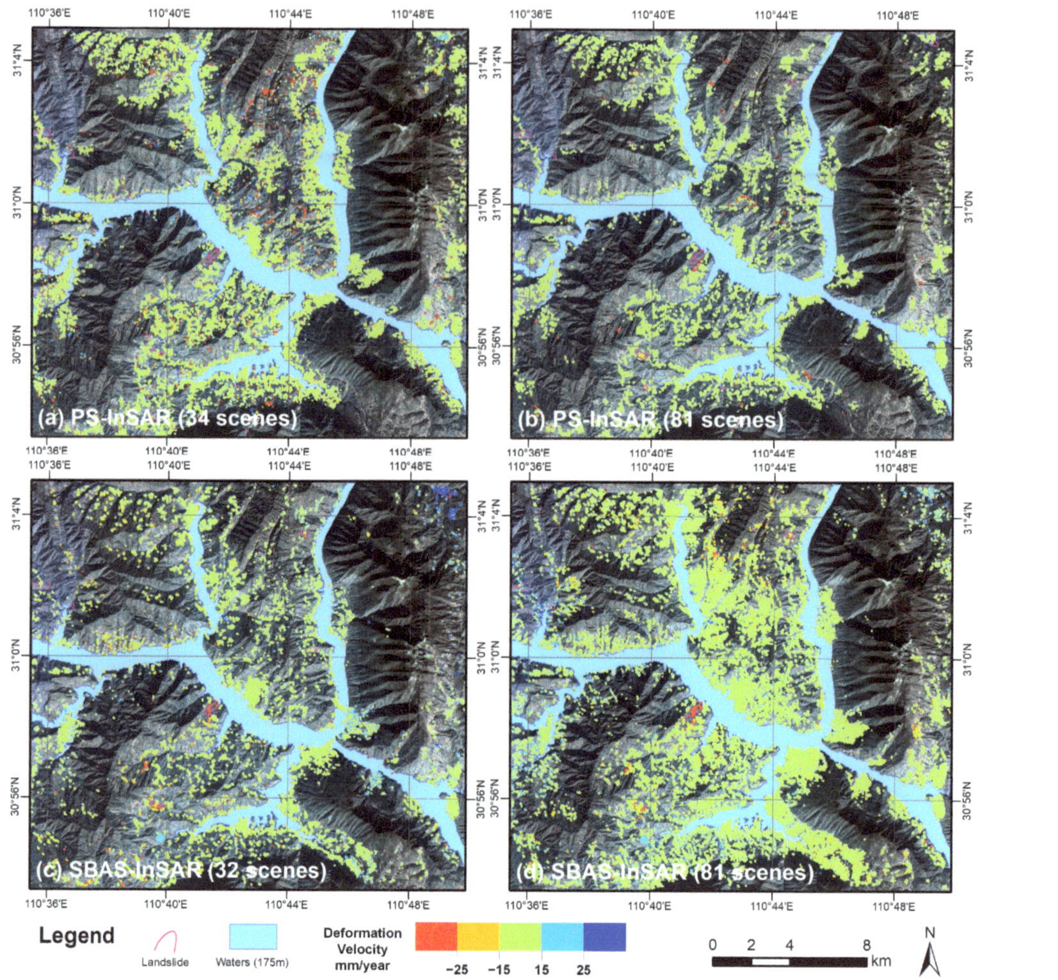

Figure 12. Comparison of Sentinel-1A time series InSAR interpretation results: (**a**) PS-InSAR (34 scenes); (**b**) PS-InSAR (81 scenes); (**c**) SBAS-InSAR (32 scenes); (**d**) SBAS-InSAR (81 scenes).

As shown in Figure 13, in the deformation area of the Lianhuatuo landslide, the PS-InSAR (34 scenes) showed that the deformation in the yellow box was positive deformation, and in the red box it is approximately 0. In the other three sets of InSAR, the deformation in the yellow box is approximately 0, and the deformation in the red box is a negative value. Although there were errors in the first set, the relative deformation of the yellow box and the red box in the four sets of time series InSAR were the same and could be verified with each other. It can be seen from the four sets of results that the two sets of PS-InSAR basically did not obtain the deformation information of the landslide. In comparison, the two sets of SBAS-InSAR both obtained the deformation information, which was highly vegetated. Among them, the deformation results of the SBAS-InSAR (81 scenes) were obviously richer and more accurate. The deformation direction here was roughly the same as the direction of the Sentinel-1A ascending orbit sensor. Therefore, the slow sliding of landslides toward the water surface between 2019 and 2021 showed negative deformation in the line of sight of InSAR deformation interpretation.

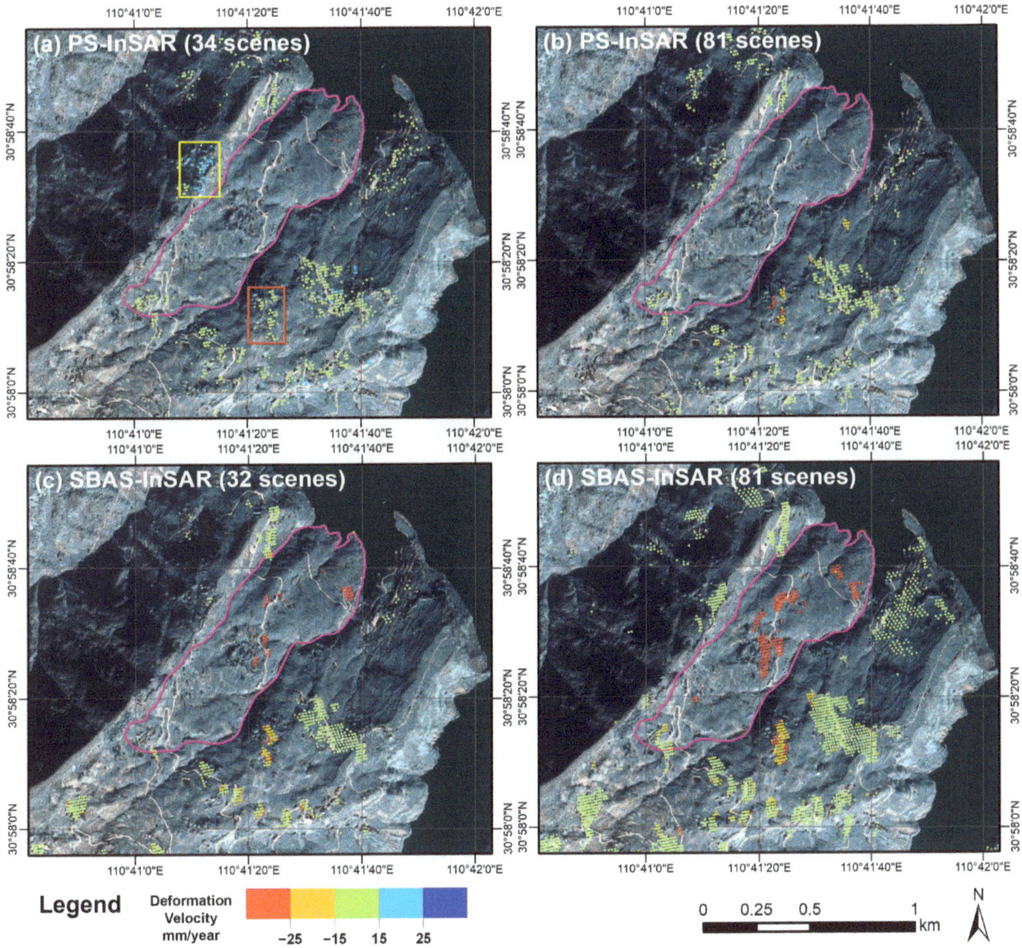

Figure 13. Comparison of Sentinel-1A time series InSAR interpretation results of Lianhuatuo landslide: (**a**) PS-InSAR (34 scenes); (**b**) PS-InSAR (81 scenes); (**c**) SBAS-InSAR (32 scenes); (**d**) SBAS-InSAR (81 scenes).

5.2.2. Different InSAR Datasets in High-Vegetation-Coverage Area

According to the previous comparison, the best detection results of the SBAS-InSAR (81 scenes) were compared with those of ALOS-2. Figure 14 is the D-InSAR deformation interpretation diagram of ALOS-2 data (abnormal deformation results in the perspective shrinkage area were eliminated).

Eight points were randomly selected for comparison. The deformations of the ALOS-2 sets and Sentinel-1A datasets basically confirm each other. Except for a certain error in P4, the deformations of the other points were close or consistent (Figure 15). Because the ALOS-2 datasets had only two scene images, multi-temporal analysis and deformation trend judgment could not be performed.

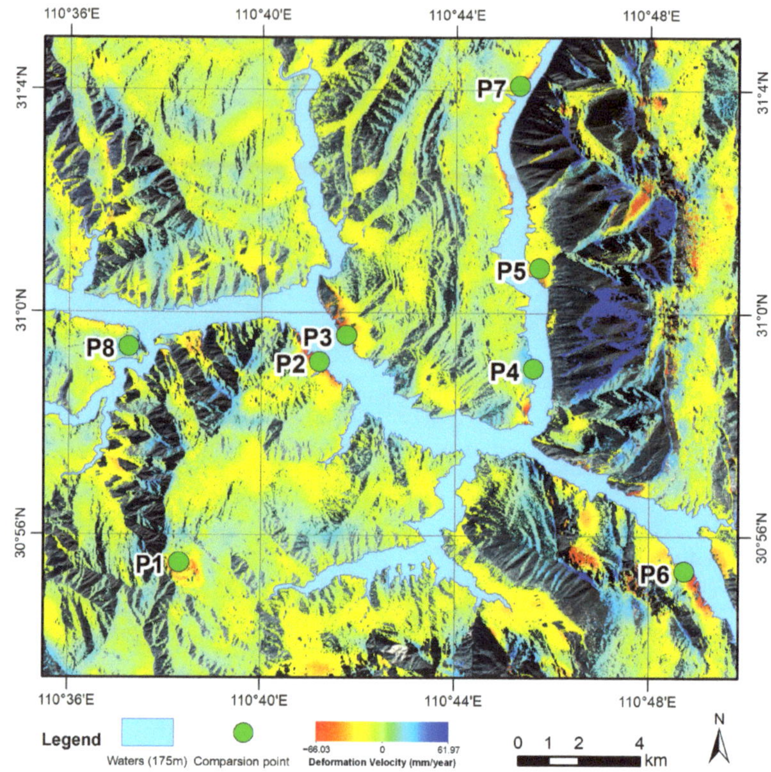

Figure 14. Deformation of ALOS-2 data by D-InSAR.

When monitoring landslides, the D-InSAR was very sensitive to landslides with deformation in the short term. Due to the high vegetation coverage in the study area, it was difficult to obtain effective PS points in many landslides with the PS-InSAR. In this study area, SBAS-InSAR was more suitable for monitoring landslide deformation. Compared with time series InSAR, D-InSAR had lower requirements for data and a shorter processing time. It is suitable for landslides with deformation in the short term but cannot conduct the deformation analysis of long-time sequences. The interpretation method should be chosen according to the time scale of the LHM. Compared with Sentinel-1A data, ALOS-2 data have a higher resolution and longer wavelength. ALOS-2 data have better monitoring results than Sentinel-1A data, but the cost is high. Obtaining the data requires customizing satellite photography plans, which may affect the time limit. When possible, using multi-view ALOS data, shortening the time baseline, and adopting time series interpretation methods can further improve the accuracy of deformation identification and the precision of dynamic LHM. Sentinel-1A data are available for free, with a 12-day spatial baseline, which is relatively stable. From the perspective of availability and economy, Sentinel-1A is more suitable for InSAR monitoring and analysis in the study area.

Therefore, the InSAR deformations used in this study were mainly SBAS-InSAR (81 scenes). In cities and towns, PS-InSAR (81 scenes) data were used for appropriate correction. In other areas, D-InSAR of the ALOS-2 datasets was supplemented as a reference.

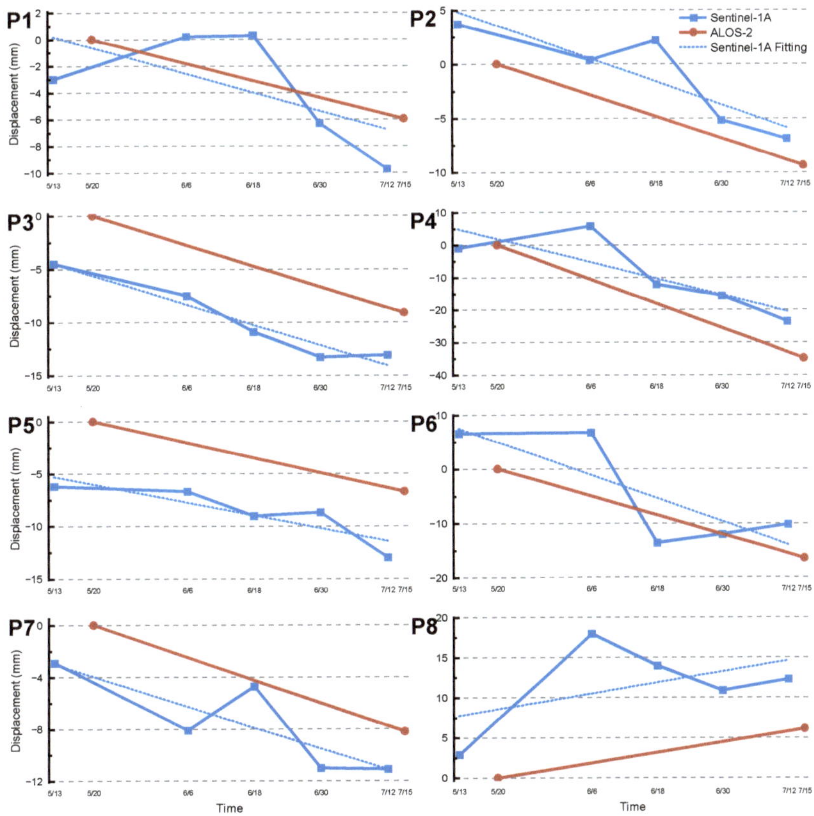

Figure 15. Interpretation result comparison of Sentinel-1A and ALOS-2.

6. Conclusions

Dynamic landslide hazard mapping incorporates surface deformation interpreted by multi-InSAR and greatly improves the accuracy. The false-positive rate for VHH was reduced by at least 60%. Taking Zigui, a high-vegetation-coverage area, as the study area, nine factors were selected to finish the landslide susceptibility to obtain the spatial probability. The rainfall exceedance probability was analyzed to calculate the time probability and define the scenarios. Dynamic landslide hazard mapping from 2019 to 2021 was finished according to the surface deformation. The research results show the following:

(1) Dynamic landslide hazard mapping can effectively reduce false-positive errors. Compared with preliminary landslide hazard mapping, while ensuring the assessment's accuracy, the areas of very-high-hazard zones in 2019, 2020, and 2021 were reduced by 97.8%, 60.4%, and 89.3%, respectively. The high-hazard zones were reduced by 87.9%, 26.6%, and 62.1%, respectively. The landslide hazard in the study area was the highest in 2020. The landslide hazard mapping of each year was in line with the actual situation verified by the field. It can provide effective data support and a reference for landslide early warning and risk management;

(2) This study compared the application effects of two SAR datasets and three InSAR interpretation methods in a high-vegetation-coverage area. It showed that the L-band SAR datasets had a better penetration ability. L-band SAR can penetrate vegetation to obtain more surface deformation information. L-band SAR datasets should be used as much as possible. Long-time series SBAS-InSAR obtained more deformation data in high-vegetation-coverage areas, and the quality of the deformation data was positively correlated with

the number of images. D-InSAR was more sensitive to short-term deformation, with a lower requirement of images and fast processing; hence, it is suitable for short-term and emergency hazard mapping.

Dynamic landslide hazard mapping has excellent indicative predicting capabilities. Geological engineers can explore its application in landslide disaster prevention and mitigation, especially landslide early warning. Due to the limitations of the study, the contribution of each factor was not independently analyzed. Further research could differentiate them to obtain a more precise dynamic LHM.

Author Contributions: Conceptualization, L.Y. and Q.X.; methodology, L.Y.; software, Q.X. and X.S.; validation, Q.X., X.S. and S.Y.; formal analysis, L.Y. and Q.X.; investigation, Q.X.; resources, D.L.; data curation, X.S.; writing—original draft preparation, L.Y. and Q.X.; writing—review and editing, E.C.; visualization, L.Y.; supervision, D.L.; project administration, D.L.; funding acquisition, Q.X. All authors have read and agreed to the published version of the manuscript.

Funding: This research was funded by the Scientific Research Project of Hubei Geological Survey, grant number KJ2024-15, and by the National Key Research and Development Program of China, grant number 2023YFC3007203.

Data Availability Statement: The Sentinel-1A datasets, ALOS-2 datasets, and precise orbit data were provided by the European Space Agency. Data such as rainfall and landslides are provided by the project that supported this article. Data are available upon request by e-mail to the authors.

Acknowledgments: We thank Runqing Ye for his help with the field work.

Conflicts of Interest: The authors declare no conflicts of interest.

References

1. Gómez, D.; García, E.F.; Aristizábal, E. Spatial and Temporal Landslide Distributions Using Global and Open Landslide Databases. *Nat. Hazards* **2023**, *117*, 25–55. [CrossRef]
2. Sassa, K. "2006 Tokyo Action Plan"—Strengthening Research and Learning on Landslides and Related Earth System Disasters for Global Risk Preparedness. *Landslides* **2006**, *3*, 361–369. [CrossRef]
3. Lateltin, O.; Haemmig, C.; Raetzo, H.; Bonnard, C. Landslide Risk Management in Switzerland. *Landslides* **2005**, *2*, 313–320. [CrossRef]
4. Varnes, D.J. *Landslide Hazard Zonation: A Review of Principles and Practice*; UNESCO: Paris, France, 1984; Available online: https://unesdoc.unesco.org/ark:/48223/pf0000063038?posInSet=1&queryId=3d365353-6a5e-44a1-89c9-1daec09ab4b9 (accessed on 28 August 2024)ISBN 978-92-3-101895-4.
5. Dhakal, A.; Amada, T.; Aniya, M. Landslide Hazard Mapping and Its Evaluation Using GIS: An Investigation of Sampling Schemes for a Grid-Cell Based Quantitative Method. *Photogramm. Eng. Remote Sens.* **2000**, *66*, 981–989.
6. Chowdhury, R.; Flentje, P. Effective Urban Landslide Hazard Assessment. *Fac. Eng. Inf. Sci.—Pap. Part A* **1998**, 1–7. Available online: https://ro.uow.edu.au/eispapers/1781 (accessed on 28 August 2024).
7. Glade, T.; Anderson, M.; Crozier, M.J. *Landslide Hazard and Risk*; John Wiley & Sons Ltd.: Chichester, UK, 2005; Available online: https://onlinelibrary.wiley.com/doi/book/10.1002/9780470012659 (accessed on 28 August 2024)ISBN 0-471-48663-9.
8. Fell, R.; Corominas, J.; Bonnard, C.; Cascini, L.; Leroi, E.; Savage, W.Z. Guidelines for Landslide Susceptibility, Hazard and Risk Zoning for Land-Use Planning. *Eng. Geol.* **2008**, *102*, 99–111. [CrossRef]
9. Goyes-Peñafiel, P.; Hernandez-Rojas, A. Landslide Susceptibility Index Based on the Integration of Logistic Regression and Weights of Evidence: A Case Study in Popayan, Colombia. *Eng. Geol.* **2021**, *280*, 105958. [CrossRef]
10. Guo, Z.; Tian, B.; He, J.; Xu, C.; Zeng, T.; Zhu, Y. Hazard Assessment for Regional Typhoon-Triggered Landslides by Using Physically-Based Model—A Case Study from Southeastern China. *Georisk Assess. Manag. Risk Eng. Syst. Geohazards* **2023**, *17*, 740–754. [CrossRef]
11. Reichenbach, P.; Rossi, M.; Malamud, B.D.; Mihir, M.; Guzzetti, F. A Review of Statistically-Based Landslide Susceptibility Models. *Earth-Sci. Rev.* **2018**, *180*, 60–91. [CrossRef]
12. Jaiswal, P.; van Westen, C.J. Estimating Temporal Probability for Landslide Initiation along Transportation Routes Based on Rainfall Thresholds. *Geomorphology* **2009**, *112*, 96–105. [CrossRef]
13. Huabin, W.; Gangjun, L.; Weiya, X.; Gonghui, W. GIS-Based Landslide Hazard Assessment: An Overview. *Prog. Phys. Geogr. Earth Environ.* **2005**, *29*, 548–567. [CrossRef]
14. Pardeshi, S.D.; Autade, S.E.; Pardeshi, S.S. Landslide Hazard Assessment: Recent Trends and Techniques. *SpringerPlus* **2013**, *2*, 523. [CrossRef]

15. Del Soldato, M.; Riquelme, A.; Bianchini, S.; Tomàs, R.; Di Martire, D.; De Vita, P.; Moretti, S.; Calcaterra, D. Multisource Data Integration to Investigate One Century of Evolution for the Agnone Landslide (Molise, Southern Italy). *Landslides* **2018**, *15*, 2113–2128. [CrossRef]
16. Hufschmidt, G.; Crozier, M.J. Evolution of Natural Risk: Analysing Changing Landslide Hazard in Wellington, Aotearoa/New Zealand. *Nat. Hazards* **2008**, *45*, 255–276. [CrossRef]
17. Rodríguez, C.E.; Bommer, J.J.; Chandler, R.J. Earthquake-Induced Landslides: 1980–1997. *Soil. Dyn. Earthq. Eng.* **1999**, *18*, 325–346. [CrossRef]
18. Zhou, C.; Cao, Y.; Yin, K.; Intrieri, E.; Catani, F.; Wu, L. Characteristic Comparison of Seepage-Driven and Buoyancy-Driven Landslides in Three Gorges Reservoir Area, China. *Eng. Geol.* **2022**, *301*, 106590. [CrossRef]
19. Guzzetti, F.; Ardizzone, F.; Cardinali, M.; Rossi, M.; Valigi, D. Landslide Volumes and Landslide Mobilization Rates in Umbria, Central Italy. *Earth Planet. Sci. Lett.* **2009**, *279*, 222–229. [CrossRef]
20. Lee, S.; Won, J.-S.; Jeon, S.W.; Park, I.; Lee, M.J. Spatial Landslide Hazard Prediction Using Rainfall Probability and a Logistic Regression Model. *Math. Geosci.* **2015**, *47*, 565–589. [CrossRef]
21. Kirschbaum, D.; Stanley, T. Satellite-Based Assessment of Rainfall-Triggered Landslide Hazard for Situational Awareness. *Earth's Future* **2018**, *6*, 505–523. [CrossRef]
22. Gariano, S.L.; Rianna, G.; Petrucci, O.; Guzzetti, F. Assessing Future Changes in the Occurrence of Rainfall-Induced Landslides at a Regional Scale. *Sci. Total Environ.* **2017**, *596–597*, 417–426. [CrossRef]
23. van Natijne, A.L.; Bogaard, T.A.; van Leijen, F.J.; Hanssen, R.F.; Lindenbergh, R.C. World-Wide InSAR Sensitivity Index for Landslide Deformation Tracking. *Int. J. Appl. Earth Obs. Geoinf.* **2022**, *111*, 102829. [CrossRef]
24. Zhou, C.; Cao, Y.; Gan, L.; Wang, Y.; Motagh, M.; Roessner, S.; Hu, X.; Yin, K. A Novel Framework for Landslide Displacement Prediction Using MT-InSAR and Machine Learning Techniques. *Eng. Geol.* **2024**, *334*, 107497. [CrossRef]
25. Zhou, C.; Cao, Y.; Hu, X.; Yin, K.; Wang, Y.; Catani, F. Enhanced Dynamic Landslide Hazard Mapping Using MT-InSAR Method in the Three Gorges Reservoir Area. *Landslides* **2022**, *19*, 1585–1597. [CrossRef]
26. Wright, T.J.; Parsons, B.E.; Lu, Z. Toward Mapping Surface Deformation in Three Dimensions Using InSAR. *Geophys. Res. Lett.* **2004**, *31*, 1–5. [CrossRef]
27. Casu, F.; Buckley, S.M.; Manzo, M.; Pepe, A.; Lanari, R. Large Scale InSAR Deformation Time Series: Phoenix and Houston Case Studies. In Proceedings of the 2005 IEEE International Geoscience and Remote Sensing Symposium, IGARSS '05, Seoul, Republic of Korea, 29 July 2005; Volume 7, pp. 5240–5243. [CrossRef]
28. Ferretti, A.; Prati, C.; Rocca, F. Analysis of Permanent Scatterers in SAR Interferometry. In Proceedings of the IGARSS 2000. IEEE 2000 International Geoscience and Remote Sensing Symposium. Taking the Pulse of the Planet: The Role of Remote Sensing in Managing the Environment. Proceedings (Cat. No.00CH37120), Honolulu, HI, USA, 24–28 July 2000; Volume 2, pp. 761–763. [CrossRef]
29. Berardino, P.; Fornaro, G.; Lanari, R.; Sansosti, E. A New Algorithm for Surface Deformation Monitoring Based on Small Baseline Differential SAR Interferograms. *IEEE Trans. Geosci. Remote Sens.* **2002**, *40*, 2375–2383. [CrossRef]
30. García-Davalillo, J.C.; Herrera, G.; Notti, D.; Strozzi, T.; Álvarez-Fernández, I. DInSAR Analysis of ALOS PALSAR Images for the Assessment of Very Slow Landslides: The Tena Valley Case Study. *Landslides* **2014**, *11*, 225–246. [CrossRef]
31. Herrera, G.; Gutiérrez, F.; García-Davalillo, J.C.; Guerrero, J.; Notti, D.; Galve, J.P.; Fernández-Merodo, J.A.; Cooksley, G. Multi-Sensor Advanced DInSAR Monitoring of Very Slow Landslides: The Tena Valley Case Study (Central Spanish Pyrenees). *Remote Sens. Environ.* **2013**, *128*, 31–43. [CrossRef]
32. Yazici, B.V.; Tunc Gormus, E. Investigating Persistent Scatterer InSAR (PSInSAR) Technique Efficiency for Landslides Mapping: A Case Study in Artvin Dam Area, in Turkey. *Geocarto Int.* **2022**, *37*, 2293–2311. [CrossRef]
33. Oliveira, S.C.; Zêzere, J.L.; Catalão, J.; Nico, G. The Contribution of PSInSAR Interferometry to Landslide Hazard in Weak Rock-Dominated Areas. *Landslides* **2015**, *12*, 703–719. [CrossRef]
34. Li, Y.; Feng, X.; Li, Y.; Jiang, W.; Yu, W. Detection and Analysis of Potential Landslides Based on SBAS-InSAR Technology in Alpine Canyon Region. *Env. Sci. Pollut. Res.* **2024**, *31*, 6492–6510. [CrossRef]
35. Dwivedi, R.S.; Rao, B.R.M.; Kushwaha, S.P.S. The Utility of Day-and-Night Observation and Cloud-Penetration Capability of ERS-1 SAR Data for Detection of Wetlands. *Geocarto Int.* **2000**, *15*, 7–12. [CrossRef]
36. Liang, J.; Dong, J.; Zhang, S.; Zhao, C.; Liu, B.; Yang, L.; Yan, S.; Ma, X. Discussion on InSAR Identification Effectivity of Potential Landslides and Factors That Influence the Effectivity. *Remote Sens.* **2022**, *14*, 1952. [CrossRef]
37. Zhang, B.; Wu, S.; Ding, X.; Wang, C.; Zhu, J.; Li, Q. Use of Multiplatform SAR Imagery in Mining Deformation Monitoring with Dense Vegetation Coverage: A Case Study in the Fengfeng Mining Area, China. *Remote Sens.* **2021**, *13*, 3091. [CrossRef]
38. Walker, W.S.; Kellndorfer, J.M.; LaPoint, E.; Hoppus, M.; Westfall, J. An Empirical InSAR-Optical Fusion Approach to Mapping Vegetation Canopy Height. *Remote Sens. Environ.* **2007**, *109*, 482–499. [CrossRef]
39. Oveisgharan, S.; Saatchi, S.S.; Hensley, S. Sensitivity of Pol-InSAR Measurements to Vegetation Parameters. *IEEE Trans. Geosci. Remote Sens.* **2015**, *53*, 6561–6572. [CrossRef]
40. Maghsoudi, Y.; van der Meer, F.; Hecker, C.; Perissin, D.; Saepuloh, A. Using PS-InSAR to Detect Surface Deformation in Geothermal Areas of West Java in Indonesia. *Int. J. Appl. Earth Obs. Geoinf.* **2018**, *64*, 386–396. [CrossRef]

41. Chen, L.; Van Westen, C.J.; Hussin, H.; Ciurean, R.L.; Turkington, T.; Chavarro-Rincon, D.; Shrestha, D.P. Integrating Expert Opinion with Modelling for Quantitative Multi-Hazard Risk Assessment in the Eastern Italian Alps. *Geomorphology* **2016**, *273*, 150–167. [CrossRef]
42. Shano, L.; Raghuvanshi, T.K.; Meten, M. Landslide Susceptibility Evaluation and Hazard Zonation Techniques—A Review. *Geoenviron. Disasters* **2020**, *7*, 18. [CrossRef]
43. Medina, V.; Hürlimann, M.; Guo, Z.; Lloret, A.; Vaunat, J. Fast Physically-Based Model for Rainfall-Induced Landslide Susceptibility Assessment at Regional Scale. *CATENA* **2021**, *201*, 105213. [CrossRef]
44. Breiman, L. Random Forests. *Mach. Learn.* **2001**, *45*, 5–32. [CrossRef]
45. Shuo, Y.; Deying, L.; Liangxuan, Y.; Yuan, H.; Mingzhe, W. Landslide Susceptibility Assessment in High and Steep Bank Slopes along Wujiang River Based on Random Forest Model. *Saf. Environ. Eng.* **2021**, *28*, 131–138. [CrossRef]
46. Kullback, S.; Leibler, R.A. On Information and Sufficiency. *Ann. Math. Stat.* **1951**, *22*, 79–86. [CrossRef]
47. Ceriani, L.; Verme, P. The Origins of the Gini Index: Extracts from Variabilità e Mutabilità (1912) by Corrado Gini. *J. Econ. Inequal.* **2012**, *10*, 421–443. [CrossRef]
48. Gumbel, E.J. *Statistics of Extremes*; Columbia University Press: New York, NY, USA, 2019; Available online: https://www.degruyter.com/document/doi/10.7312/gumb92958/html (accessed on 28 August 2024)ISBN 978-0-231-89131-8.
49. Althuwaynee, O.F.; Pradhan, B.; Ahmad, N. Estimation of Rainfall Threshold and Its Use in Landslide Hazard Mapping of Kuala Lumpur Metropolitan and Surrounding Areas. *Landslides* **2015**, *12*, 861–875. [CrossRef]
50. Gabriel, A.K.; Goldstein, R.M.; Zebker, H.A. Mapping Small Elevation Changes over Large Areas: Differential Radar Interferometry. *J. Geophys. Res. Solid. Earth* **1989**, *94*, 9183–9191. [CrossRef]
51. Chao, Z. Landslide Identification and Prediction with the Application of Time Series InSAR. Ph.D. Thesis, China University of Geosciences, Wuhan, China, 2018. Available online: https://kns.cnki.net/KCMS/detail/detail.aspx?dbcode=CDFD&dbname=CDFDLAST2019&filename=1019826173.nh&v= (accessed on 28 August 2024).
52. Hungr, O.; Leroueil, S.; Picarelli, L. The Varnes Classification of Landslide Types, an Update. *Landslides* **2014**, *11*, 167–194. [CrossRef]
53. She, X.; Li, D.; Yang, S.; Xie, X.; Sun, Y.; Zhao, W. Landslide Hazard Assessment for Wanzhou Considering the Correlation of Rainfall and Surface Deformation. *Remote Sens.* **2024**, *16*, 1587. [CrossRef]
54. Hanley, J.A.; McNeil, B.J. A Method of Comparing the Areas under Receiver Operating Characteristic Curves Derived from the Same Cases. *Radiology* **1983**, *148*, 839–843. [CrossRef]
55. Manconi, A. How Phase Aliasing Limits Systematic Space-Borne DInSAR Monitoring and Failure Forecast of Alpine Landslides. *Eng. Geol.* **2021**, *287*, 106094. [CrossRef]

remote sensing

Article

Enhancing Landslide Susceptibility Mapping by Integrating Neighboring Information in Slope Units: A Spatial Logistic Regression

Leilei Li [1], Mingzhen Jia [1], Chong Xu [2,3], Yingying Tian [4], Siyuan Ma [4] and Jintao Yang [5,*]

1 School of Earth Science and Technology, Zhengzhou University, Zhengzhou 450001, China; stonejdx@zzu.edu.cn (L.L.); jiamingzhen123@gs.zzu.edu.cn (M.J.)
2 National Institute of Natural Hazards, Ministry of Emergency Management of China, Beijing 100085, China; xc11111111@126.com
3 Key Laboratory of Compound and Chained Natural Hazards Dynamics, Ministry of Emergency Management of China, Beijing 100085, China
4 Key Laboratory of Active Tectonics and Volcano, Institute of Geology, China Earthquake Administration, Beijing 100029, China; tianyingying@ies.ac.cn (Y.T.); masiyuan@ies.ac.cn (S.M.)
5 College of Economics and Management, China Agricultural University, Beijing 100083, China
* Correspondence: yangjt@lreis.ac.cn

Abstract: Landslide susceptibility mapping (LSM) is a vital tool for proactive disaster mitigation. Although numerous studies utilize slope units (SUs) for LSM, the limited integration of adjacency information, including spatial autocorrelation, often reduces predictive accuracy. In this study, GRASS GIS was utilized to generate slope units, and a spatial logistic regression (SLR) model was developed to incorporate the adjacency information of the slope units to predict the landslide susceptibility. Then, the spatial stratification heterogeneity patterns of landslide susceptibility were analyzed using GeoDetector. The results showed that the SLR model achieved an area under the curve (AUC) of 0.89, a notable improvement of 0.26 compared to the traditional logistic regression (LR) model that does not incorporate adjacency information. This indicates that incorporating adjacency information effectively enhances LSM accuracy by mitigating spatial autocorrelation. Furthermore, lithology, PGV, and distance to the epicenter were identified as the primary factors contributing to the formation of the spatial stratification heterogeneity of landslide susceptibility. Among these, the interaction between lithology and PGV exhibits the strongest nonlinear enhancement. By integrating both mapping units and their adjacency information, this study provides a novel approach to improving the predictive accuracy of LSM. Moreover, by analyzing the driving factors of spatial stratification heterogeneity in landslide susceptibility maps, the study advances the practical utility of LSM for disaster management and mitigation.

Keywords: slope units; landslide susceptibility mapping; spatial logistic regression model; spatial stratification heterogeneity; GeoDetector

Citation: Li, L.; Jia, M.; Xu, C.; Tian, Y.; Ma, S.; Yang, J. Enhancing Landslide Susceptibility Mapping by Integrating Neighboring Information in Slope Units: A Spatial Logistic Regression. *Remote Sens.* **2024**, *16*, 4475. https://doi.org/10.3390/rs16234475

Academic Editors: Federico Raspini, Chao Zhou and Kunlong Yin

Received: 28 October 2024
Revised: 21 November 2024
Accepted: 26 November 2024
Published: 28 November 2024

1. Introduction

Landslides are the most common type of natural disaster [1,2]. Rainfall, earthquakes, water level changes, rapid snowmelt, volcanic eruptions and human activities are all trigger factors of landslides [3]. In mountainous areas, earthquakes are the main trigger factor for landslides [4]. Co-seismic landslides, as one of the secondary effects of earthquakes, frequently cause significant disasters [1,5,6]. The Wenchuan earthquake in China in 2008 triggered a large number of landslides, and nearly a quarter of the casualties were caused by the landslides [7,8]. Landslide Susceptibility Mapping (LSM) has demonstrated a strong predictive capability for potential landslide occurrences, facilitating the optimization of regional planning and construction prior to the onset of disasters [9]. However, the prediction accuracy of LSM is often influenced by the choice of mapping unit and the adaptability of the employed model [10,11]. Thus, integrating these factors is essential for generating more accurate landslide susceptibility maps.

Mapping units are fundamental to LSM, with two primary types being raster units (RUs) and slope units (SUs) [12]. RUs are commonly used due to their ease of acquisition and mapping [13]. Nonetheless, the predictive accuracy of RUs is highly dependent on resolution, and their correlation with the geological environment is relatively low [14,15]. In addition, if the elevation value cannot be determined, even if a high resolution is used, the uncertainty will be higher than for a low resolution [16]. Landslides predominantly occur in mountainous regions, characterized by valleys and ridgelines, and RUs often fail to capture topographical continuity [17]. Conversely, SUs effectively address this limitation by aligning more closely with the actual landforms, and are thereby widely accepted for LSM [18–20]. In most studies, the extraction of SUs uses the curvature watershed method and the geomorphology method, which are based on the artificial extraction of valley ridgelines. However, it is difficult to obtain SUs by these two methods. Under these circumstances, the extraction of large-scale landslide units presents the challenges of being both time-consuming and prone to errors [21–23]. Therefore, the automatic zoning of slope units is particularly important. Under such circumstances, Alvioli et al. developed a modeling framework, r.slopeunits, which divides DEM data into large watersheds based on r.watershed of GRASS GIS, and then divides the watersheds into adjacent units according to the multiple flow strategy to obtain target slope units [24]. This software can be used to automatically extract SUs that are consistent with geomorphic characteristics [24]. There have been many studies that have used this software to make reasonable divisions of slope units and have obtained good results [25]. The consideration of neighboring information in the mapping of units is another critical factor for LSM, as landslides usually happen where they have occurred before and have strong spatial continuity [26]. Spatial autocorrelation, a common characteristic of spatial data, suggests that the neighboring units of landslide-prone areas also play an indicative role in landslide susceptibility. Given the inevitable impact of spatial autocorrelation on LSM modeling, recent studies have increasingly focused on optimizing models to mitigate this effect [27]. For instance, spatial residual models and spatial logistic regression models have been employed [28,29]. The spatial residual model typically requires the consideration of numerous factors; otherwise, it is prone to significant errors. However, it is impractical to encompass all the relevant factors within a single landslide susceptibility assessment [30,31]. The spatial logistic regression model incorporates the attribute information of neighboring mapping units as influencing variables for landslide occurrences, thereby accounting for spatial autocorrelation and significantly improving susceptibility mapping accuracy [13].

Post-mapping analysis of the spatial stratification heterogeneity of susceptibility maps offers critical insights for regional landslide disaster prevention and emergency management [32]. By examining the mechanisms of underlying past landslide events, we can effectively prevent and mitigate future landslides caused by earthquakes in the same area [33,34]. The spatial stratification heterogeneity in landslide distribution results from the interplay of multiple factors rather than a single cause. Hence, analyzing the interactions between these factors is crucial for practical slope engineering applications [35]. The GeoDetector tool proves to be an effective tool for identifying the drivers of spatial stratification heterogeneity in landslide susceptibility maps, as it considers both individual influencing factors and their interactions [36]. This approach is instrumental in validating and applying LSM, offering substantial information for engineering applications.

To enhance the accuracy of landslide susceptibility maps, this study advances LSM methods and applications through three main aspects: (a) generating slope units for LSM using GRASS GIS; (b) modeling LSM by incorporating the adjacency information of neighboring units through spatial logistic regression and making comparisons with the results of the logistic regression model; and (c) analyzing the influencing factors and their interactions to understand the mechanisms behind the spatial stratification heterogeneity of landslide susceptibility.

2. Study Area and Data

2.1. Study Area

In 2013, a magnitude 5.9 earthquake struck Minxian County, Dingxi City, Gansu Province, China [6]. This earthquake triggered numerous co-seismic landslides. The study area was delineated as an elliptical region along the seismic intensity contour, encompassing the epicenter. It spans approximately 34°25′–34°43′N latitude and 103°51′–104°16′E longitude, covering an area of 927.35 km^2 with an average slope of 24.19°. The region experiences a plateau continental climate, with an average annual temperature ranging from 5.0 to 7.0 °C. Elevations in the area vary between 2192 and 3360 m, reflecting its high-altitude terrain. Influenced by the interaction of continental air masses and subtropical warm, humid air masses, coupled with topographical effects on the atmospheric uplift, the region exhibits a cold and humid climate. The Tao River serves as the primary river flowing through the study area.

The channel area of the study region consists of Quaternary sandy deposits, with Late Cenozoic sandstones distributed along both sides of the channel and Early Permian quartz sandstones covering most of the area. In the southwest and southern regions, Devonian carbonaceous shale and silty shale are present, respectively [6]. The continuous uplift and compression of the Qinghai-Tibet Plateau result in strong tectonic movements and frequent seismic activity, shaping a complex landscape characterized by numerous valleys and steep mountains.

The study area is situated in the north-central section of the North–South Seismic Belt, a region highly susceptible to earthquakes due to the presence of numerous large-scale active fault zones [37]. Figure 1 provides a comparative overview of the landslide conditions in the study area before and after the seismic event. The seismogenic structure responsible for the earthquake is a branch fault on the northern side of the Lintan–Dangchang Fault Zone, which lies between the eastern Kunlun Fault Zone and the northern edge of the western Qinling Fault Zone. This fault zone features a southward thrust with a general NWW-NW strike, a NE dip, and a dip angle of 50–70°, combined with a left-lateral strike-slip component. Variations in fault zone activity characteristics contribute to post-earthquake disasters, such as landslides and collapses [20]. To enhance the accuracy of the statistical analysis of the co-seismic landslides, AI Earth was utilized to perform a comparative analysis of pre- and post-seismic remote sensing imagery. The pre-earthquake satellite imagery, dated 6 April 2013, is shown in Figure 1a, while the post-earthquake imagery, captured on 1 August 2013, is displayed in Figure 1b.

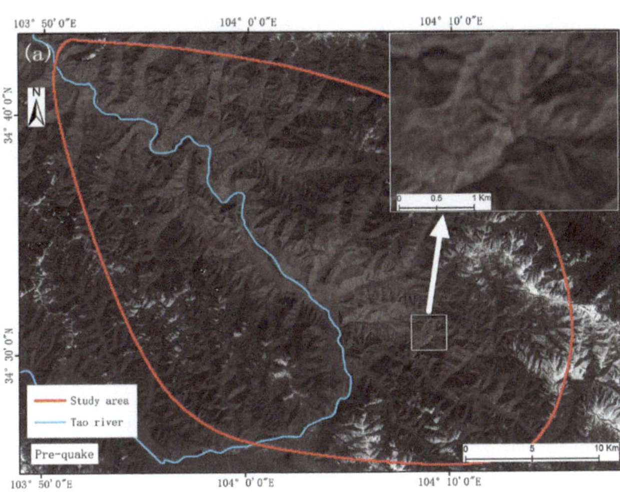

Figure 1. *Cont.*

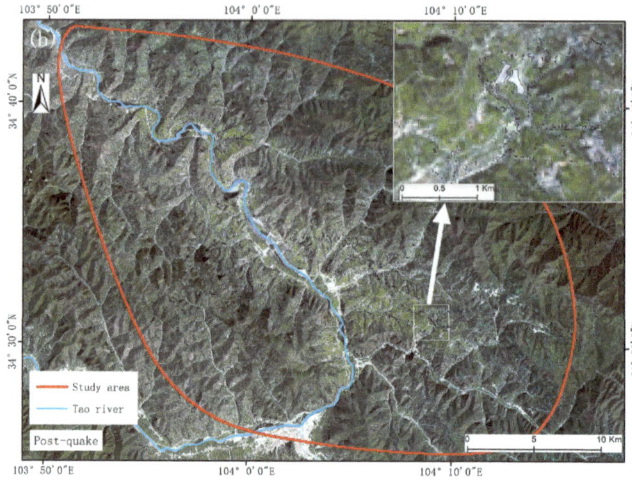

Figure 1. Maps showing the co-seismic landslides based on pre- and post-quake AI Earth images. (**a**) Pre-quake satellite image; (**b**) post-quake satellite image.

2.2. Data

Using high-resolution Google Earth images and comparing them with ZY-3 images, the landslide after the Minxian earthquake in 2013 was identified [37]. There are various types of landslides in the study area, such as falls, slides, and slumping on steep loess ridges, as well as deep continuous landslides and soil collapse [6]. In the study area, there are a total of 8526 landslides, with areas ranging from 3.68 to 51,246.19 m^2. Among them, 8236 landslides are less than 1000 m^2, accounting for 96.60% of the total number of landslides, while only 290 are large. Most of the landslides are soil landslides, with few rock landslides. These landslides are generally distributed along the Tao River, concentrated in the central and southern parts of the study area, where land use types are mostly cultivated land. The geographical location, topography, and distribution of landslides in the study area are illustrated in Figure 2.

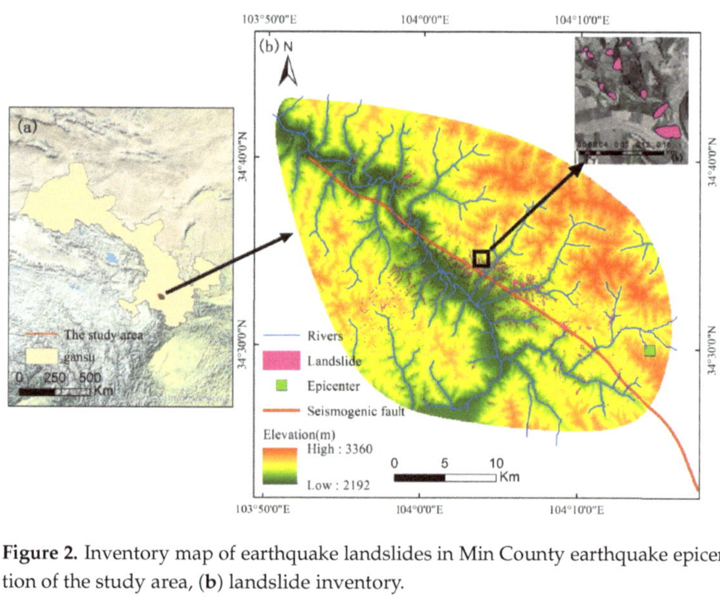

Figure 2. Inventory map of earthquake landslides in Min County earthquake epicenter area. (**a**) Location of the study area, (**b**) landslide inventory.

According to the geological conditions in the study area, eight factors were selected for LSM, including topographic and geological conditions (Figure 3). The topographic landform category includes four influencing factors: elevation, slope, profile curvature, and topographic wetness index (TWI). The elevation data have a resolution of 5 m and can be obtained from the AW3D website (https://www.aw3d.jp/en/ (accessed on 1 February 2014)) [38]. Slope, curvature, and TWI can be derived from the elevation data. The geological conditions include four influencing factors: distance to the epicenter, distance to the seismogenic fault, peak ground velocity (PGV), and lithology. The distance to the seismogenic fault and the epicenter represents the distance from the centroid of each unit to the seismogenic fault and the epicenter, respectively. Proximity to the seismogenic fault significantly increases both the probability and severity of landslides, while closer proximity to the epicenter is associated with a higher likelihood of landslide occurrence. PGV is an important indicator for describing earthquake intensity, and PGV data have a quantitative relationship with landslide density, according to the ShakeMap released by the United States Geological Survey (USGS) [39]. The lithology data come from the 1:200,000 geological map classified by stratigraphic age provided by the National Geological Data Center [6].

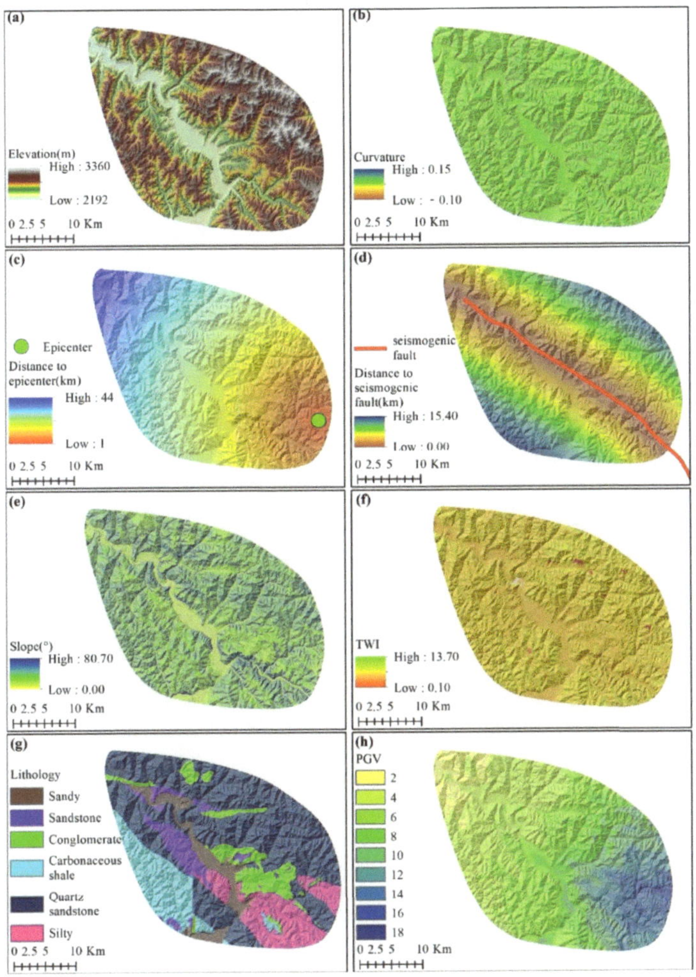

Figure 3. Maps showing distribution of influencing factors in study area. (**a**) Elevation; (**b**) curvature; (**c**) distance to epicenter; (**d**) distance to seismogenic fault; (**e**) slope; (**f**) TWI; (**g**) lithology; (**h**) PGV.

3. Method

This study evaluated landslide susceptibility by following three core steps: first, slope units were generated from elevation data using the GRASS GIS 7.8, and the attributes of each influencing factor were mapped to the slope units based on their spatial locations; second, an LR model and SLR model were employed to assess landslide susceptibility, and the results were verified and compared; and finally, the spatial stratification heterogeneity mechanisms in the landslide susceptibility map were analyzed using GeoDetector. Figure 4 depicts the comprehensive technical workflow of the study.

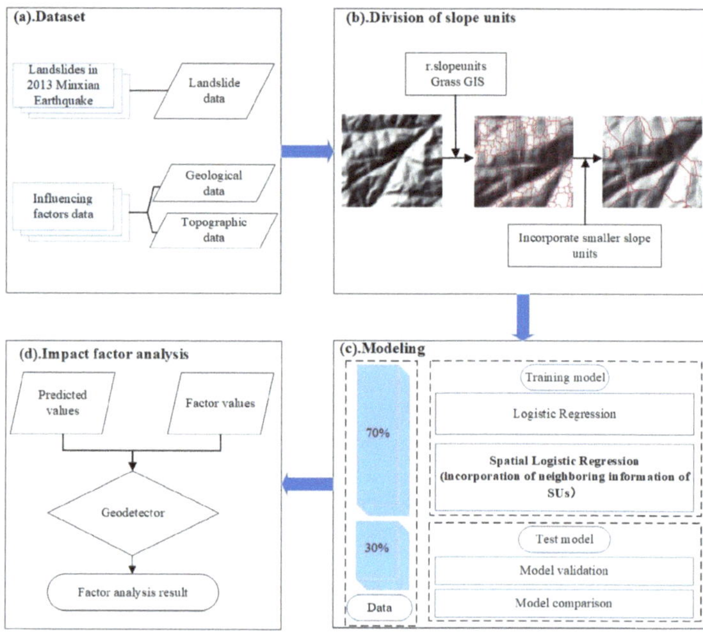

Figure 4. Workflow of the method proposed in this research to predict earthquake-induced landslide events.

3.1. Slope Unit Delineation

A slope unit (SU) refers to a topographic unit enclosed by drainage and watersheds. Several methods have been proposed for the division of slope units; among these, the r.slopeunits v1.0, based on GRASS GIS, is an automatic way to roughly divide the elevation data along the landform into SUs [40]. The parameters that need to be set include the following: the initial flow aggregation area threshold t (km^2); the reduction coefficient r, used to iteratively calculate the new value of the flow aggregation area threshold; the minimum square difference c of the terrain slope direction within a slope unit; and the minimum area a (m^2) of the SU [40]. Parameters a and c are important parameters for controlling the size of the SUs, and the time required by r.slopeunits to extract the SUs is also closely related to a and c [20].

Considering previous research and the actual shape size of the slope units in the study area, the following values were obtained, corresponding to several combinations of input parameters in the r.slopeunits software: t = 0.08 km^2, 0.1 km^2, 0.2 km^2, 0.3 km^2; a = 70,000 m^2, 90,000 m^2, 150,000 m^2; c = 0.15, 0.2; and r = 5. The slope morphology of t = 0.1 km^2, a = 70,000 m^2, c = 0.2, and r = 5 was the most consistent with that of the actual study area. To avoid excessive area differences between the slope elements, the slope units with too small an area were merged with adjacent units, and a total of 8647 SUs were generated. These units have a maximum area of 0.76 km^2 and an average area of 0.11 km^2 (Figure 5). There were 6743 units without landslides and 1904 units with landslides.

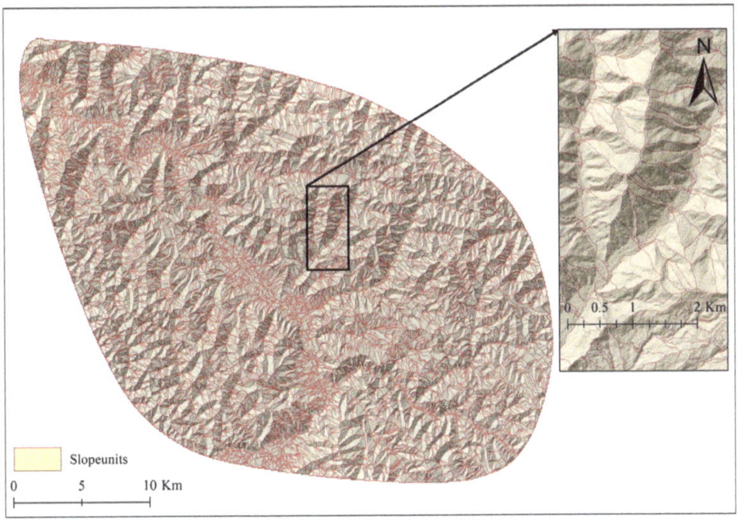

Figure 5. The slope unit map.

3.2. Spatial Logistic Regression Model

The logistic regression model can predict the probability of the landslide by using the linear regression relationship between the variables and multiple independent variables [19]. Since the variable that occurs in the landslide is a dual variable, the independent variable type can be continuous or discrete [41]. The equation is as follows:

$$Y = ln\frac{P(x)}{1 - P(x)} = \beta_0 + \beta_1 X_1 + \beta_2 X_2 + ... + \beta_i X_i + \varepsilon \ (i = 1, 2, ..., n) \quad (1)$$

$$P = \frac{e^Y}{1 + e^Y} \quad (2)$$

where X_1, X_2, ... X_i are independent variables, β_0 is the intercept coefficient, and β_1, β_2, ..., β_i are the undetermined regression coefficients corresponding to each factor. $P(x)$ is the probability of landslide occurrence, P is the prediction probability, and x is the conditional factor.

$$Y = ln\frac{P(x)}{1 - P(x)} = \beta_0 + \beta_1 X_1 + \beta_2 X_2 + ... + \beta_i X_i + \rho W Y + \varepsilon \ (i = 1, 2, ..., n) \quad (3)$$

In practical situations, the occurrence of landslides is not only related to various factors under the influence of the spatial autocorrelation effect, but also to the environment around the landslides [10]. Therefore, in addition to being influenced by the independent variable, the value of the Y variable also needs to consider the neighboring information of SUs. Spatial regression is a modification of the regression equation that incorporates the adjacency matrix [10]. Equation (3) represents the spatial autoregressive model, which considers the adjacency relationship of spatial data based on Equation (1) [27,42]. ρ is the coefficient of spatial autocorrelation of Y, ε is the error term that conforms to a normal distribution, W is the n-dimensional spatial adjacency matrix, and n is the total number of samples. Compared with Equation (1), the increase of ρWY in Equation (3) excludes the spatial structure information contained in the adjacency relationship from the residuals, reducing the possibility of obtaining biased estimates. The specific content of W is as follows:

$$W = \begin{bmatrix} 0 & f(d_{12}) & \cdots & f(d_{1j}) \\ f(d_{21}) & 0 & \cdots & f(d_{2j}) \\ \vdots & \vdots & \ddots & \vdots \\ f(d_{i1}) & f(d_{i2}) & \cdots & 0 \end{bmatrix} \quad (4)$$

$$L = y\ln\frac{e^{(\alpha+X\beta+\rho Wy)}}{1+e^{(\alpha+X\beta+\rho Wy)}} - (1-y)\ln\left(1+e^{(\alpha+X\beta+\rho Wy)}\right) \quad (5)$$

Equation (4) represents the adjacency matrix, and it is used to represent the adjacency relations in a SU, especially in graph theory and network analysis. If two mapping units are adjacent, the value of $f(d_{ij})$ is 1; otherwise, if they are not adjacent, $f(d_{ij})$ is 0. Equation (5) is the maximum likelihood function, which is used to solve the bias caused by the traditional least squares method and other parameter estimations.

3.3. Implementation of SLR

The 8647 SUs obtained above were used as mapping units, and the mean values of the independent variables of six factors, including elevation, curvature, distance to the epicenter, distance to the seismogenic fault, slope, and TWI, were assigned to the SUs. For the two discrete factors, lithology and PGV, the value with the largest proportion of landslide area was assigned to the corresponding SU. Dependent variables need to be considered during SLR modeling. If there is a landslide in the slope unit, the value of Y is 1; otherwise, it is 0. The spatial logistic regression model, as a supervised learning model, relies on training data to estimate parameters, and requires test data to verify the generalization ability and performance of the model. This process ensures that the model can have a better prediction effect in practical application. Considering the difficulty of acquiring future data sets, in this study, we adopted the conventional random sampling method for verification; 70% of the samples were used for training and 30% for testing [43]. The LR model and SLR model were solved using the free R-INLA package [44,45], and the expression for the SLR model is as follows:

$$\begin{aligned} Y = {} & (0.173 \times \text{Elevation}) + (-0.012 \times \text{Curvature}) + (-0.406 \times \text{Distance to epicenter}) \\ & + (-0.237 \times \text{Distance to seismogenic fault}) + (-0.040 \times \text{Slope}) + (0.017 \times \text{TWI}) \\ & + (-1.420 \times \text{Lithology(sandy)}) + (-1.174 \times \text{Lithology(sandstone)}) \\ & + (-1.509 \times \text{Lithology(conglomerate)}) + (-0.611 \times \text{Lithology(silty shale)}) \\ & + (-1.928 \times \text{Lithology(carbonaceous shale)}) + (-2.159 \times \text{Lithology(quartz sandstone)}) \\ & + (-1.425 \times \text{PGV (2)}) + (-1.164 \times \text{PGV (4)}) + (-1.197 \times \text{PGV (6)}) + (-1.353 \times \text{PGV (8)}) \\ & + (-1.443 \times \text{PGV (10)}) + (-1.153 \times \text{PGV (12)}) + (-1.781 \times \text{PGV (14)}) \\ & + (-1.207 \times \text{PGV (16)}) + (-0.274 \times \text{PGV (18)}) \end{aligned}$$

The regression coefficients represent the impact of influencing factors on landslide susceptibility and are determined by the dimensions of each factor. The regression coefficient of a categorical variable means the relative risk of each category. A negative regression coefficient indicates that the category is less likely to trigger the event compared to the reference category. For instance, in this study, the regression coefficient for quartz sandstone is −2.159, whereas for silty shale it is −0.611, suggesting that quartz sandstone has a lower likelihood of causing landslides compared to silty shale. The regression coefficient, therefore, reflects the relative risk. For continuous independent variables, the regression coefficient indicates the direction and magnitude of influence on the dependent variable. A positive coefficient implies that an increase in the independent variable raises the probability of landslide occurrence, whereas a negative coefficient indicates that the probability decreases as the independent variable increases.

3.4. GeoDetector

GeoDetector can identify the causes of spatial stratification heterogeneity in landslide susceptibility maps caused by influencing factors and their interactions [36]. This study used the free R package "GD" to process the data [35,46]. The predicted value of each slope unit was taken as the dependent variable Y, and the continuous variable X is categorized. The number of categories for each continuous variable was set to 6–10 categories. The "GD" package was used to determine the best classification method and classification number of each continuous factor. It calculates the contribution of each factor to landslides and their combined effect on landslides.

GeoDetector includes four parts: a risk detector, a factor detector, an ecological detector, and an interactive detector [47]. Among these, the factor detector and interactor detector are the most helpful for the impact factor analysis of this study. The role of the factor detector is to detect the spatial heterogeneity of Y and the extent to which variable X explains the spatial heterogeneity of attribute Y. The interactor detector is used to identify the interactions between variables X; that is, to assess whether two factors acting together increase or decrease the explanatory power of the dependent variable Y, or whether the effects of these factors on Y are independent [48]. The principle of GeoDetector is as follows:

$$q = 1 - \frac{1}{N\sigma^2}\sum_{i=1}^{m} N_i\sigma_i^2 \quad q \in [0,\ 1] \tag{6}$$

where m is the number of categories of the independent variable X—taking lithology as an example, m is 6; N is the total number of slope units; σ_i^2 is the variance of the i-th layer of the dependent variable Y; and σ^2 is the variance of the whole study area Y. The q value indicates the extent to which the independent variable X explains the spatial stratification heterogeneity of the dependent variable, and a higher q value indicates a greater contribution of X to the landslides.

4. Results

4.1. Model Verification and Comparison

The predictive accuracy of the model can be characterized by the receiver operating characteristic (ROC) curve and the area under the curve (AUC). The farther the ROC curve is from the diagonal, the better the differentiation of the model [49]. An AUC value between 0.7 and 0.9 indicates that the model has a high accuracy [38,50]. In addition to the AUC, the confusion matrix not only quantitatively assesses the predictive performance of the model, but also reveals more detailed information. In this study, the AUC of the LR and SLR models were 0.63 and 0.89, respectively (Figure 6). According to the confusion matrix of the two models in Table 1, the overall prediction accuracy of SLR was 85.20%, which was 17.70% higher than that of the LR model, and the recall also increased by 22.07%. The results show that the SLR model has a higher accuracy.

Table 1. The confusion matrix of LR and SLR.

Method	Landslide Occurred	Prediction		Recall	Accuracy
		Yes	No		
LR	Yes	353	218	61.82%	67.50%
	No	625	1398	69.10%	
SLR	Yes	479	92	83.89%	85.20%
	No	292	1731	85.57%	

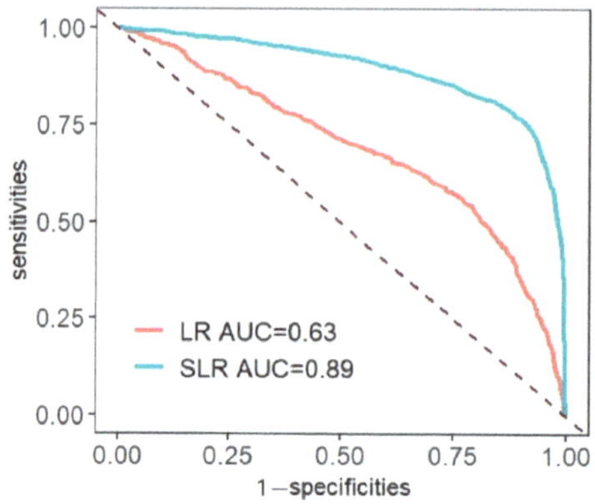

Figure 6. Comparison of the LR and SLR models by using the ROC curve. AUC is the acronym of the area under the ROC curve.

4.2. Landslide Susceptibility Map

The Seed Cell Area Index (SCAI) is an indicator for assessing the efficiency of classification, with lower SCAI values in highly sensitive areas and higher SCAI values in low-sensitive areas [51]. Table 2 shows that the SLR model has a gradually increasing sensitivity from high to low SCAI values, indicating that the classification results are reasonable, while the LR model has a large SCAI value in the high-risk area, indicating that the LR model overestimates the high-risk area. Figure 7 shows the spatial effect of the SUs' adjacent information extracted by the SLR model. The spatial effect indicates that SUs can effectively reduce the influence of spatial effects on the results. The predicted landslide risks are divided into three categories using the classification of natural break points [52]. The landslide susceptibility grade distribution maps obtained by the LR and SLR models are shown in Figure 8a and 8b, respectively. The LR model divides most areas into the same disaster level, while the SLR model is more dispersed.

Table 2. The densities of landslides among susceptibility classes of both methods.

Model	Sensitivity Level	Area Ratio	Number of Landslides	SCAI
LR	Low	46.36%	3998	1.003
	Moderate	34.21%	2990	0.989
	High	19.41%	1659	1.012
SLR	Low	50.93%	4385	1.004
	Moderate	35.13%	3041	0.999
	High	13.92%	1221	0.987

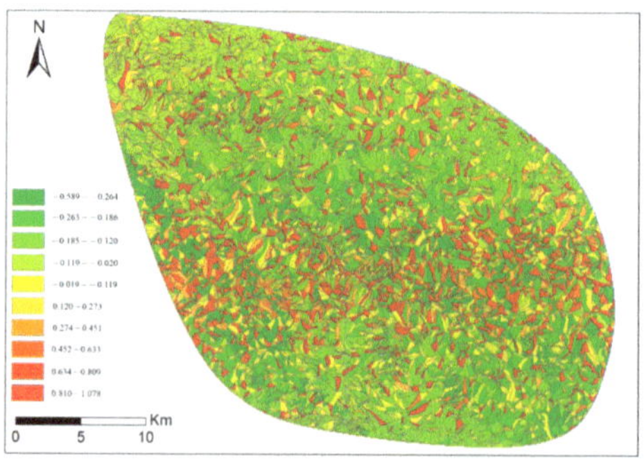

Figure 7. Map of the study area illustrating the neighboring structural effect.

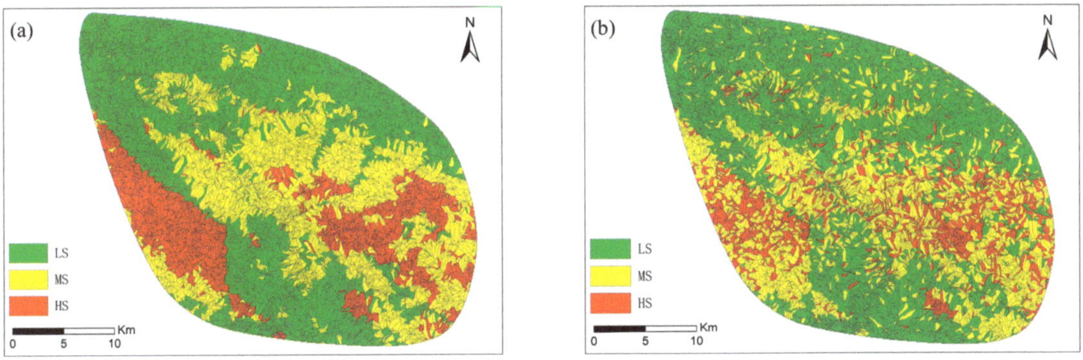

Figure 8. Landslide susceptibility map. (**a**) is the landslide susceptibility map produced by LR. (**b**) is the landslide susceptibility map produced by SLR. LS, MS, and HS, respectively, indicate three landslide susceptibility levels of low, moderate, and high.

4.3. Factor Analysis for the Spatial Pattern of the Landslide Susceptibility Map

The spatial stratification heterogeneity of the landslide susceptibility map was investigated using factor analysis. Lithology, PGV, and distance to the epicenter were identified as the three factors contributing most significantly to the spatial distribution of the landslides (Table 3), and their q values are 0.299, 0.207, and 0.175, respectively. This indicates that lithology alone can explain 29.9% of the spatial stratification heterogeneity of landslide susceptibility. This result shows that earthquakes are a trigger factor of landslides, but their main controlling factor is lithology. The interaction analysis of factors shows that, while the interactions between lithology and slope, distance to the epicenter, and PGV involve bivariate enhancement, the enhancement of other factors is nonlinear. Among these, the interaction between lithology and PGV shows the strongest nonlinear enhancement, followed by the interaction between lithology and distance to the epicenter (Figure 9). The risk detector reflects the average area of landslide occurrence in each layer of the impact factor, as well as the statistically significant difference in the average landslide area between the layers. DEM, curvature, distance from fault, slope, TWI, and PGV all show significant differences (Figure 10). Ecological detection was used to compare whether there were significant differences in the effects of the two factors on the spatial distribution of landslides. The results showed that all the factors have significant effects on landslides (Figure 11), indicating that the factors selected in this study are reasonable and there is no redundancy, which affects the accuracy of landslide susceptibility mapping [53].

Table 3. The factor detector calculated by Geodetector. (X1–X8 represent DEM, curvature, distance to epicenter, distance to fault, slope, TWI, lithology, and PGV, respectively).

Variables	X7	X8	X3	X1	X4	X5	X6	X2
q	0.299	0.207	0.175	0.030	0.016	0.013	0.002	0.001
p	<0.05	<0.05	<0.05	<0.05	<0.05	<0.05	<0.05	<0.05

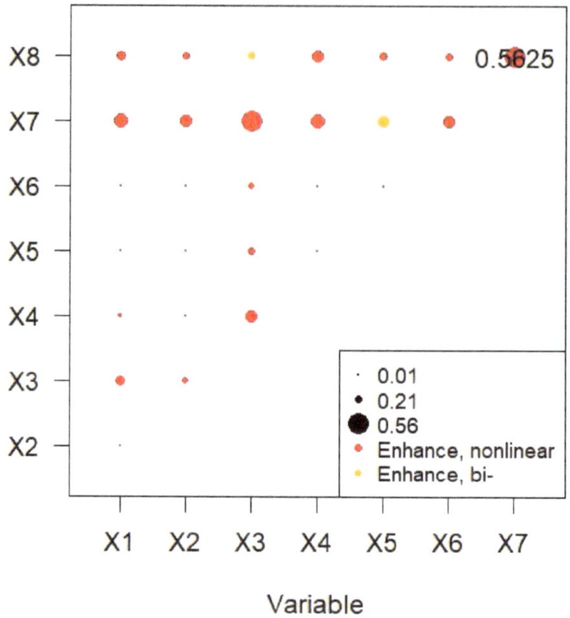

Figure 9. Interaction detector result.

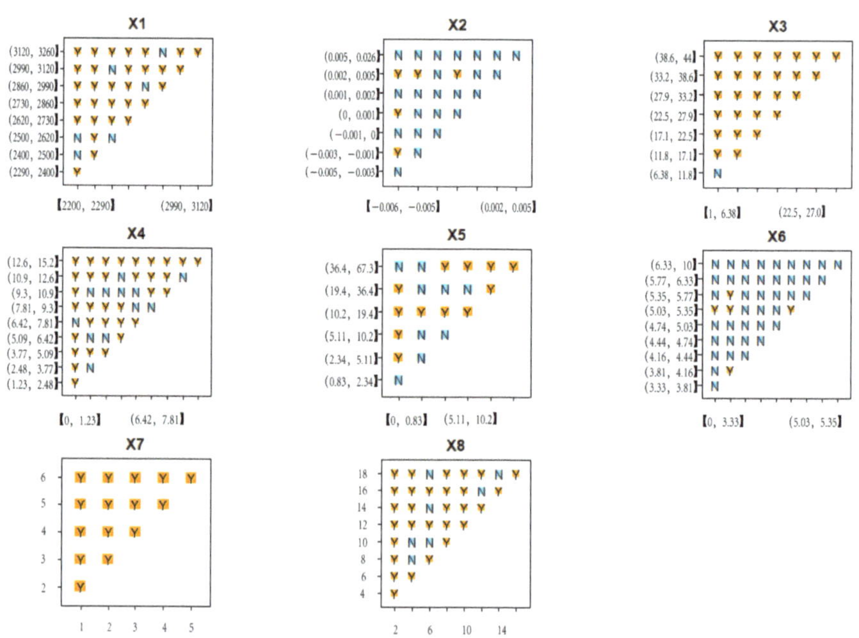

Figure 10. Risk detector result.

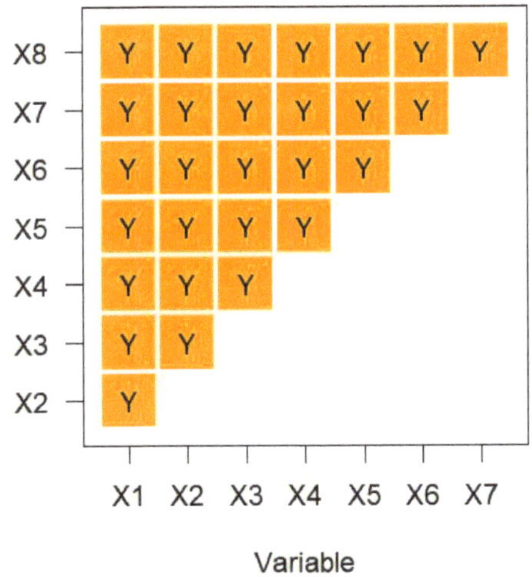

Figure 11. Ecological detector result.

5. Discussion

This study addressed the influence of the spatial autocorrelation of spatial data on LSM, by dividing slope units and including the neighboring information of SUs in susceptibility modeling. Additionally, by analyzing the contributions of the interactions among various factors to the spatial heterogeneity of landslide susceptibility, the accuracy of the susceptibility map was improved. Compared with the LR model, the AUC and confusion matrix of the SLR model had a higher accuracy, which indicated that the landslide susceptibility evaluation established by the SLR model was more consistent with the landslide occurrence rule compared with the LR model. The SLR model demonstrates objectivity, feasibility, and a strong predictive capability for the study area. This evaluation can play a certain guiding role in the prevention and management of landslide disasters in the region.

The use of SUs significantly reduces the spatial autocorrelation of landslide occurrence. The spatial effect map obtained by the SLR model is only approximately the same as the actual landslide density map (Figure 8b, while previous studies based on grid cells have shown that there is a strong correlation between the spatial effect map and the actual landslide density map [13]. This is mainly due to the consideration of the spatial similarity of adjacent units. RUs fail to consider the spatial similarity of adjacent units. SUs are extracted from RUs, and the value of each factor attribute in each SU is not particularly different, while the adjacent SUs show differences. In this study, 96% were small landslides, and one SU included many landslides, which greatly reduced the correlation of landslide occurrence among adjacent units, thereby making the spatial data more independent. For landslide susceptibility mapping, an increased independence on input data leads to reduced negative effects of spatial autocorrelation on the model. Previous studies have shown that SUs can reduce input uncertainty more effectively than RUs [54], which is consistent with the findings of this study.

The performance of LR and SLR models differs significantly in areas with high landslide incidence. The LR model predicts a series of high-risk areas in the southwest carbonaceous shale region, the central conglomerate region, and a small portion of the quartz sandstone region. This pattern arises because the LR model heavily relies on lithology factors, causing the predicted high-risk areas to align with lithological trends. However, as shown in Figure 2b, the actual landslide distribution is dispersed, and even within a single lithology area, landslides are not contiguous. This indicates that the LR model, by focusing solely on slope element attributes, erroneously classifies some low-to-medium-risk areas as high-risk. In contrast, the SLR model incorporates not only the attribute information of mapping units, but also spatial adjacency information, recognizing that the occurrence of landslides in the surrounding units greatly influences the likelihood of landslides within a slope unit. By accounting for these spatial relationships, the SLR model can distinguish three types of risk areas even within a single lithology region. This approach better reflects the heterogeneity of landslide distribution and aligns more closely with real-world conditions.

In general, the prediction accuracy of the SLR model is greatly improved compared with the LR model. The spatial model used in this study incorporates the adjacency information between mapping units. Unlike general regression models, an adjacency matrix aligned with geomorphic characteristics is included, which helps to mitigate the influence of spatial autocorrelation between adjacent units. However, the LR model is a linear model, and the effect is poor when the relationship between landslide and impact factors is not linear and is complicated [43,55]. This enhances the model's precision and the accuracy of landslide susceptibility mapping. However, the SU approach utilized in this paper inherently reduces the impact of spatial autocorrelation on modeling to some extent. To demonstrate more comprehensively the characteristics of the SLR model, it is essential to investigate the influence of the SLR model on spatial autocorrelation under RU and SU unit mapping approaches. This investigation constitutes the next step in our research plan.

GeoDetector's factor detection analysis reveals key differentiation characteristics of landslide susceptibility. Lithology, peak ground velocity (PGV), and distance to the epicenter are identified as the most significant factors, with lithology being the primary determinant of landslide instability. The study area's lithology comprises sand, sandstone, conglomerate, carbonaceous shale, quartz sandstone, and silty rock, accounting for 9%, 6%, 9%, 13%, 51%, and 12% of the slope units, respectively. Although quartz sandstone dominates in area, Figures 3g and 8 show that carbonaceous shale is mainly classified as moderate-to-high-risk, while quartz sandstone is predominantly low-risk. This demonstrates that lithology significantly influences landslide susceptibility. Carbonaceous shale, silty rock, conglomerate, and sandstone are the most landslide-prone lithologies, while soft riverine areas and quartz sandstone are less susceptible. Soft rocks, like sandy loams and carbonaceous shales, deform under seismic vibration, reducing landslide risk by altering their stress state. In contrast, hard rocks such as quartz sandstone resist sliding due to their high shear strength, even during vibrations.

Interaction detection identifies the interplay of multiple factors influencing landslide susceptibility. Figure 9 highlights that the interaction between lithology and peak ground velocity (PGV) exhibits the strongest nonlinear enhancement, followed by the interaction between lithology and distance to the epicenter. Generally, areas with hard lithology are less prone to landslides. However, high-risk zones are observed in some quartz sandstone regions near the focal point (southeast of the study area in Figure 8b). In contrast, in distant focal areas, high-risk zones are more common in regions with soft lithology. This indicates that landslide occurrence is not determined solely by individual factors such as lithology, but by the combined effect of geological conditions and seismic forces. The strong interaction between lithology and PGV suggests that landslides are heavily influenced by the combined effects of seismic dynamics and geological conditions. Similarly, the interaction between lithology and distance to the epicenter underscores the varying impact of geological and dynamic effects at different spatial scales. Specifically, as distance to the epicenter increases, PGV decreases, making soft lithology areas more susceptible to landslides, while hard lithology areas remain less prone. Conversely, closer to the epicenter, even areas with hard lithology exhibit a higher likelihood of landslides due to the amplified seismic effects.

Factor analysis and interaction analysis revealed that geological factors exert a greater influence on landslides than topographic conditions, highlighting earthquakes as the primary triggering factor. Given the challenges posed by topographic conditions, geological factors, wet climates, and the increasing frequency of extreme weather events, greater attention should be paid to the role of geology in landslide occurrence. In landslide prevention and mitigation, regular geotechnical monitoring is essential [56]. Priority should be given to shale and sandstone areas in slope engineering, where techniques such as shotcrete, bolt support, and other reinforcement methods can be applied to enhance slope shear strength and reduce the likelihood of collapse or landslides. Additionally, using ecological materials such as plant fibers and ecological bags for slope protection not only helps to prevent landslides, but also promotes slope biodiversity, providing both environmental and structural benefits.

The spatial logistic regression (SLR) model incorporates spatial information by considering the connectivity relationships between mapping units. When selecting the spatial adjacency matrix, various methods can be employed, such as the adjacency matrix and the distance matrix. In this study, the connection method used was queen adjacency within the adjacency matrix, which involves connections through common vertices and adjacent edges. Spatial samples based on the queen adjacency rule exhibit a closer correlation with their surrounding spatial units. This adjacency method is straightforward to operate and manage. However, its limitation is that it only considers the influential strength of adjacent units, assigning an influential strength of zero to non-adjacent units. This does not accurately reflect real-world conditions. Therefore, future model improvements should explore different connection methods to achieve a higher prediction accuracy that aligns more closely with actual situations.

6. Conclusions

Landslide susceptibility mapping is of great significance for landslide prediction and control. In this study, slope units were used as the mapping units for landslide susceptibility mapping, ensuring that the results aligned more closely with the actual natural environment. Furthermore, the spatial logistic regression (SLR) model was employed to incorporate the adjacency information of slope units, demonstrating a 26% higher accuracy compared to the logistic regression (LR) model, which does not account for spatial adjacency. This suggests that adjacency information can serve as a valuable new index for landslide susceptibility mapping. Additionally, the driving factors of landslide susceptibility and their interactions were analyzed using GeoDetector. The results identified lithology, peak ground velocity (PGV), and distance to the epicenter as the main factors contributing to landslides. Among these, the interaction between lithology and PGV exhibited the strongest nonlinear enhancement, followed by the interaction between lithology and the epicenter. These findings highlight the importance of considering lithological characteristics and seismic vibration conditions in slope engineering control. By delineating high-risk landslide areas, more precise engineering reinforcement measures and risk warnings can be implemented. This study not only deepens our understanding of the spatial differentiation of landslide susceptibility, but also enhances the practical application of susceptibility mapping in real-world scenarios.

Author Contributions: Conceptualization, J.Y. and L.L.; methodology, J.Y.; validation, M.J.; formal analysis, J.Y. and M.J.; investigation, S.M.; resources, C.X.; data curation, J.Y.; writing—original draft preparation, M.J.; writing—review and editing, L.L., J.Y., Y.T. and M.J.; visualization, M.J.; supervision, L.L.; project administration, L.L.; funding acquisition, L.L. All authors have read and agreed to the published version of the manuscript.

Funding: This research was funded by the National Funding Postdoctoral Fellowship Program of CPSF (GZC20233022); the China Postdoctoral Science Foundation (2024M763573); the Henan Province Fund (232102320002); and the National Institute of Natural Hazards, Ministry of Emergency Management of China (2023-JBKY-57).

Data Availability Statement: The data supporting the findings of this study are available from the corresponding author upon reasonable request.

Acknowledgments: Thanks to Jinfeng Wang, who provided the free-ware GeoDetector, and the contributors of the open-source R package INLA.

Conflicts of Interest: The authors declare no conflicts of interest.

References

1. Petley, D. Global Patterns of Loss of Life from Landslides. *Geology* **2012**, *40*, 927–930. [CrossRef]
2. Youssef, A.M.; Pourghasemi, H.R.; Pourtaghi, Z.S.; Al-Katheeri, M.M. Landslide Susceptibility Mapping Using Random Fore Boosted Regression Tree, Classification and Regression Tree, and General Linear Models and Comparison of Their Performance at Wadi Tayyah Basin, Asir Region, Saudi Arabia. *Landslides* **2016**, *13*, 839–856. [CrossRef]
3. Fang, K.; Tang, H.; Li, C.; Su, X.; An, P.; Sun, S. Centrifuge Modelling of Landslides and Landslide Hazard Mitigation: A Revie *Geosci. Front.* **2023**, *14*, 101493. [CrossRef]
4. Basharat, M. A Review of Landslides Related to the 2005 Kashmir Earthquake: Implication and Future Challenges. *Nat. Hazard* **2021**, *108*, 1–30. [CrossRef]
5. Huang, Y.; Xu, C.; Zhang, X.; Xue, C.; Wang, S. An Updated Database and Spatial Distribution of Landslides Triggered by the Milin, Tibet Mw6.4 Earthquake of 18 November 2017. *J. Earth Sci.* **2021**, *32*, 1069–1078. [CrossRef]
6. Xu, C.; Xu, X.; Shyu, J.B.H.; Zheng, W.; Min, W. Landslides Triggered by the 22 July 2013 Minxian–Zhangxian, China, Mw 5.9 Earthquake: Inventory Compiling and Spatial Distribution Analysis. *J. Asian Earth Sci.* **2014**, *92*, 125–142. [CrossRef]
7. Chong, X.U. Detailed Inventory of Landslides Triggered by the 2008 Wenchuan Earthquake and Its Comparison with Other Earthquake Events in the World. *Sci. Technol. Rev.* **2012**, *30*, 18–26. [CrossRef]
8. Zhang, S.; Zhang, L.M.; Xiang, B.; Yuan, Q. Travel Distances of Earthquake-Induced Landslides. In Proceedings of the Geo-Congress, San Diego, CA, USA, 25 February 2013; American Society of Civil Engineers: San Diego, CA, USA, 2013; pp. 991–1001.
9. Dai, F.C.; Lee, C.F.; Ngai, Y.Y. Landslide Risk Assessment and Management: An Overview. *Eng. Geol.* **2002**, *64*, 65–87. [CrossRef]
10. Erener, A.; Düzgün, H.S.B. Landslide Susceptibility Assessment: What Are the Effects of Mapping Unit and Mapping Method? *Environ. Earth Sci.* **2012**, *66*, 859–877. [CrossRef]
11. Zhang, T.; Fu, Q.; Quevedo, R.P.; Chen, T.; Luo, D.; Liu, F.; Kong, H. Landslide Susceptibility Mapping Using Novel Hybrid Model Based on Different Mapping Units. *KSCE J. Civ. Eng.* **2022**, *26*, 2888–2900. [CrossRef]
12. Reichenbach, P.; Rossi, M.; Malamud, B.D.; Mihir, M.; Guzzetti, F. A Review of Statistically-Based Landslide Susceptibility Models. *Earth-Sci. Rev.* **2018**, *180*, 60–91. [CrossRef]
13. Yang, J.; Song, C.; Yang, Y.; Xu, C.; Guo, F.; Xie, L. New Method for Landslide Susceptibility Mapping Supported by Spatial Logistic Regression and GeoDetector: A Case Study of Duwen Highway Basin, Sichuan Province, China. *Geomorphology* **2019**, *324*, 62–71. [CrossRef]
14. Arnone, E.; Francipane, A.; Scarbaci, A.; Puglisi, C.; Noto, L.V. Effect of Raster Resolution and Polygon-Conversion Algorithm on Landslide Susceptibility Mapping. *Environ. Model. Softw.* **2016**, *84*, 467–481. [CrossRef]
15. Guzzetti, F.; Galli, M.; Reichenbach, P.; Ardizzone, F.; Cardinali, M. Landslide Hazard Assessment in the Collazzone Area, Umbria, Central Italy. *Nat. Hazards Earth Syst. Sci.* **2006**, *6*, 115–131. [CrossRef]
16. Temme, A.J.A.M.; Heuvelink, G.B.M.; Schoorl, J.M.; Claessens, L. Chapter 5 Geostatistical Simulation and Error Propagation in Geomorphometry. In *Developments in Soil Science*; Elsevier: Amsterdam, The Netherlands, 2009; Volume 33, pp. 121–140. ISBN 978-0-12-374345-9.
17. Huang, F.; Cao, Y.; Li, W.; Catani, F.; Song, G.; Huang, J.; Yu, C. Uncertainties of Landslide Susceptibility Prediction: Influences of Different Study Area Scales and Mapping Unit Scales. *Int. J. Coal Sci. Technol.* **2024**, *11*, 26. [CrossRef]
18. Tanyas, H. A Global Slope Unit-Based Method for the near Real-Time Prediction of Earthquake-Induced Landslides. *Geomorphology* **2019**, *327*, 126–146. [CrossRef]
19. Sun, X.; Chen, J.; Han, X.; Bao, Y.; Zhan, J.; Peng, W. Application of a GIS-Based Slope Unit Method for Landslide Susceptibility Mapping along the Rapidly Uplifting Section of the Upper Jinsha River, South-Western China. *Bull. Eng. Geol. Environ.* **2020**, *79*, 533–549. [CrossRef]
20. Jacobs, L.; Kervyn, M.; Reichenbach, P.; Rossi, M.; Marchesini, I.; Alvioli, M.; Dewitte, O. Regional Susceptibility Assessments with Heterogeneous Landslide Information: Slope Unit- vs. Pixel-Based Approach. *Geomorphology* **2020**, *356*, 107084. [CrossRef]
21. Wang, F.; Xu, P.; Wang, C.; Wang, N.; Jiang, N. Application of a GIS-Based Slope Unit Method for Landslide Susceptibility Mapping along the Longzi River, Southeastern Tibetan Plateau, China. *ISPRS Int. J. Geo-Inf.* **2017**, *6*, 172. [CrossRef]
22. Guzzetti, F.; Carrara, A.; Cardinali, M.; Reichenbach, P. Landslide Hazard Evaluation: A Review of Current Techniques and Their Application in a Multi-Scale Study, Central Italy. *Geomorphology* **1999**, *31*, 181–216. [CrossRef]
23. Luo, W.; Liu, C.-C. Innovative Landslide Susceptibility Mapping Supported by Geomorphon and Geographical Detector Methods. *Landslides* **2018**, *15*, 465–474. [CrossRef]
24. Alvioli, M.; Marchesini, I.; Reichenbach, P.; Rossi, M.; Ardizzone, F.; Fiorucci, F.; Guzzetti, F. Automatic Delineation of Geomorphological Slope Units with r.slopeunits v1.0 and Their Optimization for Landslide Susceptibility Modeling. *Geosci. Model Dev.* **2016**, *9*, 3975–3991. [CrossRef]
25. Wu, H.; Ye, C.; Pei, X.; Oguchi, T.; He, Z.; Yang, H.; Huang, R. Incorporating Effects of Slope Units and Sliding Areas into Seismically Induced Landslide Risk Modeling in Tectonically Active Mountainous Areas. *Remote Sens.* **2024**, *16*, 3517. [CrossRef]
26. Temme, A.; Guzzetti, F.; Samia, J.; Mirus, B.B. The Future of Landslides' Past—A Framework for Assessing Consecutive Landsliding Systems. *Landslides* **2020**, *17*, 1519–1528. [CrossRef]

27. Erener, A.; Düzgün, H.S.B. Improvement of Statistical Landslide Susceptibility Mapping by Using Spatial and Global Regression Methods in the Case of More and Romsdal (Norway). *Landslides* **2010**, *7*, 55–68. [CrossRef]
28. Wu, Z.; Chen, Y.; Han, Y.; Ke, T.; Liu, Y. Identifying the Influencing Factors Controlling the Spatial Variation of Heavy Metals in Suburban Soil Using Spatial Regression Models. *Sci. Total Environ.* **2020**, *717*, 137212. [CrossRef]
29. Jones, J.P.; Foster, S.A. Testing Regression Residuals for Spatial Autocorrelation Using SAS: A Technical Note. *Geogr. Res. Forum* **1991**, *11*, 78–83.
30. Moreiras, S.M. Landslide Susceptibility Zonation in the Rio Mendoza Valley, Argentina. *Geomorphology* **2005**, *66*, 345–357. [CrossRef]
31. Jebur, M.N.; Pradhan, B.; Tehrany, M.S. Optimization of Landslide Conditioning Factors Using Very High-Resolution Airborne Laser Scanning (LiDAR) Data at Catchment Scale. *Remote Sens. Environ.* **2014**, *152*, 150–165. [CrossRef]
32. Chen, T.; Zhu, L.; Niu, R.; Trinder, C.J.; Peng, L.; Lei, T. Mapping Landslide Susceptibility at the Three Gorges Reservoir, China, Using Gradient Boosting Decision Tree, Random Forest and Information Value Models. *J. Mt. Sci.* **2020**, *17*, 670–685. [CrossRef]
33. Guzzetti, F.; Reichenbach, P.; Ardizzone, F.; Cardinali, M.; Galli, M. Estimating the Quality of Landslide Susceptibility Models. *Geomorphology* **2006**, *81*, 166–184. [CrossRef]
34. Tian, Y.; Xu, C.; Hong, H.; Zhou, Q.; Wang, D. Mapping Earthquake-Triggered Landslide Susceptibility by Use of Artificial Neural Network (ANN) Models: An Example of the 2013 Minxian (China) Mw 5.9 Event. *Geomat. Nat. Hazards Risk* **2019**, *10*, 1–25. [CrossRef]
35. Song, Y.; Wang, J.; Ge, Y.; Xu, C. An Optimal Parameters-Based Geographical Detector Model Enhances Geographic Characteristics of Explanatory Variables for Spatial Heterogeneity Analysis: Cases with Different Types of Spatial Data. *GIScience Remote Sens.* **2020**, *57*, 593–610. [CrossRef]
36. Yang, J.; Xu, C.; Jin, X. Joint Effects and Spatiotemporal Characteristics of the Driving Factors of Landslides in Earthquake Areas. *J. Earth Sci.* **2023**, *34*, 330–338. [CrossRef]
37. Tian, Y.; Xu, C.; Xu, X.; Chen, J. Detailed Inventory Mapping and Spatial Analyses to Landslides Induced by the 2013 Ms 6.6 Minxian Earthquake of China. *J. Earth Sci.* **2016**, *27*, 1016–1026. [CrossRef]
38. Ma, S.; Shao, X.; Xu, C. Landslide Susceptibility Mapping in Terms of the Slope-Unit or Raster-Unit, Which Is Better? *J. Earth Sci.* **2023**, *34*, 386–397. [CrossRef]
39. García, D.; Mah, R.T.; Johnson, K.L.; Hearne, M.G.; Marano, K.D.; Wald, D.J.; Worden, C.B. ShakeMap Atlas 2.0: An Improved Suite of Recent Historical Earthquake ShakeMaps for Global Hazard Analyses and Loss Model Calibration. In Proceedings of the World Conference on Earthquake Engineering 2012, Lisbon, Portugal, 24–28 September 2012.
40. Schlögel, R.; Marchesini, I.; Alvioli, M.; Reichenbach, P.; Rossi, M.; Malet, J.-P. Optimizing Landslide Susceptibility Zonation: Effects of DEM Spatial Resolution and Slope Unit Delineation on Logistic Regression Models. *Geomorphology* **2018**, *301*, 10–20. [CrossRef]
41. Chalkias, C.; Kalogirou, S.; Ferentinou, M. Landslide Susceptibility, Peloponnese Peninsula in South Greece. *J. Maps* **2014**, *10*, 211–222. [CrossRef]
42. Gamerman, D.; Moreira, A.R.B. Multivariate Spatial Regression Models. *J. Multivar. Anal.* **2004**, *91*, 262–281. [CrossRef]
43. Cao, J.; Zhang, Z.; Wang, C.; Liu, J.; Zhang, L. Susceptibility Assessment of Landslides Triggered by Earthquakes in the Western Sichuan Plateau. *CATENA* **2019**, *175*, 63–76. [CrossRef]
44. Bakka, H.; Rue, H.; Fuglstad, G.-A.; Riebler, A.; Bolin, D.; Illian, J.; Krainski, E.; Simpson, D.; Lindgren, F. Spatial Modeling with R-INLA: A Review. *WIREs Comput. Stat.* **2018**, *10*, e1443. [CrossRef]
45. Lombardo, L.; Opitz, T.; Huser, R. Numerical Recipes for Landslide Spatial Prediction Using R-INLA. In *Spatial Modeling in GIS and R for Earth and Environmental Sciences*; Elsevier: Amsterdam, The Netherlands, 2019; pp. 55–83. ISBN 978-0-12-815226-3.
46. Wang, J.-F.; Zhang, T.-L.; Fu, B.-J. A Measure of Spatial Stratified Heterogeneity. *Ecol. Indic.* **2016**, *67*, 250–256. [CrossRef]
47. Wang, J.; Li, X.; Christakos, G.; Liao, Y.; Zhang, T.; Gu, X.; Zheng, X. Geographical Detectors-Based Health Risk Assessment and Its Application in the Neural Tube Defects Study of the Heshun Region, China. *Int. J. Geogr. Inf. Sci.* **2010**, *24*, 107–127. [CrossRef]
48. Wang, J.-F.; Hu, Y. Environmental Health Risk Detection with GeogDetector. *Environ. Model. Softw.* **2012**, *33*, 114–115. [CrossRef]
49. Beguería, S. Validation and Evaluation of Predictive Models in Hazard Assessment and Risk Management. *Nat. Hazards* **2006**, *37*, 315–329. [CrossRef]
50. Cantarino, I.; Carrion, M.A.; Goerlich, F.; Martinez Ibañez, V. A ROC Analysis-Based Classification Method for Landslide Susceptibility Maps. *Landslides* **2019**, *16*, 265–282. [CrossRef]
51. Yilmaz, C.; Topal, T.; Süzen, M.L. GIS-Based Landslide Susceptibility Mapping Using Bivariate Statistical Analysis in Devrek (Zonguldak-Turkey). *Environ. Earth Sci.* **2012**, *65*, 2161–2178. [CrossRef]
52. Pourghasemi, H.R.; Moradi, H.R.; Fatemi Aghda, S.M. Landslide Susceptibility Mapping by Binary Logistic Regression, Analytical Hierarchy Process, and Statistical Index Models and Assessment of Their Performances. *Nat. Hazards* **2013**, *69*, 749–779. [CrossRef]
53. Zhou, X.; Wen, H.; Zhang, Y.; Xu, J.; Zhang, W. Landslide Susceptibility Mapping Using Hybrid Random Forest with GeoDetector and RFE for Factor Optimization. *Geosci. Front.* **2021**, *12*, 101211. [CrossRef]
54. Bornaetxea, T.; Rossi, M.; Marchesini, I.; Alvioli, M. Effective Surveyed Area and Its Role in Statistical Landslide Susceptibility Assessments. *Nat. Hazards Earth Syst. Sci.* **2018**, *18*, 2455–2469. [CrossRef]

55. Shao, X.; Ma, S.; Xu, C.; Zhang, P.; Wen, B.; Tian, Y.; Zhou, Q.; Cui, Y. Planet Image-Based Inventorying and Machine Learning-Based Susceptibility Mapping for the Landslides Triggered by the 2018 Mw6.6 Tomakomai, Japan Earthquake. *Remote Sens.* **2019**, *11*, 978. [CrossRef]
56. Confuorto, P.; Di Martire, D.; Infante, D.; Novellino, A.; Papa, R.; Calcaterra, D.; Ramondini, M. Monitoring of Remedial Works Performance on Landslide-Affected Areas through Ground- and Satellite-Based Techniques. *CATENA* **2019**, *178*, 77–89. [CrossRef]

MDPI AG
Grosspeteranlage 5
4052 Basel
Switzerland
Tel.: +41 61 683 77 34

Remote Sensing Editorial Office
E-mail: remotesensing@mdpi.com
www.mdpi.com/journal/remotesensing

www.ingramcontent.com/pod-product-compliance
Lightning Source LLC
Chambersburg PA
CBHW041401210926
51850CB00013B/3151